Life in the Solar System and Beyond

Springer
London
Berlin
Heidelberg
New York
Hong Kong
Milan
Paris
Tokyo

Barrie W. Jones

Life in the Solar System and Beyond

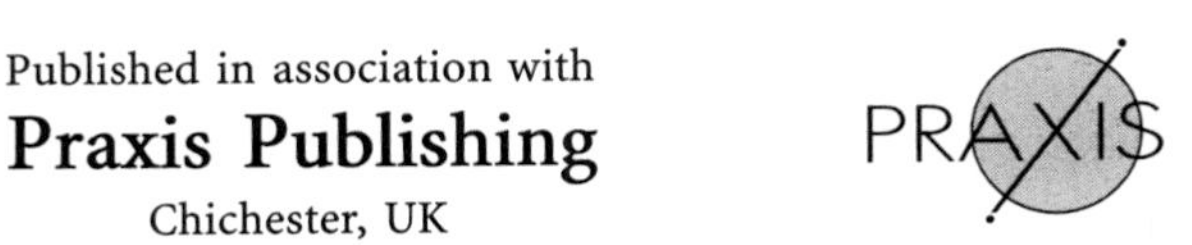

Professor Barrie W. Jones
Physics and Astronomy Department
The Open University
Milton Keynes
UK

SPRINGER–PRAXIS BOOKS IN ASTRONOMY AND SPACE SCIENCES
SUBJECT *ADVISORY EDITOR*: John Mason B.Sc., M.Sc., Ph.D.

ISBN 1-85233-101-1 Springer-Verlag Berlin Heidelberg New York

British Library Cataloguing in Publication Data
Jones, Barrie William, 1941–
Life in the Solar System and Beyond. –
(Springer-Praxis books in astronomy and space sciences)
1. Life on other planets 2. Exobiology
I. Title
576.839

ISBN 1852331011

A catalogue record for this book is available from the Library of Congress

Printed in the USA

Cover design: Jim Wilkie
Project management: Originator, Gt Yarmouth, Norfolk, UK

Printed on acid-free paper

Contents

Preface

Life in the Solar System and beyond is a subject that includes the study of life on Earth, but particularly the possibility of life elsewhere. It is of enormous interest and widespread significance, not least because of the progress that has been made in recent decades in our understanding of life on Earth, and of the potential for life existing elsewhere. At present, with no firm evidence as yet for extraterrestrial life, the study of life beyond the Earth is concerned with the location and study of possible habits, not only within the Solar System but outside it too. It is a wide-ranging subject, embracing aspects of all the core areas of science – astronomy, planetary science, chemistry, biology, and physics. This breadth is a challenge for authors and readers alike. Another challenge is the rapid pace of development of the subject, and it is hoped that the websites included in Resources at the end of the book will help us all keep up to date.

I have written a broad introduction to 'life in the Universe', a subject that is variously known as bioastronomy and astrobiology, though increasingly as the latter. The text is aimed at people with some background in science, though not much, given the breadth of science that astrobiology draws on. The readership could include science undergraduates, graduates with little previous knowledge of the field, and the interested non-specialist, including the amateur astronomer. A basic knowledge of atoms and the chemical elements will help, as will familiarity with graphs and with simple algebraic expressions and equations.

In studying the book it will benefit you to absorb the summaries at the end of each chapter – these are intended to encapsulate the main points to carry forward. There are also a few end of chapter questions. These are not comprehensive in their coverage, but are intended to illustrate how various topics can be progressed – they require more than recall. Full answers are given at the back of the book to all the questions set.

The chapters are arranged as follows. Chapter 1 is a broad introduction to the cosmos, with an emphasis on where we might find life out there. In Chapters 2 and 3 we discuss life on Earth, the one place we know to be inhabited, which provides us with an essential guide in our search. Chapter 4 is a brief tour of the Solar System, leading us in Chapters 5 and 6 to two promising potential habitats, Mars and Europa. Each of these worlds, and their exploration for life now or in the past, has separate chapters devoted to them. In Chapter 7 we meet the fate of life in the Solar System, which gives us extra reason to consider life further afield. Chapter 8 focuses on the types of stars that might host habitable planets, and where in the Galaxy these might be concentrated. Chapters 9 and 10 describe the instruments and techniques being employed to discover planets of other stars (exoplanetary systems), and those that will be employed in the near future. In Chapter 11 a summary is given of the known exoplanetary systems, and an outline of the sort of systems we expect to discover soon, particularly those containing habitable planets. Chapter 12 describes how we will attempt to find life on these planets, and Chapter 13, the final chapter, brings us to the search for extraterrestrial intelligence, to where we wonder whether we are alone.

Acknowledgements

I am very grateful to John Mason and Nick Sleep, who each read the whole book and provided many careful comments, to which I hope I've responded adequately. Irene Ridge, Hugh Jones, and Julian Hiscox read various groups of chapters and helped me to clarify the text in important ways. Many people helped with short pieces of text or with particular figures (credits given in captions). Those who made a substantial contribution to text development include (in alphabetical order): Coryn Bailer-Jones, David Erskine, Scott Gaudi, Andreas Glindemann, Alan Penny, Dave Rothery, Peter Sheldon, Peter Skelton, Peter Wizinowich, and Ian Wright.

Clive Horwood of Praxis has been a helpful and patient publisher, and my wife Anne Jones has withstood well my hermit-like habits, now happily ended.

Nick Sleep and Anne Jones provided essential help with the final preparation of the material for the publisher. Nick Sleep subsequently helped me check the proofs.

To my family
and
any extraterrestrials who can read

Abbreviations

ADP	adenosine diphosphate
ATP	adenosine triphosphate
ALMA	Atacama Large Millimetre Array
ATA	Allen Telescope Array
BETA	Billion Channel Extraterrestrial Assay
CETI	communicating with extraterrestrial intelligence
DNA	deoxyribonucleic acid
E–K	Edgeworth–Kuiper
ELT	extremely large telescope
ETI	extraterrestrial intelligence
EIRP	equivalent isotropically radiated power
GMST	global mean surface temperature
GCMS	gas-chromatograph mass-spectrometer
HZ	habitable zone
HST	Hubble Space Telescope
H-R	Hertzsprung-Russell
IMF	initial mass function
NIR	near-infrared
OWL	Overwhelmingly Large Telescope
OGLE	Optical Gravitational Lensing Experiment
OSETI	optical search for extraterrestrial intelligence
PAHs	polycyclic aromatic hydrocarbons
psf	point-spread function
PRIMA	Phase Referenced Imaging and Microarcsecond Astrometry
RNA	ribonucleic acid
SETI	search for extraterrestrial intelligence
snr	signal-to-noise ratio
SIM	Space Interferometry Mission

STARE	Stellar Astrophysics and Research on Exoplanets
SERENDIP	Search for Extraterrestrial Radio Emissions from Nearby Developed Independent Populations
TPF	Terrestrial Planet Finder
UMa	Ursae Majoris
UV	ultraviolet
VLT	Very Large Telescope
ZAMS	zero age main sequence

Figures

Tables

Plates (between pages 150 and 151)

1

The cosmos

There are many and varied places beyond the Earth where we might find life. In this chapter you are provided with a context for them, for where they are located in relation to the Earth, and for their physical nature. Liquid water is a prime indicator of potential habitats. Within the Solar System Mars is a good candidate, now a cool, dry, and dusty planet, but perhaps warmer and wetter early in its history. We also locate Europa, a world covered in ice, beneath which there probably lies an ocean where life might exist today. Beyond the Solar System, a significant proportion of the nearby stars are known to have their own planetary systems, and it is possible that some of these might harbour life. Further away there are many more stars; our Galaxy alone contains about two hundred thousand million, and there are billions of galaxies. However, the chances of finding evidence of life are greater when nearer the target, and therefore we will spend most of this chapter exploring the Solar System and the nearby stars – our cosmic neighbourhood.

1.1 THE SOLAR SYSTEM

1.1.1 The orbits of the planets

The Solar System consists of the Sun, the planets, and a large number of smaller bodies. Figure 1.1 shows the orbits of the planets, which are roughly circular and which all lie approximately in the same plane. The orbit of the outermost planet, Pluto, is the least circular, and it is also the most tilted with respect to the other orbits. Pluto's orbit is shown in Figure 1.2, and we can use it to define the 'vital statistics' of all the planetary orbits.

Each orbit is an *ellipse*, which has the shape of a circle viewed obliquely, the more oblique the view the more elongated the appearance. For an ellipse the longest diameter is called the major axis. At one end of this axis the planet is closest to the

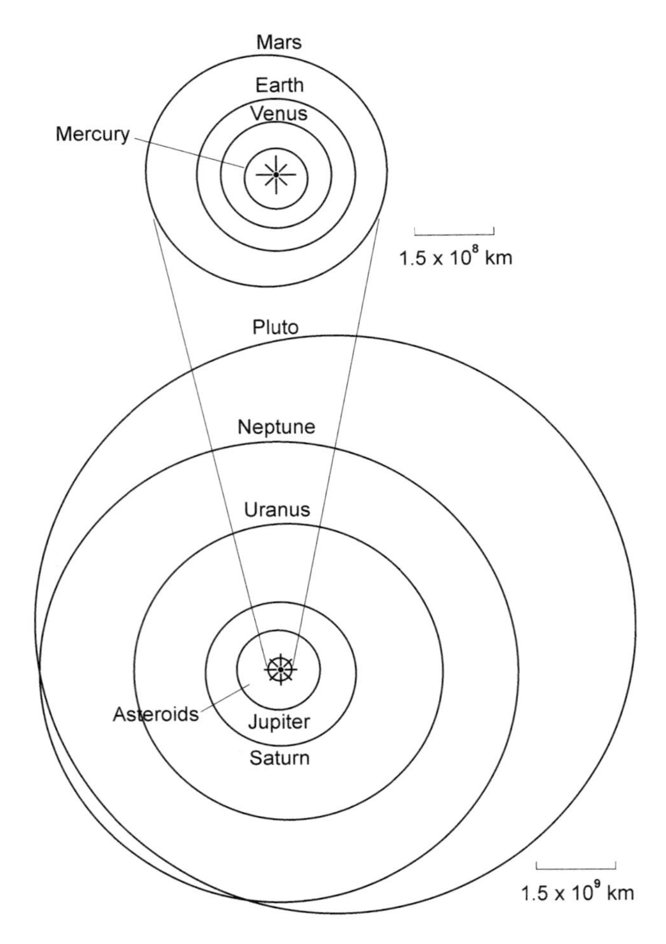

Figure 1.1 The orbits of the planets around the Sun, as they would appear from a distant viewpoint perpendicular to Earth's orbit, with the North Pole of the Earth pointing towards us.

Sun; this is the point of *perihelion* (Figure 1.2). At the other end the planet is furthest from the Sun; this is *aphelion*. Half of the major axis is the *semimajor axis* and it is given the symbol a. On the major axis, and displaced equal distances to each side of its midpoint (C), are the two *foci* (F) of the ellipse. You can see in Figure 1.2 that the Sun is at one of the foci, and that the other is empty. The less elongated an ellipse the nearer are its foci to the midpoint, the limiting case being a circle, where both foci are at the midpoint. Therefore the distance FC in Figure 1.2 is a measure of the departure from circular form. This departure is specified by the *eccentricity* $e = \mathrm{FC}/a$. It is also the case that $e = (1 - b^2/a^2)^{1/2}$ where b is the semiminor axis, which is half of the shortest diameter of an ellipse. For Pluto's orbit $e = 0.249$. The edgewise view of Pluto's orbit in Figure 1.2 shows the orbital inclination i with

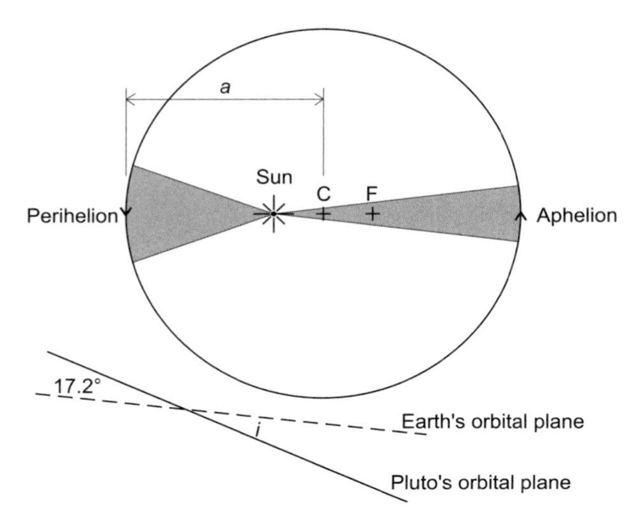

Figure 1.2 The orbit of Pluto, face-on and edge-on. The shaded areas in the face-on view are swept out in equal time intervals.

respect to the plane of the Earth's orbit, the ecliptic plane, which is taken as the reference plane in the Solar System. For Pluto's orbit $i = 17.2°$.

The quantities a, e, and i are three of what are called the elements of a planetary orbit. Other elements, which further specify the orientation of an orbit with respect to that of the Earth's orbit, will not concern us. The value of a for the Earth's orbit is 1.496×10^{11} m, and this is a convenient unit of distance in the Solar System and within other planetary systems; it is called the *astronomical unit*, symbol AU. (Because the semimajor axis of the Earth varies, albeit very slightly, the AU is now defined exactly as $1.495\,978\,707 \times 10^{11}$ m, not quite the same as the present-day value of a for the Earth.) For Pluto $a = 39.4$ AU.

The planets move around their orbits in the same direction, anticlockwise as viewed from above the Earth's North Pole – this is called the prograde direction. They move fastest near to perihelion because the gravitational pull of the Sun is greatest there, and they move slowest at aphelion. The line from the planet to the Sun sweeps out equal areas in equal time intervals, as illustrated in Figure 1.2. The time to go around an orbit once is called the *orbital period*, P. For the Earth it is one year, whereas for Pluto it is 248 years. The relationship between P and a is very simple, P is proportional to $a^{3/2}$:

$$P = ka^{3/2} \tag{1.1}$$

where k is the constant of proportionality. That P increases as a increases is not surprising – the orbit is bigger. However, this alone would make P proportional to a. The extra sensitivity to a is because the speed of the planet in its orbit decreases as a increases. In the Solar System, if P is measured in years and a in AU then the constant of proportionality is 1, and so $P = a^{3/2}$.

We have now encountered three important laws of planetary motion. These are called *Kepler's laws* after the German astronomer Johannes Kepler, who announced the first two in 1609 and the third in 1619. The modern form of the laws is as follows.

First law Each planet moves around the Sun in an ellipse, with the Sun at one of the foci of the ellipse.

Second law As the planet moves around its orbit, the straight line from the planet to the Sun sweeps out equal areas in equal intervals of time.

Third law $P = ka^{3/2}$, where (in the Solar System) $k = 1$ if P is measured in years and a in AU.

After their discovery, it was shown that these laws can be accounted for by fundamental physical theories, specifically Newton's laws of motion and law of gravity, developed by the English scientist Isaac Newton, in the second half of the 17th century – see the physics texts in Resources.

As well as planets, the Solar System also contains smaller bodies. Several thousand asteroids are known to orbit the Sun, mainly between the orbits of Mars and Jupiter. The asteroids are rocky and of the order of a few kilometres in radius. By contrast, comets are mixtures of small rocky particles and icy materials, mainly water ice. Comets originate as small icy–rocky bodies in two regions. The Edgeworth–Kuiper (E–K) belt consists of bodies in orbits of fairly low inclination. It extends outwards from 34 AU to at least 50 AU, perhaps far enough to merge with the Oort cloud. This cloud is a thick spherical shell of icy–rocky bodies, extending from about 10^3 AU (perhaps 10^4 AU) to the very edge of the Solar System at about 10^5 AU.

Whereas E–K objects are close enough to be seen with telescopes as tiny dots, Oort objects are much too far away. The existence of the Oort cloud is inferred from those of its members that are perturbed so that they travel through the inner Solar System, were they develop huge, spectacular tails. These tails consist of gases evaporated from icy materials by the heat of the Sun, and of dust particles entrained in this gas flow (Plate 1). Other comets are members of the E–K belt, likewise perturbed so that they pass through the inner Solar System.

Table 1.1 gives the orbital elements a, e, and i, and the orbital periods P of the planets and the largest asteroid (Ceres). Most planets also have satellites, and these are described in Section 1.1.4.

Table 1.1. The orbital elements a, e, i, and orbital periods P of the planets and the largest asteroid Ceres (2003).

	Mercury	Venus	Earth	Mars	Ceres	Jupiter	Saturn	Uranus	Neptune	Pluto
a (AU)	0.387	0.723	1.000	1.524	2.766	5.202	9.581	19.13	29.95	39.4
e	0.206	0.0067	0.0167	0.0936	0.0794	0.0490	0.0574	0.0499	0.0096	0.249
i (°)	7.00	3.39	0.00	1.85	10.6	1.30	2.49	0.77	1.77	17.2
P (years)	0.241	0.615	1.000	1.881	4.600	11.86	29.65	83.67	163.9	247.7

1.1.2 The Sun as a body

Though planets and large satellites are the sort of bodies where we expect to find extraterrestrial life, their surfaces would not be habitable without the heat and light from the star they orbit – the Sun in the case of the Solar System. What is it that sustains the Sun's energy output and how long can we expect it to last?

Figure 1.3 shows a cross section through the Sun as it is today. The outer part has much the same composition as the whole Sun did when it was born from interstellar gas and dust 4600 million years ago. In proportions by mass, this is 71% hydrogen, 27% helium, and 2% for all the other chemical elements, the so-called *heavy elements*. This is in stark contrast to the Earth, where the heavy elements account for over 99% of the mass. The core of the Sun had the initial composition, but it has since been enriched in helium at the expense of hydrogen through nuclear reactions. This is the process that sustains the Sun's energy output.

Nuclear reactions result from the collisions between atomic nuclei. The nucleus accounts for nearly all of the mass of an atom, and it has a positive electric charge that equals the negative electric charge of the electrons that surround the nucleus. The nuclear charge is carried by protons. A chemical element is defined by the number of protons in its atomic nucleus. Hydrogen (H) has one proton, and helium (He) has two. Iron, for example, has 26. The nucleus can also contain

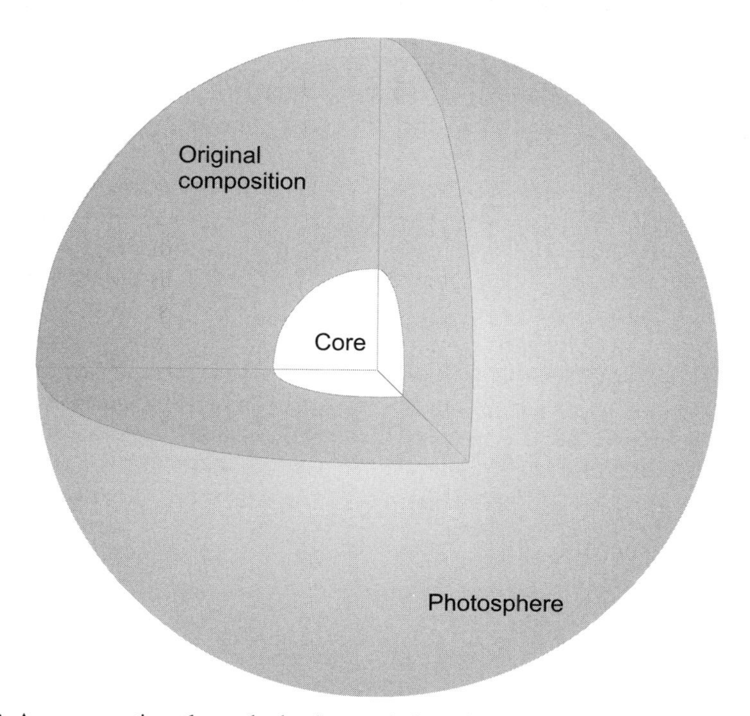

Figure 1.3 A cross section through the Sun as it is today, showing the core where fusion of hydrogen to helium liberates energy that sustains the Sun's energy output.

neutrons, with very nearly the same mass as a proton, but no electric charge. For a given number of protons there can be various numbers of neutrons, and each particular number defines a different isotope of the element. Most of the hydrogen in the Sun, and indeed in the Universe, has no neutrons. Therefore the nucleus is the simplest possible, a single proton, and is denoted by 'p' for proton, or by ^{1}H, the '1' indicating that the sum of the number of protons and neutrons is 1. About 0.01% of hydrogen nuclei have one neutron, so are denoted ^{2}H. Unusually, this isotope has a name, deuterium, and is sometimes given the symbol D. Another hydrogen isotope, tritium, has two neutrons, but it is unstable and breaks up. The common isotope of helium is ^{4}He, and so has two neutrons.

In the solar interior the temperatures are so great that the collision speeds of atoms are sufficiently high to strip off the atomic electrons. The atomic nuclei are therefore moving about in a 'sea' of electrons. Such a medium is called a plasma. The nuclei collide, and if they do so with sufficient energy then a nuclear reaction will occur. Because the energy is derived from the random thermal motions they are called thermonuclear reactions. The energy required is less the fewer the number of protons in the nucleus because the electrostatic repulsion between nuclei is then smaller. This gives an advantage to nuclei of hydrogen, as does their great abundance in the Sun. It is therefore thermonuclear reactions involving hydrogen that are important. In the Sun the dominant reactions are shown in pictogram form in Figure 1.4. Their net effect is the (thermonuclear) fusion of four ^{1}H nuclei into one ^{4}He nucleus:

$$4\,^1\mathrm{H} \rightarrow\, ^4\mathrm{He} + 2\mathrm{e}^+ + 2\nu_\mathrm{e} + 2\gamma \tag{1.2}$$

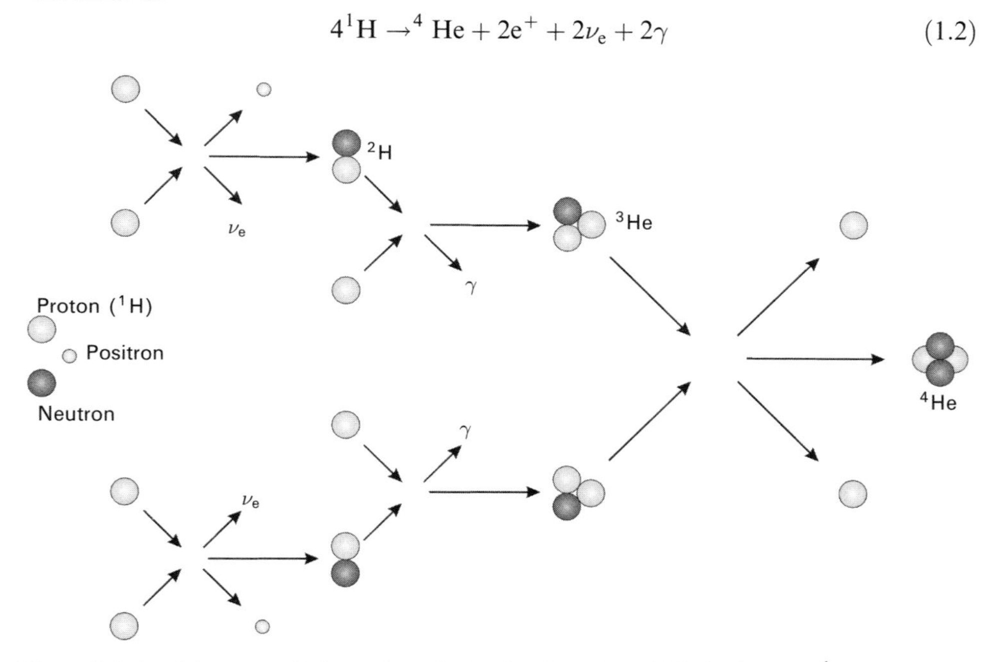

Figure 1.4 A pictogram of the ppI cycle in the Sun, in which hydrogen (^{1}H) undergoes thermonuclear fusion to helium (^{4}He), liberating energy.

The other particles involved are positrons e^+ (electrons with positive charge), neutrinos ν_e (very low-mass, unreactive particles), and gamma rays γ (short wavelength electromagnetic radiation).

The particular set of reactions in Figure 1.4 is called the ppI cycle – 'pp' because it starts with a collision between two protons (p). The 'I' indicates that there are other pp cycles. There are in fact two, ppII and ppIII; and these also convert hydrogen into helium, but most of the Sun's energy output is due to the ppI cycle. The energy comes from the conversion of mass to energy, as in Einstein's famous equation $E = mc^2$, where c is the speed of light, $2.998 \times 10^8\,\mathrm{m\,s^{-1}}$. The sum of the masses on the left of Reaction (1.2) is slightly greater than on the right. This energy appears as the kinetic energy of the electron and the neutrino, and particularly as the energy in the gamma rays.

The ppI cycle is confined to the central core of the Sun (Figure 1.3). Temperatures increase with depth throughout the Sun and the rate at which the cycle proceeds increases rapidly as temperature rises. There is thus a rapid transition from a very low reaction rate just outside the core to a far higher rate just inside it. The temperature at the surface of the core is about $8 \times 10^6\,\mathrm{K}$ (kelvin). The ppI cycle sustains the temperature gradient that, in turn, sustains the pressure gradient that supports the Sun against the inward force of gravity. The Sun is thus in mechanical equilibrium, neither expanding nor contracting. It is also in radiative equilibrium, the rate of energy generation in the core equalling the rate at which the Sun loses energy to space. This is almost entirely in the form of electromagnetic radiation, though a small proportion is carried by neutrinos and other particles. The electromagnetic radiation that escapes to space, though its source is in the core, comes from a thin layer that we see as the surface of the Sun, called the photosphere.

Most of the radiation from the photosphere is the familiar visible light that we see, plus infrared radiation, as shown in Figure 1.5, which is a solar spectrum omitting fine detail. The smooth curve is the spectrum of an ideal thermal source called a *black body*. The shape of a black body spectrum depends only on its temperature, and the curve in Figure 1.5 is for a black body at 5780 K. This is a good fit to the solar spectrum, and so we can regard 5780 K as a representative temperature of the photosphere – the local temperature varies from place to place. The exact value is chosen because the power radiated by unit area of a black body at 5780 K then equals the power radiated by unit area of the Sun. This means that 5780 K is the *effective temperature* of the Sun. The total power radiated by the Sun, from its whole surface, over all wavelengths, is called its *luminosity*, and it has a value of $3.85 \times 10^{26}\,\mathrm{W}$ (watts).

The Sun settled into its hydrogen-fusing equilibrium 4600 million years ago, when its contraction from interstellar gas and dust was halted by the onset of thermonuclear fusion of hydrogen in its core. The equilibrium is not perfect, and over this time the Sun's luminosity has gradually increased from about 70% of its present value. The temperature of the photosphere has also increased. Additionally, there are shorter term, though far smaller fluctuations, such as in the 11-year cycle of solar activity that has been traced back for hundreds of years. In the future the Sun

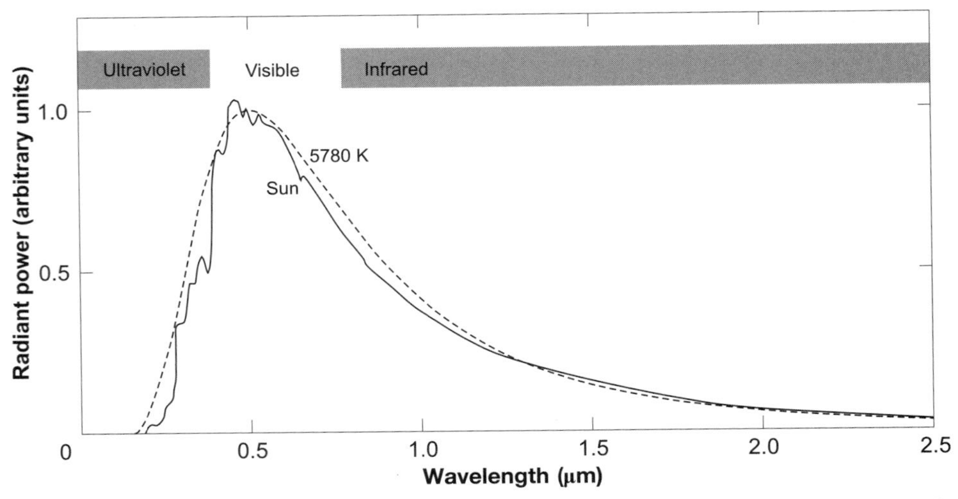

Figure 1.5 A simplified solar spectrum, with the spectrum of a black body at 5780 K for comparison. Note that 1 μm = 10^{-6} m.

will continue to evolve slowly. In about 6000 million years the pace of change will quicken. This will be when the core runs out of hydrogen, thus bringing to an end the *main sequence phase* – the phase of thermonuclear fusion of hydrogen in the core. A star in this phase of its life is called a main sequence star. The Sun is thus nearly half way through its main sequence lifetime. What will happen afterwards is a subject for Chapter 7, where you will also meet the consequences for life in the Solar System of the Sun's continuing evolution.

1.1.3 The planets as bodies

The planets are intriguingly different from each other. This is at once apparent from their external appearances, as illustrated in Plates 2–9. They also differ greatly in size, as shown in Figure 1.6, with the Sun for comparison. The inner four planets, Mercury, Venus, Earth, and Mars, are called the *terrestrial planets*. This is because their composition is dominated by rocky materials (which include metallic iron). Volatile substances, such as water and the gases that make up any atmospheres, are important, but constitute only a tiny fraction of each terrestrial planet's mass. By contrast, the *giant planets* Jupiter and Saturn are dominated by hydrogen and helium, though there is a few-fold enrichment in heavy elements compared to the Sun's 2%, more so in Saturn. Uranus and Neptune are also called giant planets, though they are intermediate in size between the terrestrial planets and Jupiter, and water makes up well over half the mass of each, with hydrogen and helium contributing only the order of 10%. Consequently, they are sometimes called water planets. Pluto, a small world, is icy in composition – water ice, methane ice, and other ices. We will now pay a brief visit to each of these worlds.

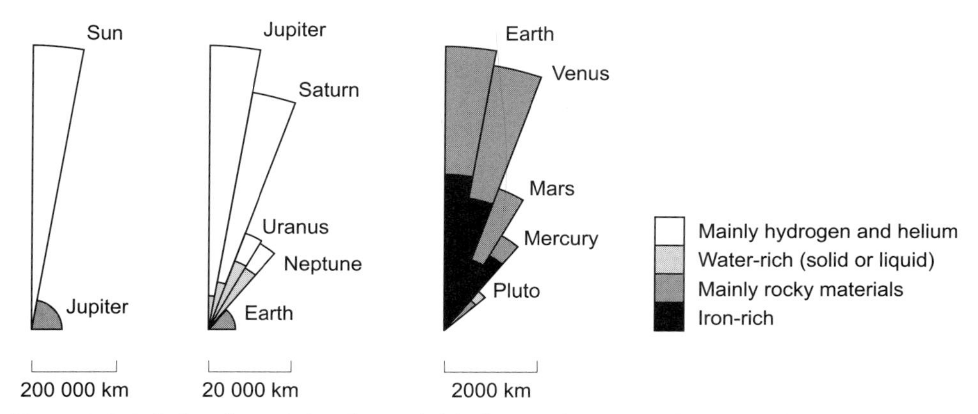

Figure 1.6 Relative sizes of the planets (with the Sun for comparison). The internal structures are indicated.

Mercury (Plate 2) has a surface scarred by craters caused by impacts from space. A surface becomes ever more cratered with the passage of time unless they are removed by geological processes (such as volcanism) or by weathering. Therefore, the abundance of impact craters on the surface of Mercury is evidence of a lack of geological activity and weathering. Lack of weathering is consistent with the near absence of an atmosphere on Mercury today, and the near certainty that it has not had much by way of an atmosphere in the past. Its proximity to the Sun means that the Sun-facing surface can reach temperatures as high as about 700 K, but its thin atmosphere and slow rotation mean that the night-time surface can become as cold as about 90 K. To obtain these values on the more familiar Celsius temperature scale we subtract 273, thus obtaining 427°C and −183°C. Planetary temperatures will be given in °C and stellar temperatures in K.

Venus (Plate 3) is permanently shrouded in cloud, not the familiar water clouds of Earth, but clouds consisting largely of sulphuric acid (H_2SO_4) droplets. These clouds lie high in a thick, dry atmosphere of carbon dioxide (CO_2) which for a long time has sustained a huge greenhouse effect that raises the average surface temperature to 460°C. Moreover, the thick atmosphere and cloud cover ensure that there is no temperate zone at the poles. The surface has few impact craters, but there are many volcanoes and widespread evidence of volcanic activity. Such widespread volcanism, along with atmospheric weathering, has been responsible for obliterating impact craters.

Earth (Plate 4) is the only terrestrial planet to have abundant liquid water at its surface. About 70% of the surface is covered in oceans. Clouds of water particles, either as liquid droplets or ice, typically cover about half the surface. The atmosphere is much less massive than that of Venus – the *column mass*, which is the mass per unit area of the surface, is $1.03 \times 10^4\ \mathrm{kg\,m^{-2}}$ compared to $102 \times 10^4\ \mathrm{kg\,m^{-2}}$ for Venus. Though there is water vapour in the Earth's atmosphere, the main constituents are oxygen (O_2) and nitrogen (N_2), with CO_2 little more than a trace. The Earth is very active geologically, with extensive volcanism, and with a process that is unique in the Solar System – plate tectonics. The Earth's rocky surface is divided into a few dozen plates. These are in motion with respect to each other, being created at some plate boundaries, sliding past each other at other boundaries, and being destroyed at boundaries where one plate dives beneath another. This geological activity, plus atmospheric weathering, has ensured that evidence of impact craters is rare.

Mars (Plate 5) resembles Venus in one way – it too has an atmosphere consisting largely of CO_2 – but in many other ways it is a very different planet, and is different from the Earth too. Its atmosphere has a column mass of only $0.015 \times 10^4\,\mathrm{kg\,m^{-2}}$. This thin atmosphere, the sparse cloud cover, and its distance from the Sun have made Mars a cold world. On a good day at the equator the temperature can reach about 10°C, but it plunges to −100°C or even lower at night. The surface divides into two roughly equal areas. The northerly hemisphere shows considerable evidence of geological activity, perhaps confined to the past, and a corresponding low density of impact craters. The southerly hemisphere is much more impact scarred, displays little evidence of geological activity, and bears features that suggest that Mars was probably a much warmer, wetter place early in its history. Today, water has only been seen in solid form, in the polar caps (Plate 5), in frosts that form here and there at night (Plate 15), and as thin clouds of ice crystals, though it is probably present as liquid beneath the surface. We visit Mars properly in Chapter 5, because it offers one of the better prospects for finding extraterrestrial life.

Jupiter (Plate 6) is a very different world from any terrestrial planet. It has no proper surface at all. As we go deeper, the atmosphere, dominated by hydrogen and helium in roughly solar proportions, gets denser (and hotter) until we are in a global ocean of hot hydrogen and helium. The enrichment of heavy elements has probably led to their concentration near the centre, as a hot core of liquid water and liquid rock. From the outside we see the top of the uppermost layer of clouds, with occasional glimpses to cloud layers of ammonium hydrosulphide (NH_4SH) and water deeper down. The uppermost clouds are icy particles of ammonia (NH_3), coloured in some unknown way, and formed into bands and whorls by atmospheric circulation. The largest whorl in Plate 6 is the Great Red Spot, and it is at least 100 years old, probably much older. Saturn (Plate 7), other than its extensive ring system of small particles, is rather like Jupiter, though it is somewhat smaller, somewhat more enriched in heavy elements, and the cloud features are subdued by atmospheric haze.

Uranus (Plate 8) and Neptune (Plate 9) also have atmospheres rich in hydrogen and helium, but these planets overall are *not* dominated by these elements. They probably have surfaces marking the transition to the dominant constituents, which are liquid water with substantial proportions of NH_3 and methane (CH_4), but any surfaces are hidden by atmospheric haze and cloud. Note that though water, NH_3, and CH_4 are called icy materials, this is the name of a chemical group, and does not imply that they are necessarily present as solids (i.e., as ices). Indeed, the interiors of Uranus and Neptune are far too hot for icy materials to freeze.

Pluto differs from all the other planets in its small size (Figure 1.6) and in its composition. Most of its volume is probably water ice, but there are also rocky materials, probably concentrated into a core. It is so far from the Sun that its surface is very cold, never exceeding about −210°C, and covered by the more volatile ices such as N_2, CH_4, and carbon monoxide (CO). Pluto resembles some of the large satellites in the Solar System more than it resembles the other planets. It is increasingly regarded as the largest known member of the E–K belt.

Table 1.2 lists some of the properties of the Sun and planets. Note that the radii of the giant planets are to the level in the atmosphere at the equator where the pressure is 10^5 Pa (pascals). This is close to atmospheric pressure at the Earth's surface.

1.1.4 The large satellites

A satellite is a body in orbit around a planet, rather than in its own orbit around the Sun. There are many satellites in the Solar System, and almost all of them orbit the giant planets. These giant planets account for all but one of the seven large satellites – the exception is our Moon. Figure 1.7 shows the sizes of the seven large satellites in comparison to the smallest three planets, Mars, Mercury, and Pluto. Io, Europa, Ganymede, and Callisto are the Galilean satellites of Jupiter, named after the Italian astronomer Galileo Galilei, who discovered them in 1610 when he made some of the very first observations of the heavens with the then recently invented telescope. Titan belongs to Saturn, and Triton to Neptune. The largest of all the remaining satellites is also shown, which is Uranus's satellite Titania, and the largest asteroid, Ceres. This highlights a considerable step down in size. By contrast there is no such step down from planets to satellites. Indeed, the two largest satellites, Jupiter's Ganymede and Saturn's Titan, are larger than

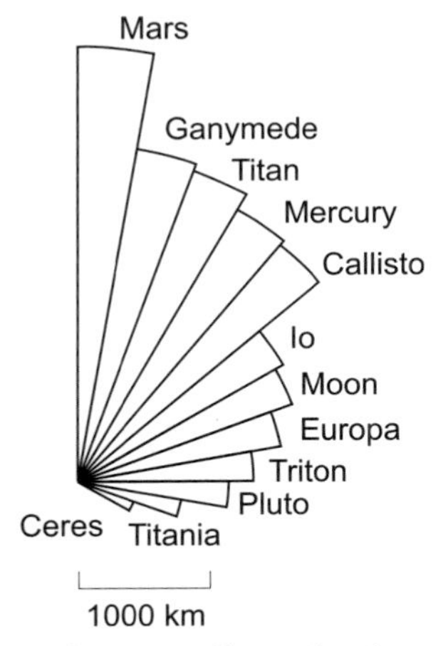

Figure 1.7 Relative sizes of the seven large satellites, the three smallest planets, the largest remaining satellite, Titania (of Uranus), and the largest asteroid, Ceres. Io, Europa, Ganymede, and Callisto are the Galilean satellites of Jupiter, Titan belongs to Saturn, and Triton to Neptune.

Table 1.2. Some properties of the Sun, the planets, and the largest asteroid Ceres.

	Sun	Mercury	Venus	Earth	Mars	Ceres	Jupiter	Saturn	Uranus	Neptune	Pluto
Radius* (km)	696 000	2440	6052	6378	3397	502	71 490	60 270	25 560	24 765	1150
Mass† (Earth)	333 000	0.0553	0.815	1	0.107	0.00018	318	95.2	14.5	17.2	0.0025
Main materials	Hydrogen + helium	Rocky‡ + iron	Rocky + iron	Rocky + iron	Rocky + iron	Rocky + iron	Hydrogen + helium	Hydrogen + helium	Icy‡	Icy	Icy + rocky

* Radius at the equator.
† The mass of the Earth is 5.974×10^{24} kg.
‡ 'Rocky' and 'icy' are labels for groups of materials, and do not imply that they are solid or liquid.

Mercury, and all seven are larger than Pluto. If the large satellites were in their own orbits around the Sun they would be regarded as planets – they are certainly *planetary bodies* with various icy–rocky or rocky compositions. Plates 10–13 show Io, Europa, and Titan.

Among the satellites, as you will see in Chapter 6, Europa offers the best prospect for finding life elsewhere in the Solar System today. Europa is a rocky body, though this is not the impression gained from the outside (Plate 11). The whole of Europa is covered in a veneer of water ice probably underlain by liquid water. It is this possibility of a widespread ocean that makes Europa a good prospect for life.

1.2 THE ORIGIN OF THE SOLAR SYSTEM

The age of the Solar System has been firmly established from radiometric dating of meteorites as a little less than 4600 million years, almost exactly one-third of the 13 700 million year age of the Universe. There is much observational evidence to support the view that the Solar System was formed from a fragment of an interstellar cloud of gas and dust. Figure 1.8 shows an artist's impression of such a fragment at the stage when a central condensation has been heated by gravitational energy to the point where its contraction has been slowed and it is close to becoming a main sequence star – it is a protostar. It is surrounded by the rest of the fragment, as an extended disc, perhaps a few percent of the mass of the protostar. Most astronomers believe that it was from such a disc, called the *solar nebula*, that the planets in the Solar System were formed. Here is the widely accepted theory.

1.2.1 The nebular theory

The composition of the solar nebula was the same as that of the protoSun, which is the same as the Sun outside its core today – by mass, 71% hydrogen, 27% helium, and 2% heavy elements (Section 1.1.2). Most of the nebula was gas, consisting almost entirely of molecular hydrogen (H_2) and monatomic helium. About 1% by mass of the nebula was in fine dust, much of it the size of cigarette smoke particles. The composition of the dust depended on the local temperature. Far from the protoSun where it was cold, water was a prominent component. Nearer to the protoSun, the water was in the gas phase, and so the dust was dominated by silicates (compounds of silicon, oxygen, and metals), metallic iron, and the heavier hydrocarbons (compounds of carbon and hydrogen). Closer still, only certain silicates and iron remained.

The dust settled under gravity to form a thin sheet of order 10^4 km thickness around the protoSun, enveloped in the disc of the solar nebula, as shown edgewise in Figure 1.9. The dust particles were now much more likely to collide than when they were more thinly dispersed, and so they grew into bodies 0.1–10 km across, called *planetesimals*. These were massive enough to gravitationally attract other

Figure 1.8 An artist's impression of the solar nebula at an early stage, where the Sun at the centre is a protostar. This is an oblique view – the disc is really circular.
Julian Baum (Take 27 Ltd).

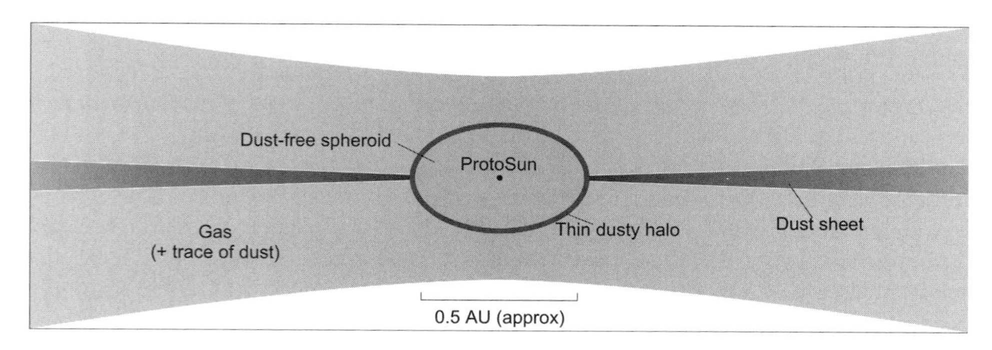

Figure 1.9 Edgewise view of the central part of the sheet of dust surrounding the protoSun, with the predominantly gaseous solar nebula enveloping the sheet.

planetesimals over considerable distances, thus modifying their orbits. This led to the acquisition of smaller planetesimals by larger ones, a process called accretion. In the inner Solar System, within about 4 AU of the Sun, the outcome was a hundred or so *embryos*, each with a mass the order of a few percent that of the Earth (i.e., the order of a few times the mass of the Moon). There were also many leftover planetesimals. By now, from the formation of the solar nebula around the protoSun, only a few hundred thousand years had elapsed. The process of putting embryos together to make the terrestrial planets was very much slower, and must have been influenced by developments further out.

By the time the embryos were appearing in the terrestrial zone, Jupiter and Saturn were well on the way to forming. In this 4–10 AU region, the models indicate that only a few embryos would have formed, and that most, even all of them would be the seed of a giant planet. This is because the embryo mass here was typically the order of ten times the mass of the Earth, and this was sufficient for the embryo to gravitationally capture gas from the solar nebula and thus grow to giant size (to 100 or more Earth masses). The gas was predominantly hydrogen and helium, and so the giant planets acquired their observed composition.

The embryos, now called *kernels*, grew so massive for two reasons. First, beyond about 4 AU it was cool enough for the abundant icy substance water to condense as ice – this boundary is called the *ice line*. The appearance of ice moderated the decrease in column mass of condensates with increasing distance from the protoSun. Second, the 'feeding zone' of an embryo was much wider, so it gathered a lot more material. Therefore very few embryos formed, but each was a massive kernel.

It is estimated that Jupiter acquired much of its mass in at most a few million years. The gravitational field of Jupiter (and to a lesser extent Saturn) then aided further terrestrial planet growth by scattering planetesimals into the terrestrial zone, and by increasing the eccentricities of the embryos and planetesimals there so that collisions were frequent. Collision led to fragmentation, but out to about 1 AU (the Earth) most of the fragments reassembled. Further out, near Mars, the collisions were more violent, and as a consequence the growth of Mars was stunted, hence its low mass (Table 1.2). Beyond about 2 AU the collisions were so disruptive that most of the material was removed, and we were left with the asteroids, which don't amount to much more than a few percent of the Moon's mass in total. Overall, it took a few tens of millions of years for the terrestrial planets to acquire 99% of their mass.

The formation of the giant planets was halted by what is called the T Tauri phase of the protoSun, when there was a violent outflow of gas and a high level of ultraviolet radiation. It immediately preceded the main sequence phase and lasted about ten million years, but very early in the T Tauri phase it would have dissipated the solar nebula and thus halted the acquisition of nebula gas by the kernels of the giant planets. The solar nebula had lasted a few million years.

This relatively short lifetime for the solar nebula makes our understanding of the formation of Uranus and Neptune problematical. Kernel formation must have slowed dramatically with increasing distance from the protoSun, and at the present distance of Uranus and Neptune it would have taken many millions of years. It is possible that the kernels formed in time to capture a small amount of nebula gas, thus explaining why Uranus and Neptune consist predominantly of icy materials, and are much less massive than Jupiter and Saturn, but this requires critical timing which can seem contrived.

One way around this problem is planetary migration. Uranus and Neptune could have formed closer to the Sun, and while they were acquiring nebular gas were gravitationally perturbed by Jupiter or Saturn into more distant orbits. Another way to achieve such orbital migration is through interaction with planetesimals. There is evidence in the E–K belt for the planetesimal-induced migration of Neptune, and you will see in Chapter 11 that there is evidence of giant migration in other planetary systems, so it does offer a plausible solution.

An alternative theory of giant planet formation

Instead of the two-step process for forming giants, inherent in the kernel formation and gas capture theory, there is also a one-step theory, in which the giant planets form through gravitational instabilities in the solar nebula beyond about 5 AU. This results in the gravitational contraction of a fragment of the nebula to form a protoplanet, that further contracts to giant planet dimensions. The composition is initially an unbiased sample of the nebula, and therefore is the same as that of the young Sun, but planetesimals are subsequently captured, giving some degree of enrichment in the heavy elements.

The formation of giant planets in this way would have occurred early in the life of the solar nebula, and so there would have been a protoJupiter plenty early enough to influence planetary formation in the terrestrial zone in the manner outlined above. In support of this theory it has been found that gravitational instabilities occur readily within the nebula, though there is much detailed work to be done to see if this is really a way in which the giant planets could have formed.

1.2.2 Pluto, comets, and satellites

Regardless of how the giant planets formed, there were surely many icy planetesimals in the giant region. If so then the gravitational fields of the giants would have captured some of them, and scattered others far outwards, where those retained in the Solar System would have constituted the Oort cloud.

The E–K objects, which lie beyond the giant region but not nearly as far as the Oort cloud (Section 1.1.1), are readily explained as a consequence of the slow rate of accretion at such large distances from the protoSun, even slower than in the Uranus–Neptune region. It is quite possible that accretion did not proceed even as far as lunar-size embryos, but stopped with a range of small icy embryos and icy planetesimals. This is exactly what we see today, with Pluto as the largest icy embryo, and a variety of E–K objects, some about half the radius of Pluto, others much smaller.

The satellites of the giant planets were probably acquired in two ways. First, the majority were acquired from a circumplanetary disc of material in a manner somewhat analogous to the formation of the terrestrial planets around the protoSun. The remainder could have been acquired by capture of passing planetesimals, icy or rocky. The two tiny satellites of Mars, Phobos and Deimos, are probably captured asteroids. The Moon seems to have been created in a different way, through the collision between a Mars-sized embryo and the Earth when the latter was nearly fully formed. Some of the debris from this huge impact would have remained in orbit around the Earth, and accreted to form the Moon. Many of the chemical differences and similarities between the Earth and the Moon can be explained by this giant impact theory, as can the comparatively large mass of the Moon – the Earth is only 81 times more massive. Charon, the comparatively large icy satellite of Pluto might have been acquired in a similar way, or perhaps by capture.

1.2.3 The acquisition of volatile substances by the terrestrial planets

Though the terrestrial planets consist almost entirely of rocky materials and metallic iron, life would not exist on the Earth or anywhere else if there were not also traces of volatile substances, notably the water on which all terrestrial life depends, but also other volatiles that constitute the atmospheres, such as CO_2 and N_2. These volatiles arrived in two ways. First, the dust from which the bulk of these planets formed, contained volatiles that later outgassed onto their surfaces when the interiors became hot. The dust might have borne only a tiny trace of volatiles, but in any case there seems to have been an abundant second supply towards the end of planetary formation, and subsequently.

This second supply was by the volatile-rich and icy planetesimals from the outer Solar System that are thought to have swept through the inner Solar System because of gravitational scattering by the giant planets, particularly Jupiter. The delivery was particularly effective when the swarm of rocky planetesimals in the inner Solar System had thinned. The bombardment was concentrated in the first 700 million years or so of Earth history (i.e. from 4600 to 3900 million years ago). This is called the *heavy bombardment*, and it must have peppered the other terrestrial planets too, and also the Moon. Indeed, it is from the radiometrically dated surfaces on the heavily cratered Moon that we get the timings – the Earth has been too effectively resurfaced by geological activity for any such ancient terrain to survive. During the heavy bombardment, as well as volatile-rich bodies, leftover rocky planetesimals would also have arrived. But it is the former that brought a significant proportion of the terrestrial planets' volatiles, and thus the possibility of life.

1.2.4 The origin of the heavy elements

It should be clear from the preceding sections that the heavy elements are of prime importance in the formation of planets. In the case of the terrestrial planets and icy–rocky bodies they dominate the composition, and have endowed these planets with the volatile substances essential for life. Moreover, the biochemicals that are the basis for all life on Earth have carbon, oxygen, nitrogen, and other heavy elements, as essential and prominent ingredients, as you will see in Chapter 2. Yet when the Universe was very young, it consisted almost entirely of H_2 and He, with far less than 2% of heavy elements, so little in fact that terrestrial planets and icy–rocky bodies could not have formed anywhere, and perhaps even the formation of giant planets was inhibited, if kernel formation was an essential stage in the process.

The essential enrichment of the Universe in heavy elements has resulted largely from the death of stars at least several times the mass of the Sun. After their main sequence lifetimes stars support further thermonuclear reactions that build heavy elements. The more massive the star the greater the variety and quantity of elements produced. But these elements need to reach the interstellar medium so that they could have been incorporated in the solar nebula, and in other stellar nebulae that would give rise to other planetary systems. They reach the interstellar medium through huge explosions, called supernovae, that terminate the lives of massive stars (Chapter 7). During these explosions further heavy elements are produced. Massive stars have short lives, measured in a few million years. Therefore, the Universe was soon enriched in heavy elements. When the Sun was born, 4600 million years ago, the Universe was about two-thirds of its present age, and the interstellar cloud that gave birth to it had consequently been enriched by many generations of massive stars.

1.3 BEYOND THE SOLAR SYSTEM

Beyond the Solar System we look to other planetary systems – to *exoplanetary systems* – as potential abodes of life. A significant number of exoplanetary systems is known, and we shall examine them in Chapter 11. More remain to be discovered. A star lies at the heart of each system, so here we will look briefly at the stars and at their distribution in space.

1.3.1 The stars

The stars as bodies

The stars are like the Sun in that they start their main sequence lives with compositions dominated by H and He in the approximate mass ratio 3 : 1, and with thermonuclear fusion of hydrogen in their cores sustaining their luminosity. The predominance of hydrogen and helium is because these elements dominate the whole Universe, including the interstellar clouds from which stars are born.

At the start of their main sequence lives, stars differ from each other in two main ways. First, they differ slightly in composition. There are slight differences in the H : He ratio, with much larger differences in the proportion of heavy elements. This proportion, called the *metallicity* of the star, can be as low as 0.05 of one percent to as high as a few percent. The Sun's metallicity of 2% is thus fairly high – it must have been born from a comparatively well-endowed interstellar cloud. The metallicity has an influence on the effective temperature and luminosity, and on the evolution of the star. Second, stars also differ in mass. Measured masses range from about 100 times the solar mass ($M_\odot$) down to 0.08 $M_\odot$. The mass has a huge effect on the effective temperature, luminosity, and evolution of the star, as discussed in Chapter 8. Luminous objects with masses less than 0.08 $M_\odot$ exist, but the interior temperatures never become high enough for sustained thermonuclear reactions, so they are not classified as stars. Such objects are called brown dwarfs.

The evolution of a star is also greatly influenced if it has a close stellar companion, within a few stellar radii. Such close binaries, as they are called, are fairly rare, but other binary systems are much less rare, including those that are still sufficiently close to be seen as a single point with the unaided eye. Less frequent are systems in which there are three or more stars. Overall, a high proportion of stars are in binary, triple, or higher multiple systems, in which the stars are in orbit around each other. In the solar neighbourhood the proportion is about 70%, but in any case, multiple systems account for over half of the points of light you see in the sky.

The stars in space

The nearest star, Proxima Centauri, is 2.67×10^5 AU from the Sun. This is a huge distance even compared to the 40 AU orbit of the outermost planet Pluto, though if the Oort cloud extends to about 10^5 AU then this is much of the way there. Figure 1.10 shows the distances to the Sun's 12 nearest neighbours, in *light years*. This is the distance travelled by light (or by any other electromagnetic radiation) in a vacuum in one year (near enough the distance travelled through space). One light year (ly) is 9.46×10^{15} m, which is 6.32×10^4 AU. Proxima Centauri, is thus 4.22 ly from the Sun. Another measure of distance is the *parsec* (pc). This is the distance to an object that appears to shift in position by one *arcsec* (= second of arc, 1/3600°) when our viewpoint shifts by 1 AU, as illustrated in Figure 1.11. This apparent shift is called the trigonometric parallax. One parsec = 3.26 ly.

Note that there are several multiple star systems in Figure 1.10. These are the triple system Proxima Centauri/Alpha Centauri A/Alpha Centauri B, and the binary systems Sirius A/B, and L726-8 A/B. Only one of the stars in Figure 1.10 is as yet known to have a planetary system – Epsilon Eridani.

If the space density of star systems were independent of distance, then the number of stars lying within a sphere centred on the Sun would be proportional to the volume of the sphere. Therefore the number within a distance d of the Sun

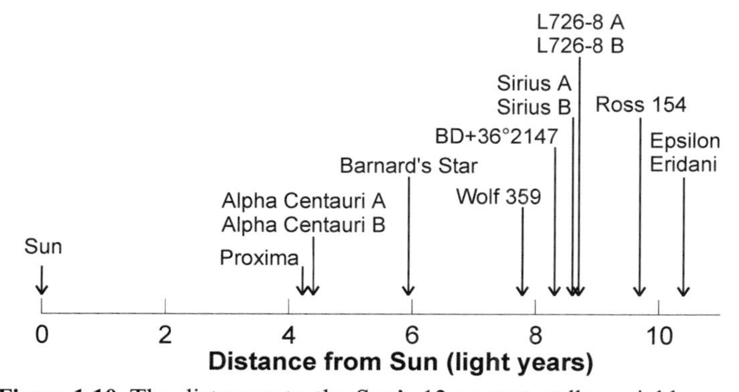

Figure 1.10 The distances to the Sun's 12 nearest stellar neighbours.

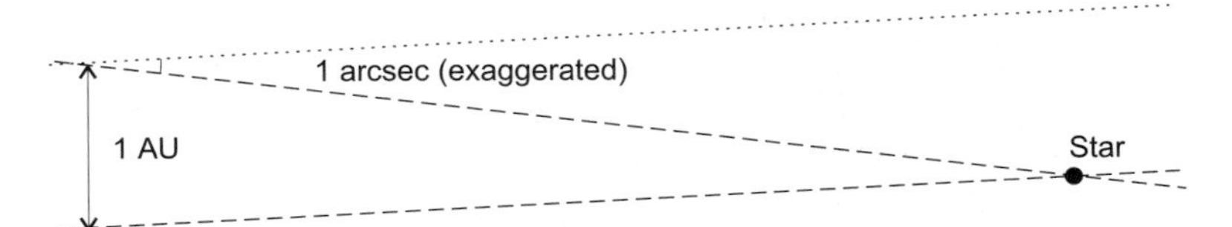

Figure 1.11 The definition of the parsec, as the distance to an object whose trigonometric parallax is one arcsec on a baseline of 1 AU.

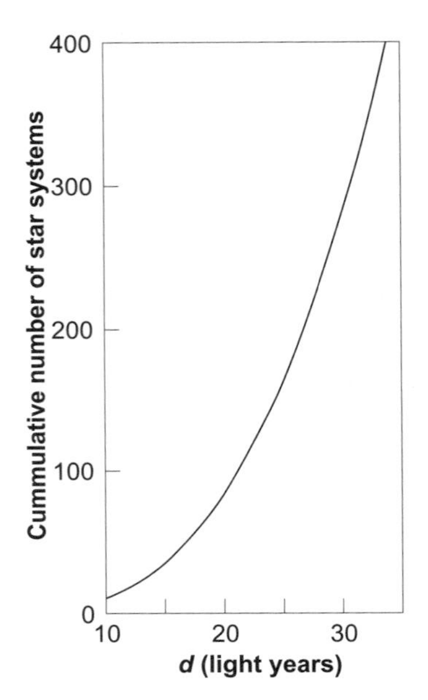

Figure 1.12 The number of star systems that would lie within various distances d from the Sun if the space density were independent of distance. This is approximately the case out to a few hundred light years.

would be proportional to d^3, as shown in Figure 1.12. The observed numbers are not greatly different from those shown, and so the space density of star systems in our cosmic neighbourhood does not change much with distance, at least up to 30 ly, and in fact to a few hundred light years. There is also not much concentration in particular directions. But if we look beyond several hundred light years, we would notice that in some directions the number of stars is less than in other directions. This is because we are then encountering the structure of our Galaxy.

1.3.2 The Galaxy (and others)

The Sun is one of about two hundred thousand million stars that make up the *Galaxy*. From extensive observations made from Earth it is clear that this vast

Figure 1.13 A face-on view of a spiral galaxy rather like ours. This is NGC1232.
European Southern Observatory.

assemblage of stars has a form that, face-on is something like that in Figure 1.13. The stars, and tenuous interstellar gas and dust, are concentrated into a disc highlighted by spiral arms. In our Galaxy the disc is about 100 000 ly in diameter and most stars are in a thin sheet about 1000 ly thick – roughly the same ratio of diameter to thickness as a compact disc. This sheet is called the thin disc. It is enclosed in a thick disc about 4000 ly thick. The spiral arms are delineated by a high space density of particularly luminous stars and luminous interstellar clouds. Elsewhere in the disc the space density of the stars and interstellar clouds is no less – it is just that they are not as bright. At its centre the disc has a bulge called the nuclear bulge, also full of stars and interstellar matter. The bulge is shown circular in the face-on view in Figure 1.13, though it might be slightly elongated into a bar. The disc is enveloped in the halo (not visible), a roughly spherical volume in which interstellar matter is particularly tenuous and the space density of stars is low. About 1% of the halo stars and some in the bulge are concentrated into globular clusters (Figure 1.14), each one containing the order of a million stars. There are clusters in the

Figure 1.14 A globular cluster of several hundred thousand stars. This is M80, about 28 000 ly from the Sun.

The Hubble Heritage Team (AURA/STScI/NASA).

disc too, but the great majority are much more open and irregular in structure, and contain the order of 1000 stars – these are called open clusters.

The Sun is located near the edge of a spiral arm, roughly half way from the centre of the Galaxy to the edge of the disc. Nearly all the exoplanetary systems so far discovered lie within a few hundred light years of the Sun, in our cosmic backyard, though much of the region on our side of the nuclear bulge is accessible to our instruments, or to those we will soon have.

Beyond our Galaxy there are many more. We have available the observable Universe. This is a sphere centred on us with a radius equal to the speed of light times the age of the Universe, 13 700 million years, so the sphere has a radius of 13 700 million light years. We cannot see any further because there has not yet been enough time for light to reach us from such distances. But within the observable Universe there are hundreds of billions of galaxies, so there is plenty of space to search for extraterrestrial life.

1.4 SUMMARY

- The Sun is a huge body that had an initial composition by mass of 71% hydrogen, 27% helium, and 2% all the other chemical elements (the heavy elements). It is a main sequence star (i.e. sustained by the thermonuclear fusion of hydrogen in its core, where hydrogen is being converted into helium, mainly through the ppI cycle).

- The Earth is one of nine planets that orbit the Sun in roughly circular orbits, approximately in the same plane. The Earth is one of the four terrestrial planets, the others being Mercury, Venus, and Mars. They are characterised by a rocky/iron composition and locations in the inner Solar System.
- Of the remaining planets, Jupiter and Saturn are giants, dominated by hydrogen and helium. Uranus and Neptune are lesser giants, dominated by water. Pluto, a small world, is dominated by water ice plus rocky materials.
- The only place we know to harbour life in the Universe is our own Earth. Elsewhere, liquid water is a prime indicator of potential habitats. Mars is a good candidate, for surface life in the past, and perhaps for life beneath its surface today. The Galilean satellite Europa is the only other good candidate in the Solar System. This rocky world is covered by a shell of ice, but beneath it there may be a widespread ocean of liquid water.
- The Solar System is widely thought to have originated 4600 million years ago from the solar nebula. The terrestrial planets formed from the dust that settled to form a thin sheet in the nebular disc. The giant planets might have formed in two steps, with kernels of icy and rocky materials that became massive enough to capture nebula gas. Alternatively, they might have formed in one step, via gravitational instability in the disc. The terrestrial planets acquired volatiles through outgassing of the material that forms their bulk, and via volatile-rich planetesimals that arrived mainly during the heavy bombardment, 4600–3900 million years ago.
- The heavy elements in the Solar System, essential for terrestrial planets and for life, were produced by massive stars, particularly by the supernovae that terminate such stars' lives.
- Beyond the Solar System we look to exoplanetary systems for potential habitats. Nearly all the systems so far discovered lie in our cosmic neighbourhood. This occupies a position near the edge of one of the spiral arms of our Galaxy, about half way from the centre to the edge of the Galactic disc. Most of the Galaxy is thus unexplored for exoplanets, as are the hundreds of billions of other galaxies in the observable Universe.

1.5 QUESTIONS

Answers are given at the back of the book.

Question 1.1

Suppose that the Earth has its present semimajor axis, but an orbital eccentricity of 0.9.

(i) Calculate the perihelion and aphelion distances of this Earth.
(ii) Comment on the habitability of the surface of this Earth.

Question 1.2

By making reasonable suppositions about the orbital stability of an 'Earth' at 1 AU, compare the habitability of binary systems consisting of two solar-type stars, where the stars have various spacings, 0.1 AU, and 1 AU, 100 AU. What implications might your conclusions have for the habitability of globular clusters?

2

Life on Earth

The Earth is the only place where we know that life exists. We must use terrestrial organisms as an indicator of what it is we are looking for. This will guide us towards the extraterrestrial locations that have the greatest chance of being inhabited. In this chapter we look at life on Earth today. In the next chapter we consider how it has evolved from the earliest organisms, and how these earliest organisms originated. But first we look briefly at the Earth itself.

2.1 THE EARTH

2.1.1 The Earth's interior

The interior of the Earth is better known than that of any other planetary body. This is because of the wealth of gravitational, seismic, and other data. Figure 2.1 shows a segment of the Earth reaching down to its centre, where the temperature is about 4000°C. The main compositional division is between the iron-rich core and the silicate-rich mantle and crust. The core extends a little over half way to the Earth's surface, but accounts for only one-sixth of the volume. It is divided into an inner core that is solid, and an outer core that is liquid. The outer core is actually slightly cooler than the inner core, so it might surprise you that it, and not the inner core is liquid. This is partly because of the increase in pressure with depth, and partly because of a slight compositional difference – whereas the inner core is almost entirely made up of iron plus a few percent nickel, the outer core contains a few percent of additional constituents, perhaps iron sulphide (FeS) or even iron hydrides (FeH_x). The melting point of the outer core is lower for both reasons. The outer core is the source of the Earth's magnetic field. This is because it is an electrical conductor, a liquid, and in motion due to a combination of heat from the inner core plus the Earth's rotation.

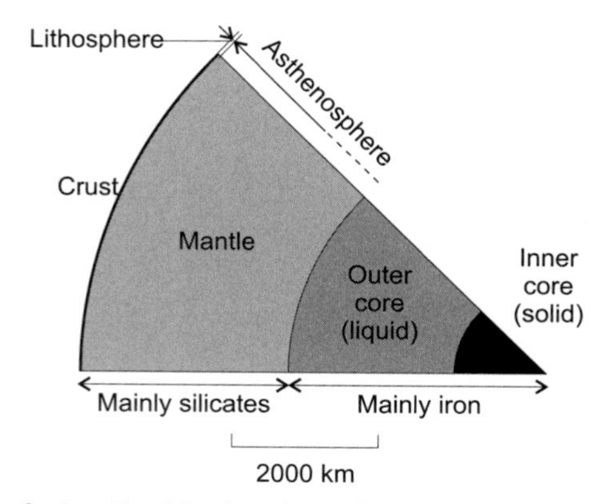

Figure 2.1 A model of the Earth's interior, showing the main divisions by chemical composition.

The Earth's interior was originally heated and partially melted by the gravitational energy released as planetesimals and embryos came together, and by the decay of short-lived radioisotopes, notably one of aluminium, ^{26}Al, perhaps provided by a supernova in the region of the Solar System's birth. Any downward separation (differentiation) of the iron core would have provided further heating. Subsequently, and in addition to residual primordial heat, long-lived radioisotopes, particularly of uranium (^{235}U, ^{238}U), potassium (^{40}K), and thorium (^{232}Th), have sustained the high internal temperatures. A source of heat that could be crucial to keeping the outer core moving, is the growth of the inner core at the expense of the outer core. Without this, the Earth might have no magnetic field. This would make the Earth's surface less habitable – the magnetic field deflects energetic charged particles from space.

Outside the core silicates predominate (compounds of one or more metallic elements with the abundant elements silicon and oxygen). The mantle reaches from the core almost to the Earth's surface, and is fairly homogeneous in composition. The upper mantle consists almost entirely of pyroxene ($(Ca,Fe,Mg)_2Si_2O_6$) and olivine ($(Mg,Fe)_2SiO_4$), and though in the lower mantle the crystalline form is different, the chemical mix is much the same. At a typical depth of about 10 km the crust is encountered, also dominated by silicates, but substantially different from those in the mantle. In particular, the crustal silicates are richer in calcium and aluminium, and poorer in iron and magnesium. The crust has been derived from the mantle by partial melting, and is less dense than the mantle.

2.1.2 The Earth's crust, lithosphere, and plate tectonics

The crust itself has a compositional division into oceanic and continental forms (Figure 2.2). As these names imply, they are predominantly found respectively under the oceans and making up the continents. However, it is only a coincidence that the volume of the oceans raises sea level approximately to where the division occurs. The mix of silicates that makes up oceanic crust is loosely referred to as

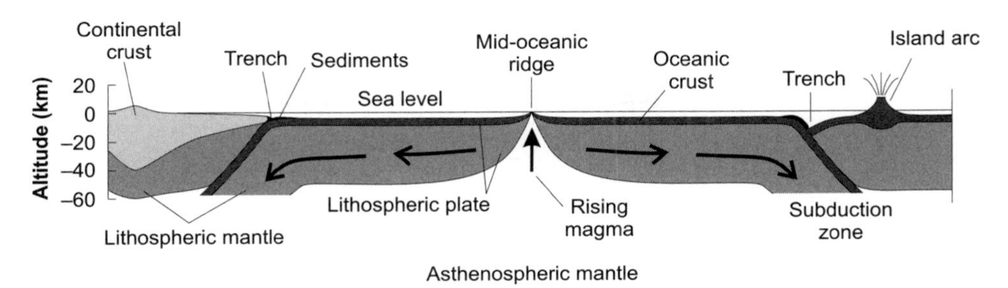

Figure 2.2 A section through the Earth's crust and upper mantle, showing the main features associated with plate tectonics.

basalt, or of basaltic composition. Overall, these silicates are richer in magnesium and iron than those that constitute continental crust, though they are still short of the proportions found in the mantle.

The partial melting that produces the crust and that promotes the creation of the two crustal types is driven by *plate tectonics*, briefly outlined in Section 1.1.3. The crust is fairly rigid, and so too is the upper few tens of kilometres of the mantle. Together, these constitute the lithosphere (Figure 2.2). The lithosphere is underlain by a much less rigid layer, extending from about 50 km downwards, perhaps to the core, called the asthenosphere. The lithosphere is divided into a few dozen plates, of various sizes, and these move over the asthenosphere. At some plate boundaries the plates are moving apart, with fresh oceanic crust appearing from partial melting of the mantle. At others they are just sliding sideways, and at others the plates push together, with one of them being reincorporated into the mantle, at what is called a subduction zone. If one of the plates at a subduction zone carries continental crust then the partial melting will create more continental crust. There are thus great cycles of plate motion, on timescales of tens of millions of years, where the continental crust is trundled around the globe, deforming, assembling, breaking up, and growing in volume. Volcanism, and other forms of geological activity, are thus maintained at a high level.

The immediate cause of plate motion is convection in the mantle. Most of the mantle is solid, so this is solid-state convection, facilitated by the high pressures and temperatures in the mantle, and by the plasticity of a partially molten zone at the top of the asthenosphere. Convection is sustained today by the heat from long-lived radioisotopes plus residual primordial heat. For the plates to move in response to convection it is essential that the continental crust is flexible, otherwise the plates would jam. It might also be necessary for water to be present in the crust – compounded in rocks, it lowers their melting points.

2.1.3 Atmosphere, oceans, and biosphere

The Earth's atmosphere has the composition shown in Table 2.1, where the number fraction of each component is given – this is the fraction of all the molecules in the

Table 2.1. The main constituents of the Earth's atmosphere.

	N_2	O_2	H_2O	Ar	CO_2
Number fraction	0.78	0.21	0.01 (mean)	0.0093	0.000 345
Partial pressure (Pa)	0.79×10^5	0.21×10^5	10^3	940	34.9

atmosphere contributed by the component. The mean pressure at sea level is 1.013×10^5 pascal (Pa), which is 1013 millibars, or 1.013 bars. This is the sum of the partial pressures of each component, also given in Table 2.1. The partial pressure is proportional to the number fraction. For example, the partial pressure of O_2 at sea level is $0.21 \times 1.013 \times 10^5$ Pa $= 0.21 \times 10^5$ Pa. Atmospheric pressure declines rapidly with altitude, so that at the modest altitude of 2 km it is already as low as 7.95×10^4 Pa, and at 20 km it is only 5.5×10^3 Pa – for all its importance to life the atmosphere is just a thin veneer. The oceans are intimately coupled with the atmosphere, chemically and physically, and account for most of the water above the base of the crust. There are traces in the mantle, that, because of the mantle's huge volume, could exceed the quantity in the oceans.

The global mean surface temperature (GMST) is 15°C, and most of the time in most places it is between 0°C and 35°C. The GMST is the outcome of the balance between the energy gains by the surface of a planet, and its energy losses. If you imagine the energy gains being switched on then the temperature of the surface rises until the losses balance the gains. The losses consist of infrared radiation emitted by the surface by virtue of its temperature, plus the heat convected away through the atmosphere. The gains consist of the absorbed fraction of solar radiation (the fraction not reflected back to space by the surface, clouds, and atmosphere), plus the absorbed fraction of the infrared radiation emitted by the atmosphere. The GMST would be lower but for the *greenhouse effect*. This is the name given to the phenomenon whereby the surface temperature of a planet is raised because the atmosphere absorbs some of the infrared radiation emitted by the surface and reradiates a proportion back to the surface. Water vapour and CO_2 are responsible for nearly all of the Earth's greenhouse effect, without which the GMST would be −18°C, and much or all of the surface would be uninhabitable.

The atmosphere also shields life on Earth from damaging ultraviolet (UV) radiation from the Sun. This it does through the presence of a trace of ozone (O_3) high in the atmosphere. The ozone is derived from O_2 by the action of UV radiation.

Life on Earth constitutes the *biosphere*, the assemblage of all things living and their remains. Plate 14 shows a typical scene of life on Earth. It is neither herds on the African plains, nor a rain forest, nor an ocean teeming with fish, but single celled creatures, unicellular creatures, 1–100 μm (10^{-6}–10^{-4} m) across. The *cell* is the basic unit of life, an enclosed environment within which the processes of life are conducted. The membrane that encloses the cell regulates the exchange of substances between the cell and its environment. There is a great variety of unicellular creatures, with bacteria as a very large group. One type of bacterium is shown in Plate 14, which also shows a group of non-bacterial single cells. There are also multicellular creatures – plants, and animals being particularly familiar. The human body contains about 10^{13} cells, aggregated into various organs – heart, liver, and so on.

Although there are different types of cell, all contain the same sorts of chemical compounds. We shall first outline what sort of compounds these are, then go on to consider how they are organised within cells, and then we shall look at the fundamental processes of life.

2.2 THE CHEMICALS OF LIFE

Table 2.2 shows the composition of a typical mammalian cell in terms of the broad types of chemical compounds it contains. Though cells differ somewhat in the proportions of their different compounds, Table 2.2 gives the correct impression that any living cell consists mainly of water, with proteins as another prominent component. Except for water and inorganic ions, the cell is made up of organic compounds. These are carbon compounds containing hydrogen. They were originally named because of their association with life, though many organic compounds are abiological. Life on Earth is characterised by a special selection of organic compounds, mostly with large complex molecules, and by water. Moreover, the water has to be liquid over at least part of the cell's life cycle, and therefore *carbon–liquid water life* is an apt name for life on Earth. Water is needed as a solvent, a transport medium, and as a participant in biochemical reactions.

In terms of chemical elements, life is dominated by C, H, O, and N. In an organism the number fractions have representative values 63% H, 28% O, 7% C, and 2% N. At significant fractions of a percent are calcium and phosphorus, and as traces there are many other elements.

2.2.1 Proteins and nucleic acids

Proteins are large, complex organic compounds. About 100 000 different proteins have been identified in terrestrial organisms, and between them they fulfil a wide range of functions – structural, transport, storage, and catalysis of biochemical reactions. They are made up of many units, where each unit is one of about 20 amino acids. Figure 2.3(a) shows the general molecular plan of an *amino acid*. They differ in the nature of 'R'. The simplest is glycine, in which 'R' is a hydrogen atom H. Figure 2.3(b) illustrates the joining of just two amino acids. In

Table 2.2. Chemical composition of a typical mammalian cell.

	Water	Proteins	Lipids	Polysaccharides	Nucleic acids	Small organic molecules	Inorganic ions
Mass (% of total)	70	18	5	2	1	3	1

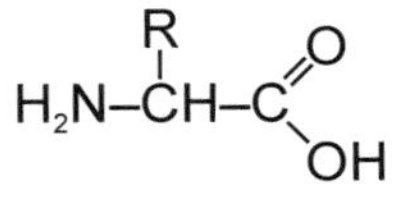

$$H_2N\text{–}\overset{R}{\overset{|}{C}}H\text{–}C\begin{smallmatrix}\nearrow O\\ \searrow OH\end{smallmatrix}$$

(a)

$$H_2N\text{–}\overset{R_1}{\overset{|}{C}}H\text{–}\overset{O}{\overset{\|}{C}}\text{–}\underset{\underset{H}{|}}{N}\text{–}\overset{R_2}{\overset{|}{C}}H\text{–}C\begin{smallmatrix}\nearrow O\\ \searrow OH\end{smallmatrix}$$

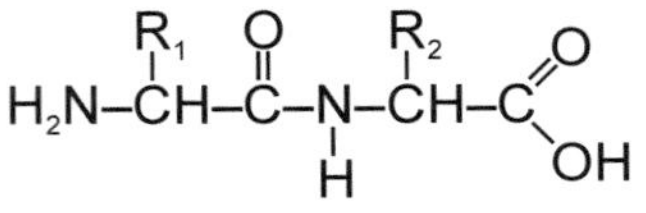

(b)

Figure 2.3 (a) Amino acids, the building blocks of proteins. They differ in the nature of 'R'. (b) Two amino acids joined together.

the joining, a molecule of water would have been liberated. (Note that models like that in Figure 2.3 label the locations of the constituent atoms, and show which atoms are bonded together. The actual atoms are large enough to touch each other, and the bond directions are not necessarily all in the same plane.) A *protein* consists of about 50–1000 amino acids joined together in a string. In some proteins several strings are intertwined but they remain string-like; these are called fibrous proteins, and their function is structural. In other proteins a single string is wound to form a roughly spherical shape. Many of these catalyse biochemical reactions. Such catalysts are called *enzymes*, and without them biochemical reactions would occur far too slowly to sustain life, or would not outpace abiological reaction rates sufficiently to allow life to find a niche. Nearly all enzymes are proteins. The precise geometrical shape of an enzyme determines which biochemical reactions it catalyses. This shape in turn depends on the particular sequence of amino acids that make up the protein. Note that with an 'alphabet' of about 20 letters and a 'word' that is 50–1000 'letters' long, the 100 000 different biological proteins are a *very* small fraction of those that could in principle exist.

The *nucleic acids* are also large, complex organic compounds, and comprise *ribonucleic acid* (*RNA*), and *deoxyribonucleic acid* (*DNA*). These are at the heart of protein synthesis, and are central to the processes by which organisms reproduce themselves. The basic unit of RNA is called a *nucleotide*, and there are four types. Each consists of a molecule of a sugar (ribose), a phosphate group (which contains the element phosphorus), and one of four different organic molecules called bases, as shown in Figure 2.4(a). The base labels A, C, G, and U stand for adenine, cytosine, guanine, and uracyl. Each base is a compound of C, H, and N, and, in some cases, O. To make RNA the units can be joined in any order, as illustrated in Figure 2.4(b). RNA consists of hundreds to tens of thousands of nucleotides in a string, and so a huge variety of RNA sequences is possible. The string might also be folded.

DNA also consists of four nucleotides. These are the same as in RNA except

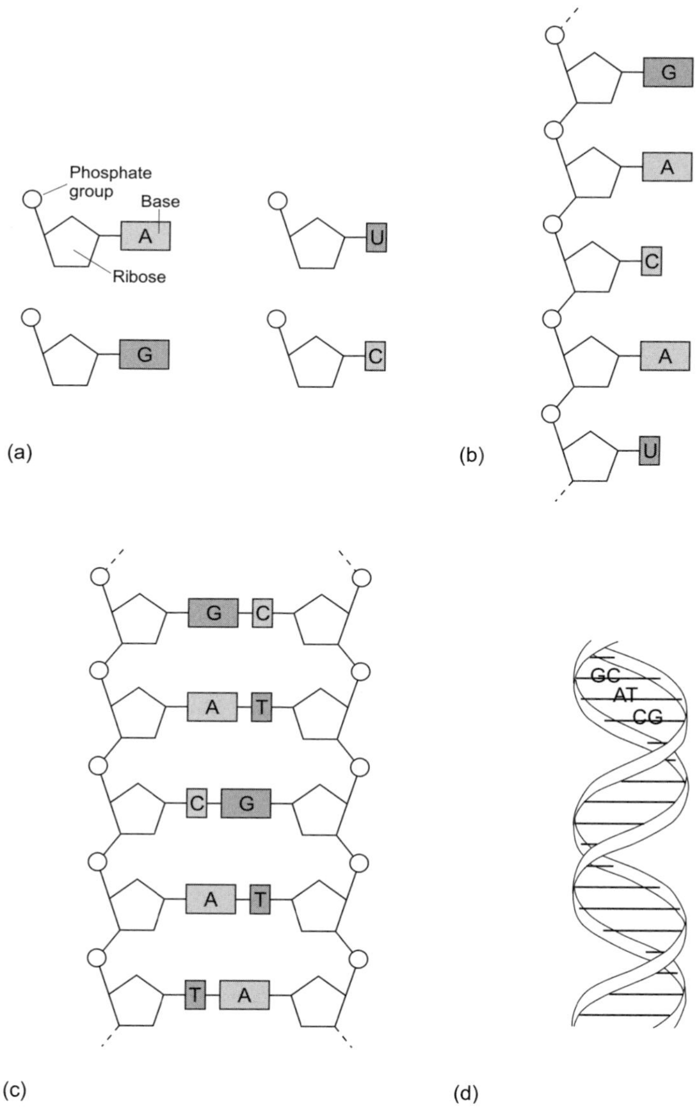

(a)

(b)

(c)

(d)

Figure 2.4 The four basic units (nucleotides) of RNA. (b) A short RNA sequence. (c) A short segment of a DNA molecule. (d) The double helix of DNA.

that uracyl is substituted by another base, called thymine (T). Another difference is that in the spine the sugar is deoxyribose rather than ribose. A third difference is illustrated in Figure 2.4(c). There are two strands joined by the bases to form a 'ladder' with 'rungs'. Each rung is either A–T (or T–A), or C–G (or G–C). The rungs can occur in any order, and so, with thousands of rungs, there is a huge number of possible base-pair sequences in a molecule of DNA. The ladder is twisted to form the famous double helix structure of DNA (Figure 2.4(d)). The double helix is not straight but curled up, to give a complex three dimensional structure.

2.2.2 Polysaccharides, lipids, and small molecules

Though polysaccharides are another class of large molecules, they are somewhat simpler than proteins and nucleic acids in that each molecule consists either of hundreds of identical units, or hundreds of just a few different units. Polysaccharide means 'many sugars', because each unit is a sugar molecule. There are various sugars, but all of them contain C, H, and O, with common structural features in the molecules that distinguish them from other C, H, and O compounds. Some sugars also contain other elements, such as N, P, and S. The simplest sugar is glucose, with the chemical formula $C_6H_{12}O_6$. Other well known sugars are fructose and sucrose. The term carbohydrate is in common use – this term embraces sugars and polysaccharides. Sugars are far more soluble in water than are most polysaccharides.

In joining sugars together to form a polysaccharide (with the ejection of a water molecule per pair of sugars joined), a great variety of outcomes is possible because of the variety of sugars and the various ways of linking them together. Cellulose, the rigid supporting framework of the cell walls in all plants, is a single string of glucose molecules, whereas starch, the major energy store of plants, though again consisting only of glucose units, is branched in most of its forms.

Unlike proteins, nucleic acids, and polysaccharides, a lipid is not a polymer but a relatively small molecule. For example, the lipid glyceril tristearate has the semistructural formula $(C_{17}H_{35}\text{–}COOCH_2)\text{–}(C_{17}H_{35}COOCH)\text{–}(C_{17}H_{35}\text{–}COOCH_2)$, small by biological standards! The simplest lipids are the fats and oils, the distinction being that oils are liquid at room temperatures whereas fats are solid. Care has to be taken to distinguish the lipids from mineral oils. The latter include compounds of hydrogen and carbon (hydrocarbons), such as paraffin oil, and are not found in cells. Lipids and mineral oils share the property of insolubility in water.

The small organic molecules (Table 2.2) include *adenosine triphosphate* (*ATP*) and *adenosine diphosphate* (*ADP*), where P represents a phosphate group. ATP and ADP play a central role in energy storage and transfer in the cell. There are also many other organic molecules present in small quantities, some of them occurring only in certain types of cell.

Our inventory of chemical compounds in the cell is completed by inorganic ions. Simple inorganic ions result when common salt is dissolved in water. A proportion of the NaCl molecules dissociate to give the positive ion Na^+ and the negative ion Cl^-. There are many other types of ion present.

2.3 THE CELL

The chemicals of life are contained within cells. All life forms on Earth consist of one or more cells. The cell is bounded by a membrane that divides the environment

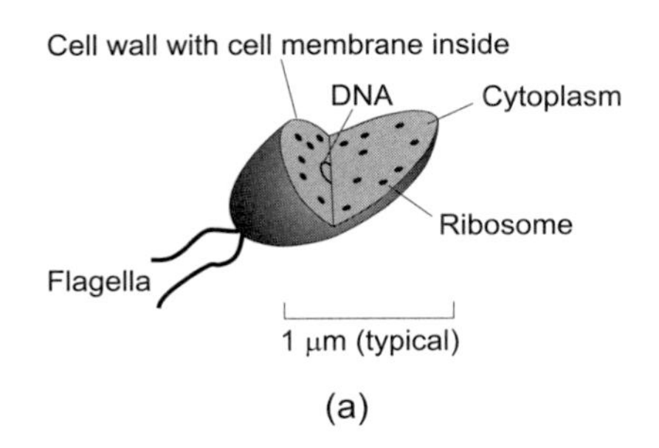

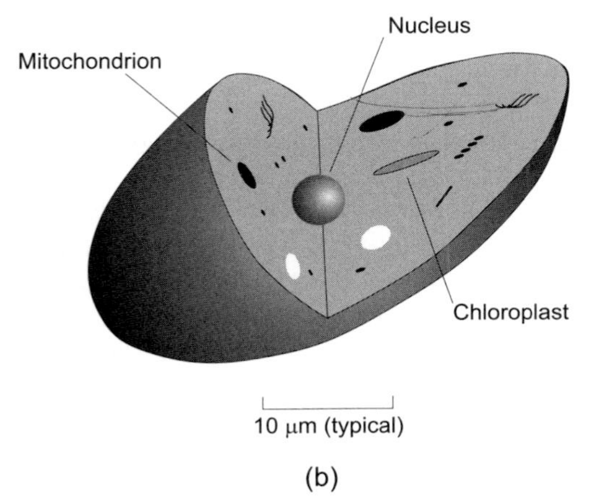

Figure 2.5 The essential components of (a) a prokaryotic cell (b) a eukaryotic cell.
Adapted from Figure 12.4 in *Earth: evolution of a habitable world* by J. I. Lunine, CUP, 1999.

within the cell from the environment outside it, and regulates the transfer of substances between the two. There are two basic types of cell, the prokaryotic cell and the eukaryotic cell. In unicellular organisms the cell can be of either type, whereas almost all multicellular organisms consist of eukaryotic cells. Few prokaryotes are multicellular.

Figure 2.5(a) shows the essential components of the *prokaryotic cell.* The membrane consists of proteins and a class of lipids called phospholipids, and little else. Outside the membrane the great majority of prokaryotes have a cell wall that provides a measure of rigidity. This wall consists of strings of 20–40 amino acids (too short to be classified as proteins) and polysaccharides. The membrane encloses cytosol, a saltwater medium containing proteins. Floating in the cytosol is DNA and small particles called ribosomes that contain RNA. Most prokaryotic cells can move by means of protein strands attached outside the cell wall – these are called flagella.

Figure 2.5(b) shows the essential components of a *eukaryotic cell.* It is typically much larger than the prokaryotic cell, 10–100 μm instead of about 1 μm. It is also much more complex. Of particular note are the various structures inside the cell called organelles. One such is the nucleus, where much (but not all) of the cell's DNA is housed. Another is the mitochondrion, inside which cell respiration occurs (Section 2.4.2). In green plants there are chloroplasts, which are the sites of photosynthesis (Section 2.4.2). Mitochondria and chloroplasts also contain DNA. There are various other organelles and other structures that will not concern us.

In multicellular creatures different cells take on different functions. For example, in animals a nerve cell carries out a different function from a muscle cell or a liver cell. This specialisation comes about during the growth of the organism from a single cell. Most plants cells have a rigid cell wall made of cellulose fibres held together by a glue. By contrast cells in animals have no wall and generally change shape readily.

2.4 THE FUNDAMENTAL PROCESSES OF LIFE

Any living organism on Earth is involved in three processes. First, biosynthesis, in which small organic molecules are constructed, and then combined to form the complex molecules that make up the cell. Second, reproduction, in which cells make copies of themselves, so that life is sustained from one generation to the next. Third, catabolism, in which molecules are broken down into smaller ones.

Each of these processes requires energy transfer. In this respect life is like every other process in the Universe – nothing can happen unless energy is transferred from one place to another, often changing from one form to another. Life also requires the availability of the chemical elements that constitute the various compounds in Table 2.2, but it is energy that reorganises these elements.

2.4.1 Chemical energy

Organisms utilise chemical energy. This is energy that is either stored or released when the configuration of electrons in atoms or molecules is changed. Such changes occur in chemical reactions. A simple chemical reaction is:

$$\mathrm{O} + \mathrm{H} \rightarrow \mathrm{OH} \tag{2.1}$$

In this reaction the electron in the hydrogen atom on the left becomes shared between the two atoms in OH. Overall, the hydrogen has lost sole possession of its electron and the oxygen has gained. The hydrogen is an electron donor, and the oxygen is an electron acceptor. This is an example of a *redox reaction*. These reactions are particularly important because they involve a lot of energy. In Reaction (2.1), the electron transfer lowers the energy of the electron and this energy is released in the form of kinetic energy (energy of motion) of the OH and infrared radiation. A reaction such as this with a net release of energy is called exothermic. Note that a third body needs to be involved in the reaction, to carry away some of the vibrational energy of the newly formed OH, or it will at once dissociate. The reverse reaction:

$$\mathrm{OH} \rightarrow \mathrm{O} + \mathrm{H} \tag{2.2}$$

requires energy to be given to the OH to raise the electron energy in the products O and H. A reaction such as this that results in an increase in the energy stored is called endothermic.

The name 'redox' is a contraction of 'reduction–oxidation'. If an atom or molecule gains electrons it is said to be reduced, with the process of gaining an electron called reduction. If it loses electrons it is said to be oxidised, and the process of loss is called oxidation. Thus, in a chemical reaction involving electron transfer, reduction and oxidation both occur, and hence it is called a redox reaction. In Reaction (2.1) the hydrogen is oxidised and the oxygen is reduced. For the substances involved in redox reactions the following terms are equivalent: electron donor = reducing agent = fuel; electron acceptor = oxidising agent = oxidant.

The terms oxidation and reduction arose many years ago to denote, respectively, the addition of oxygen and the removal of oxygen, but the terms have now been broadened and expressed in terms of electron transfer. There is certainly no need for oxygen to be involved. Thus the reaction:

$$\mathrm{Na} + \mathrm{Cl} \rightarrow \mathrm{NaCl} \tag{2.3}$$

is a redox reaction in which sodium shares an electron with chlorine to form a molecule of common salt.

2.4.2 Energy for the cell

Energy stores and energy release

There is a large variety of ways in which organisms store and release energy, some more widespread than others. First consider how energy is released from its store in the cell. Energy is stored in the form of sugars, polysaccharides, certain lipids, and, rarely, proteins. Energy is derived from these molecules when they are converted to smaller molecules (catabolism). Central to the supply of a cell's energy is ATP (Section 2.2.2). An ATP molecule gives up energy when it is converted to the closely related molecule ADP. Conversely, when energy is given to ADP it turns into ATP. Note that ATP is only a temporary store of energy – it acts as a molecular link between energy released by the breakdown of molecules in a store, and energy required for some process. The energy required to convert one molecule of ADP to ATP – 5.1×10^{-20} joules (J) – is a useful unit of energy currency in a cell, and this can be expressed as the number of molecules of ATP produced.

In almost every organism catabolism happens through a process called respiration. Aerobic respiration uses oxygen, the second most abundant component of the Earth's atmosphere, and also present in the Earth's rivers, lakes, and oceans. Most eukaryotes use oxygen in a chemical reaction that converts the organic materials in the energy store to water and carbon dioxide, releasing energy that forms ATP from ADP. In the case of a store of glucose, a large series of chemical reactions can be summarised as:

$$C_6H_{12}O_6 + 6O_2 \rightarrow 6CO_2 + 6H_2O \tag{2.4}$$

This is a redox reaction in which the fuel is glucose and the oxidant is oxygen.

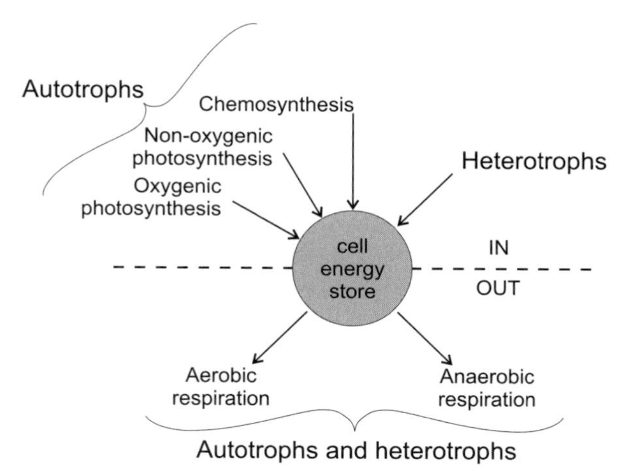

Figure 2.6 Methods of energy storage and release in cells.

There is an alternative to aerobic respiration, and it is essential for organisms intolerant of oxygen. This is called (naturally) anaerobic respiration and it is carried out today in various ways by a variety of unicellular organisms. A familiar example is the fermentation that produces alcohol through the activity of yeasts, which are colonies of unicellular eukaryotes belonging to the kingdom of fungi. Fermentation results in the formation of ATP from ADP through the energy released in the conversion of glucose into ethanol ('alcohol') and carbon dioxide. The overall effect of a number of stages is:

$$C_6H_{12}O_6 \rightarrow 2CO_2 + 2C_2H_5OH \tag{2.5}$$

However, per molecule of glucose, only 2 ATP molecules are produced, whereas with aerobic respiration 36 ATP molecules are produced.

Organisms that require oxygen in some way are called aerobes. The rest are anaerobes, and they divide into those for which oxygen is toxic and those that will use aerobic respiration when oxygen is available. Figure 2.6 displays these two forms of respiration. This Figure also summarises various ways in which energy stores in cells are created, to which we now turn.

Creating energy stores

Where do the sugars, polysaccharides, lipids, and, rarely, proteins, that constitute energy stores, come from? For some unicellular organisms, and many multicellular organisms including animals, the answer is 'food' (i.e. organic materials from other organisms, ingested or absorbed, though there might be some processing preceding storage). Humans are particularly omnivorous, obtaining organic matter from a great variety of sources. Organisms that rely on such ready-made organic material are called heterotrophs ('other-feeders'). Others synthesise them from inorganic compounds. These are the autotrophs ('self-feeders') and they lie at the base of the food chains.

Green plants are autotrophs that manufacture organic compounds through the process of *photosynthesis*, in which CO_2 and water are used to make glucose. The CO_2 is the source of carbon. Though the process is complicated, involving several stages, the overall effect in green plants can be summarised as:

$$6CO_2 + 6H_2O \rightarrow C_6H_{12}O_6 + 6O_2 \tag{2.6}$$

where the CO_2 has been reduced to form glucose, the electrons coming from H_2O which is oxidised to form oxygen. Oxygen is a by-product, so this is called *oxygenic photosynthesis*. The products (on the right-hand side of the reaction) have more energy than the reactants (on the left-hand side), and therefore some energy source is needed to promote this reaction. This is solar radiation.

Solar radiation, or any other form of electromagnetic radiation, travels through space like a wave, with a wavelength λ and a frequency f, linked by the equation $c = f\lambda$, where c is the speed of the wave, in this case the speed of light. But when electromagnetic radiation interacts with matter, it does so as if it is a stream of particles called *photons*. The energy e of a photon is proportional to the frequency of the wave, and so:

$$e = hf = h\frac{c}{\lambda} \tag{2.7}$$

where h is a universal constant called Planck's constant (see physics texts in Resources). In green plants a molecule called chlorophyll absorbs two photons in the wavelength range 0.4–0.7 μm, and this initiates a complex sequence of events. An important intermediate stage is the conversion of ADP to ATP and the production of a substance called $NADPH_2$ (details of which will not concern us). These two substances are then used to produce glucose and thus build up the energy store. Other substances are produced too, such as amino acids.

Some unicellular organisms also photosynthesise broadly in the above manner. Some prokaryotes do it differently. They perform photosynthesis without generating oxygen by using molecules such as hydrogen sulphide (H_2S) as the electron donor in place of H_2O. This particular example can be summarised schematically as follows:

$$2H_2S + CO_2 + (C) \rightarrow 2\text{‘}CH_2O\text{’} + 2S \tag{2.8}$$

CH_2O is the simplest carbohydrate, and is shown in quotes because the actual carbohydrates produced are more complicated. This is less efficient than oxygenic photosynthesis in that it captures a smaller proportion of the energy of the available solar radiation, though it can be utilised by organisms that are intolerant of oxygen. Nevertheless, biosynthesis and respiration benefited greatly from the appearance of oxygen in the Earth's atmosphere, because it made more energy available.

Some autotrophic prokaryotes create energy stores using chemical reactions that do not involve photosynthesis. This is called *chemosynthesis*. In chemosynthesis the organism takes advantage of substances in its environment that can be made to react exothermically within the cell. The energy released is used to make ATP, etc. Even in clear water, at depths greater than only about 100 m, there is insufficient sunlight for photosynthesis and therefore only chemosynthesis is possible. The same is true in underground caves and crevices. Of particular importance, because of their energy yield, are redox reactions. One example among many is the pair of dissolved gases CO_2 and H_2 that emerge from volcanic vents on the ocean floor. At temperatures of about 400°C these gases remain intact, but at lower temperatures they react as follows:

$$CO_2 + 4H_2 \rightarrow CH_4 + 2H_2O \quad (2.9)$$

However, the reaction rate is extremely low, and so they still persist, allowing certain cells the opportunity to absorb them. Inside the cell the reaction is greatly speeded by the action of enzymes (Section 2.2.1). The overall effect is the production of a carbohydrate energy store 'CH_2O', again using CO_2 as the carbon source:

$$2CO_2 + 6H_2 \rightarrow \text{'}CH_2O\text{'} + CH_4 + 3H_2O \quad (2.10)$$

The appearance of CH_4 in this process gives it the name *methanogenesis*, and the organisms that perform it are called methanogens.

Figure 2.6 summarises these various ways in which energy stores are created in cells. Let's now consider the main processes for which this energy is required.

2.4.3 Protein synthesis

After water, and excluding cell walls in some organisms, proteins make the largest contribution to the mass of a cell (Table 2.2). They also dominate in the breadth of their functions, though only a very few protein molecules can make copies of themselves. Therefore, to form a protein something else is almost invariably needed, and that is the DNA that is found in a cell. The DNA contains the instructions for making all the cell proteins, and is therefore a 'blueprint' of the cell – it is where all the genetic information resides. Protein synthesis, in most cells, consumes more energy than any other biosynthetic process.

Consider the typical prokaryotic cell. There are two main steps in protein synthesis. First, part of the DNA molecule is disrupted by an enzyme that breaks the weak bonds between the base pairs that constitute the rungs on the ladder. A particular enzyme will break the bonds from a particular start site to a particular stop site. The sequence of bases between these extremities constitutes a *gene*. The enzyme builds RNA on just one of the two strands of disrupted DNA, as shown in Figure 2.7(a), using RNA bases floating in the cell. Note that a particular form of RNA is produced in which the base sequence mirrors that on the DNA strand in accord with the rules that A(DNA) pairs with U(RNA), C(DNA) with G(RNA), G(RNA) with C(RNA), and T(DNA) with A(RNA) (Figure 2.7(b)). When this assembly is complete the enzyme and the RNA leave the DNA, which reassembles. The RNA so formed is called messenger RNA (mRNA).

The second step is the assembly of a particular set of amino acids – a particular protein – corresponding to the particular mRNA base sequence. This assembly takes place within the ribosomes (Figure 2.7(a) and (b)). The bases strung out along the mRNA, taken three at a time, specify the sequence of amino acids in the protein, and thus specify the type of protein. Each triplet of bases is called a codon, and it will correspond to just one of the 20 or so different amino acids. A form of RNA called

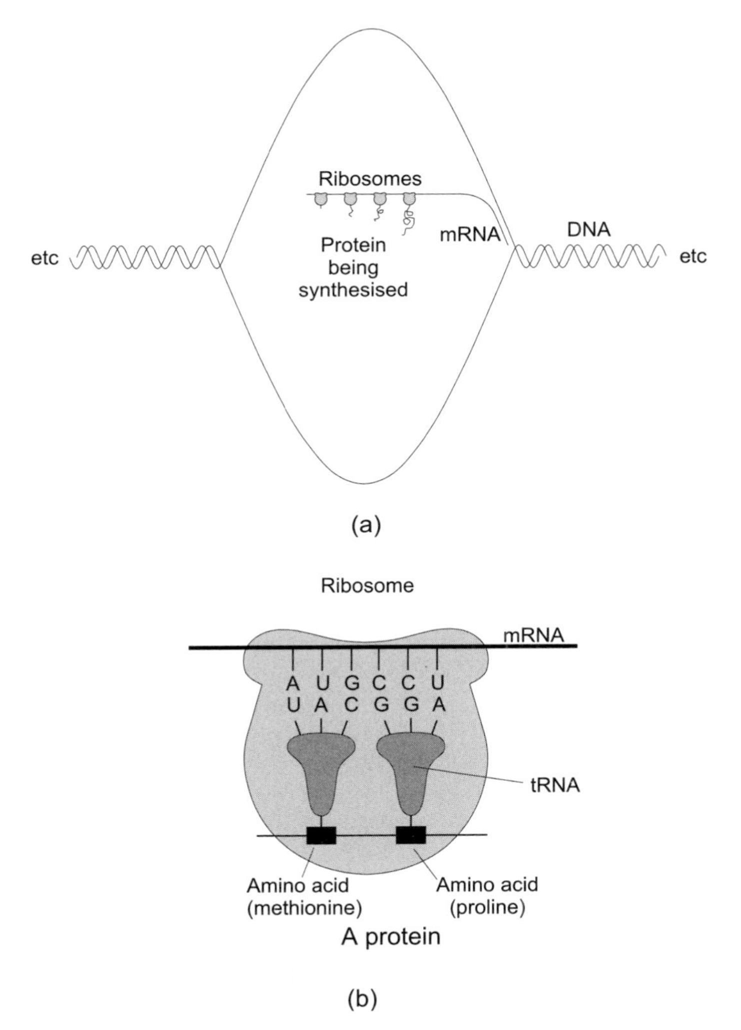

Figure 2.7 (a) A protein being formed by ribosomes, acting on instructions from mRNA that is derived from DNA. (b) A close-up of the action of a ribosome.

transfer RNA (tRNA) mediates between a codon on mRNA and an amino acid. There is one form of tRNA for each amino acid. During protein synthesis a ribosome contains a segment of the mRNA and up to two tRNA molecules that it has temporarily incorporated from the cytosol to match the mRNA segment currently inside it. The process starts with a tRNA molecule attaching itself to the mRNA at its first codon, and this tRNA attracts the corresponding amino acid floating in the cell. Another tRNA molecule then attaches to the mRNA at the next codon – a different form of tRNA unless the codon is the same as the first one. The corresponding amino acid is then attached to the first one. The first tRNA molecule then departs, a third arrives, and so on until the whole protein is synthesised.

The ribosomes act on the mRNA until the protein is complete. This whole process is repeated for other genes, until all the necessary proteins have been synthesised. A ribosome itself consists of proteins, and molecules of a third form of RNA called ribosomal RNA (rRNA). It might be the rRNA that catalyses the assembly, not the ribosomal proteins.

In eukaryotic cells the basic biochemical processes of protein synthesis involving DNA and RNA are broadly the same as in the prokaryotic cell, if rather more complicated. In heterotrophs, whether prokaryotes or eukaryotes, most of the amino acids have to come from food that traverses the cell membrane, though animals can make some of them (within the mitochondria – Section 2.3). In autotrophs they are synthesised within the cell.

2.4.4 Reproduction and evolution

All organisms have limited lifespans. It is therefore essential for the survival of a particular type of organism that it reproduces itself. In all organisms today the information for making each of its many and varied proteins is contained entirely within its DNA. Therefore, the information for making a copy of itself is also contained within the DNA. It is thus not surprising that the persistence of an organism from one generation to the next requires the passing on of DNA.

Figure 2.8 shows the main stages in prokaryotic cell division, in a highly simplified way. In stage 1 the DNA replicates in a manner to be outlined shortly. The cell then has two identical DNA molecules. Each DNA molecule migrates to

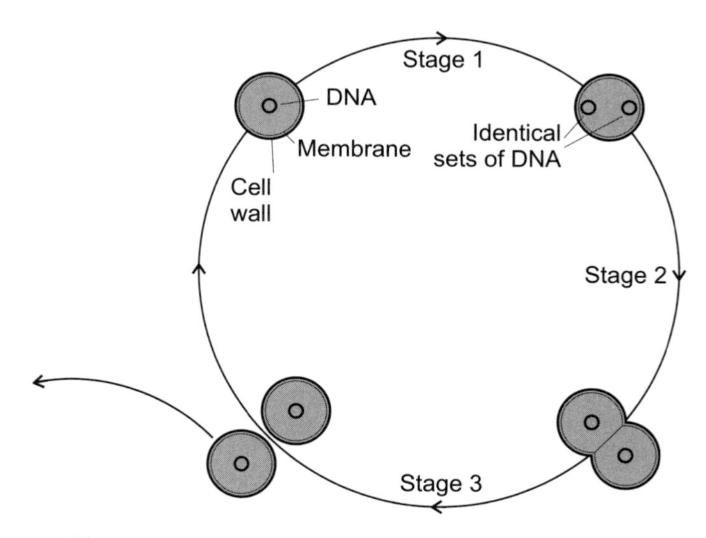

Figure 2.8 A simple view of cell division in prokaryotes.

opposite ends of the cell. Stage 2 is when the cell develops a 'waist' traversed by a new membrane, dividing the cell into halves. In the final stage the compound cell separates at the waist to yield two (nominally) identical cells. Each of these new cells can undergo further doubling.

DNA replication is triggered in ways that are beyond the scope of this book. The outcome is that certain enzymes gradually separate the DNA molecule into its two strands, as shown in Figure 2.9. This leaves two sets of exposed, unpaired bases. Each exposed base attracts its complementary nucleotides from the free nucleotides that are present in abundance in the cell. These free nucleotides will have been manufactured through enzyme action. To link them to form new DNA another enzyme is required. One complete DNA molecule thus builds up from each of the single strands of the original molecule. The base sequence is identical in both molecules.

The whole process of DNA copying is driven by enzymes, so we have enzymes causing DNA to be made, and the enzymes, being proteins, require DNA for their synthesis. When it comes to considering the origin of life in Chapter 3 we shall have to consider carefully this chicken and egg situation.

In cell division the copying of the DNA is not always perfect. The 'children' are therefore not identical to the 'parent'. If the 'children' are viable this will lead to an increase in variety. This can also result from a prokaryotic cell transferring some of its DNA to another prokaryotic cell, possibly of a substantially different type, with incorporation of the DNA into the recipient's DNA. This transfer occurs in a number of ways. Another source of variation is damage to the DNA caused by certain chemicals in the environment, by solar UV radiation, and by energetic charged particles and gamma radiation from radioactive materials on Earth and from space. Yet another cause is processes internal to the cell that cause segments of DNA to be moved from one location to another. We can group these DNA changes together as *mutations*.

Cell division also occurs in eukaryotic cells, though by different processes from that in prokaryotes. One process (called mitosis) enables unicellular eukaryotes to reproduce asexually, and enables multicellular organisms to grow and replace dysfunctional or lost cells. However, the process of reproducing a new organism can be different. In plants and animals this is usually by sexual reproduction, in which half the DNA for the new organism comes from special cells in the female, and the other half from special cells in the male (in animals, egg and sperm respectively). This means that the DNA in the offspring is not identical to that in either parent, though it is sufficiently similar that we recognise the offspring as belonging to the same type of organism as its parents – to the same species. The cell division that results in this mixing of DNA from two parents in eukaryotes is called meiosis. Sexual reproduction clearly promotes variation in the DNA of the offspring. Further variation in eukaryotes arises from DNA mutations.

Thus, in eukaryotes as well as in prokaryotes, DNA in any type of organism is not an invariable entity passed on from one generation to the next. In the great majority of cases the outcome of DNA changes has either no discernible effect, no important effect, or a damaging effect. But some descendents will be more suited to

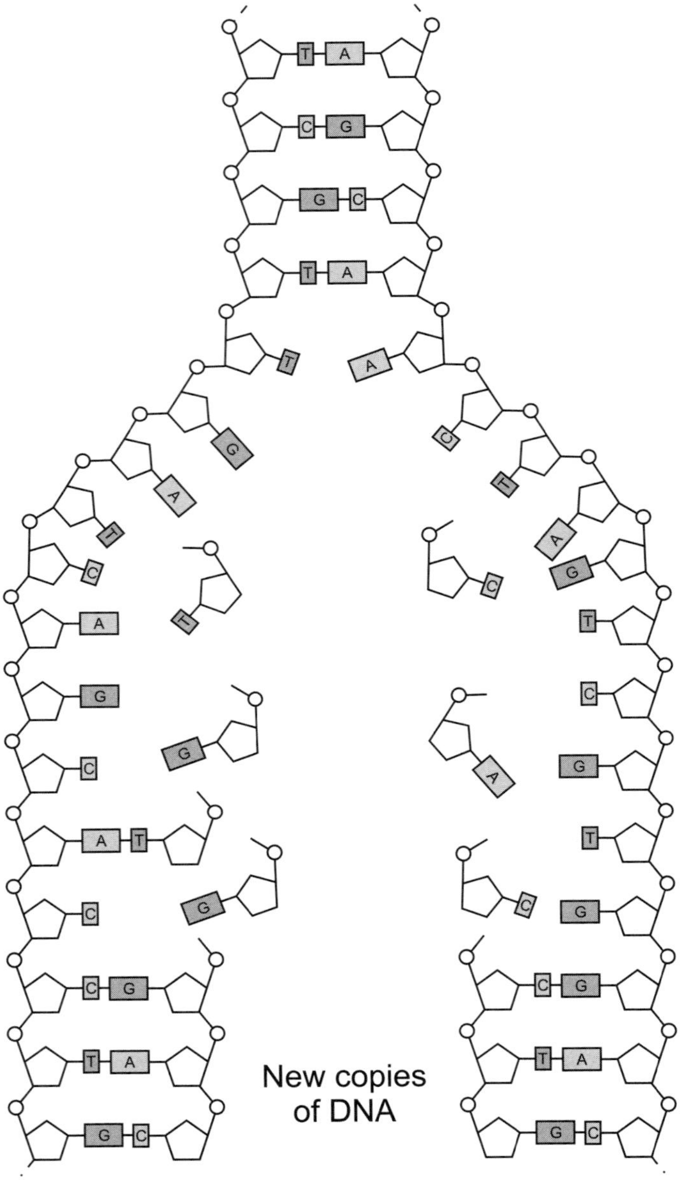

Figure 2.9 A simple view of DNA replication.

the environment than others, and it is these that will have a better chance of survival and reproduction. It is this that has enabled life to evolve from its earliest forms. It has driven evolution in a direction that has benefited the survival of species and the emergence of new, viable species. Evolution occurs because there is variety in the offspring. The inherited characteristics that promote survival to reproduce will therefore spread. This is the basis of Darwin's theory of *evolution by natural selection*, and it leads to a variety of species each adapted to its own environment.

2.5 DIVERSITY OF HABITATS

Life is ubiquitous on the Earth's surface and in the Earth's oceans. It is also present in the crust of the Earth, not just in underground caves and sediments but inside the rocks themselves. Life thus inhabits a wide diversity of habitats, though the most extreme of these are occupied mainly by prokaryotes. But in spite of this diversity, liquid water is common to all habitats. In order to survive, an organism requires access to water and to conditions under which this water can be liquid within its cells. No organism can survive without liquid water during at least part of its life cycle. For the great majority of organisms the water has to be available externally as liquid, rather than as ice or vapour. Consider non-extreme habitats first.

2.5.1 Non-extreme habitats

The most familiar non-extreme habitat is of course the Earth's surface, exposed to the Earth's oxygen-rich atmosphere (Table 2.1), where the mean pressure at sea level is 1.013×10^5 Pa and the global mean surface temperature is 15°C (most of the time, in most places, it is between 0°C and 35°C).

You or I, unprotected by clothing or dwelling places, could only survive under a very narrow range of conditions. Even if there were an ample supply of food and liquid water, we would die if exposed for a few days at any temperature outside the approximate range 5–45°C. Few animals or plants can live at sustained temperatures outside the range 0–45°C, and none at all outside the approximate range −20°C to 50°C. Other conditions to be met for survival by some organisms is the partial pressure of O_2 in the local atmosphere. For humans, a partial pressure of about 0.04×10^5 Pa is the lower limit for survival – about one-fifth that at sea level, and encountered at about 12 km altitudes. However, there are many habitats that have no oxygen, such as lake sediments, where we find anaerobes (Section 2.4.2). Indeed, early in the history of life on Earth there were only anaerobes – there was too little oxygen to sustain aerobes, as you will see in Section 3.2.3. Therefore, we will not regard the absence of oxygen as an extreme environment.

The salinity, and the alkalinity–acidity of the environment also impose constraints. If you are not familiar with the pH ('pea–aitch') measure of alkalinity–acidity, you need only note that pH = 7.0 is neutral, like pure water, that greater values are alkaline, and smaller values are acidic. Thus, washing soda solution, as normally used, is a weak alkali with a pH of about 10.5, and vinegar, as bought from the supermarket, is a weak acid with a pH of 2–3. In the vast majority of cells the internal pH is 7.7, very mildly alkaline, and this has to be sustained whatever the pH of the environment. Likewise, the salinity inside the cell has to be sustained at about one-third that of sea water. Salinity is normally measured in terms of NaCl content, though other salts are present in smaller, well-defined proportions. In the cell the proportion is roughly 0.85% NaCl by mass. Also, for all life, exposure to UV and ionising radiation must be below some limits, otherwise too much damage occurs to DNA and other molecules.

2.5.2 Extreme habitats

Table 2.3 shows the extreme environmental conditions under which various terrestrial organisms live – survival in some dormant state occurs under even wider conditions. A particular organism will not necessarily survive across the whole of a range. Instead, each range tends to be occupied only by certain organisms. At most of the extreme values only unicellular organisms are found, and most of these are prokaryotes. These do far better than plants and animals in surviving extreme environments, partly because plants and animals consist of many different types of cell, with specialised functions, and many types of organs and large-scale structures. Keeping this working is possible in most cases only under moderate conditions. Organisms living at the extremes are called *extremophiles*, 'lovers of extremes', though this is a chauvinistic term, because we regard the conditions under which we live as non-extreme.

The extremes in Table 2.3 are striking. If water freezes or boils then the cell is disrupted, so how can some organisms survive at −18°C and others at 123°C? The answer is that the familiar freezing and boiling points, 0°C and 100°C, apply to pure water at a pressure of 1.013×10^5 Pa. Figure 2.10 shows the ranges of temperature and pressure over which pure water exists in stable form as a liquid, solid, and gas. At pressures greater than 1.013×10^5 Pa the boiling point is raised. It is indeed the case that the hyperthermophiles ('lovers of extreme heat') are found at the high pressures deep in the oceans. Given sufficient pressure, the upper limit for carbon–liquid water life could in principle be as high as 160°C, the temperature at which essential carbon compounds cannot avoid being broken down.

Table 2.3. Some extreme environmental conditions under which terrestrial organisms live. Most are prokaryotes.

Parameter	Limit(s)	Type of organism
Temperature	−18°C to 15°C	Psychrophiles
	60–80°C	Thermophiles
	80–123°C	Hyperthermophiles
Pressure	610 Pa to $> 10^5$ Pa	(No special name)
	up to 1.3×10^8 Pa	Piezophiles (barophiles)
Salinity	15–37.5% NaCl	Halophiles
pH	0.7–4	Acidophiles
	8–12.5	Alkalophiles

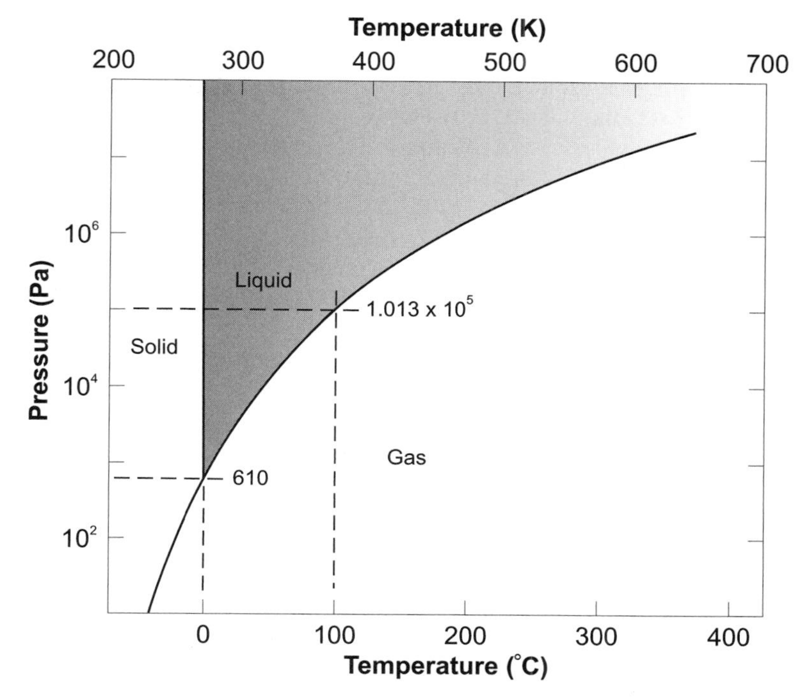

Figure 2.10 The phase diagram of pure water, showing the ranges of temperature and pressure over which pure water exists in stable form as a liquid, solid, and gas.

At the other extreme of temperature, psychrophiles ('lovers of cold') must avoid three threats. Most obviously, the water in the cell must not freeze. Pressure does not help – Figure 2.10 shows that it hardly affects the freezing point. But if substances are dissolved in water then the freezing point can be significantly lowered (and the boiling point is again raised). Certain proteins lower the freezing point of water in the cells of psychrophiles. Second, the rates of biochemical reactions must not become too low. These decline exponentially as temperature falls, and in non-psychrophiles are extremely slow below about 10°C. Psychrophiles avoid this problem via special enzymes, not found in other cells, that promote reactions at low temperatures. Finally, the cell membrane must not become too rigid to function as a regulator of what substances pass through it. This is avoided by the incorporation of special lipids into the membrane that keep it flexible.

The lower end of the pressure range in Table 2.3, 610 Pa, is determined by the requirement to have liquid water. The minimum pressure for pure water is 610 Pa – at lower pressures water can only exist as a solid or a gas (Figure 2.10). The upper end of the pressure range is the hydrostatic pressure in the deepest ocean where life has so far been found (the Mariana Trench, Western Pacific Ocean). For unicellular creatures the true limit could be far higher. Unicellular organisms living at great depths do not die when brought up to the surface, nor do those living at the surface when taken down. Multicellular creatures perish, particularly if the transition is rapid.

The range of salinities in Table 2.3 is to be measured against the 2.95% average for seawater. The pH range extends from strongly acidic to strongly alkaline. An extreme not listed in Table 2.3 is exposure to radiation, in particular UV radiation, energetic charged particles, and gamma radiation. These cause damage to DNA (Section 2.4.4). Some prokaryotes can stand levels of radiation that would quickly be lethal to us, such as those inside nuclear reactors!

Extreme conditions are found in extreme habitats. Let's look at two habitats of particular interest to the search for extraterrestrial life.

Hydrothermal vents

Hydrothermal vents, also called black smokers, are found in regions where new oceanic crust is being created at plate boundaries (Section 2.1.2). The vents are protuberances typically a metre across and a few metres tall, pouring out water at temperatures up to 400°C, much of it is recycled oceanic water that has percolated into the crustal rocks. The water contains dissolved gases, including CO_2, H_2, CH_4, and H_2S. Where the hot water meets the cool oxygenated water of the oceans there are chemical reactions in which solid sulphides form, colouring the water black and building the protuberances. Figure 2.11 shows a vent and Figure 2.12 shows some of the organisms that are found in their vicinities.

Figure 2.11 A hydrothermal vent (black smoker) about 1 m tall, pouring out water at temperatures up to 400°C, and containing dissolved gases, including CO_2, H_2, CH_4, and H_2S.
D. Foster, © Woods Hole Oceanographic Institution.

Figure 2.12 Life near a hydrothermal vent. Tubeworms, spider crabs, clams, shrimps, and unicellular creatures are among the organisms found at such vents.
F. Grassle, © Woods Hole Oceanographic Institution.

The organisms seen in Figure 2.12 are tubeworms and spider crabs. Animals such as mussels and worms can also be found along with unicellular creatures. At such depths all these creatures are piezophiles. Near the vent there are hyperthermophiles. These are prokaryotes, many of which are also anaerobes, the vent water being almost devoid of oxygen. Further from the vent the temperature drops to about 3°C, and psychrophiles live there. Thus, over distances of a few tens of metres we pass from the domain of organisms adapted to hot conditions to those adapted to cold conditions.

A food supply for heterotrophs here (and elsewhere in the deep oceans) is detritus sinking from surface waters, but there is a supplementary base to the food chain, provided by autotrophic prokaryotes, which can end up as food. No sunlight penetrates to these depths and so they do not photosynthesise. Instead, they perform chemosynthesis. However, it is not clear that these chemosynthesisers are truly independent of photosynthesis. For example, there are methanogens that use the

CO_2 and H_2 dissolved in the vent water (Reaction 2.9). The CO_2 comes partly from the break-up of carbonates that have been taken down into the Earth's interior, and partly from the oxidation of crust and mantle carbon. Much of the carbonate comes from the shells of sea creatures that have relied partly on photosynthesising organisms for food. Moreover, the oxidation of carbon could rely in part on the oxygen in the oceans, itself the result of oxygenic photosynthesis (Reaction 2.6). There might also be a biological component to the H_2, even though much of this could be produced abiologically, for example through the oxidation by water of iron in new crustal rocks. If the chemosynthesisers at hydrothermal vents are not independent of the rest of the biosphere, where might we find some that are? One possibility is in the crustal rocks.

Crustal rocks

Perhaps the most surprising type of habitat on Earth is within solid bodies – within ice or rock. In the cold deserts of Antarctica, such as the Dry Valleys of Victoria Land, there are rocky outcrops of sandstones and quartzites within which there are colonies of unicellular organisms of great variety. These are all within a centimetre or so of the surface, a depth to which sunlight can penetrate and so photosynthesis can occur. Water is also present, despite the dry environment, trapped in pores in the sandstone. Moreover, even though the surrounding air temperatures are usually below 0°C, the temperature a few millimetres into the rocks can reach 10°C. We thus have life living under not very extreme conditions; the surprise is that such conditions can be found in rocks within a much more extreme environment.

Even more surprising is the presence of dormant unicellular organisms in ice recovered from depths up to 2.5 km in Antarctica. Most astonishing of all is the discovery of unicellular creatures up to several kilometres deep in the Earth's crustal rocks. Much of the biosphere today, perhaps most of it by mass, is in unicellular form, mostly prokaryotic, well below the surface of the Earth. Water-filled pore spaces and cracks in rocks are plenty big enough for single cells. The pores are not isolated and so the crust is rather like a sponge through which unicellular life and its watery environment can pass. The water contains chemical reactants, and a variety of chemosynthetic autotrophic processes occur, including methanogenesis. Enzymes promote redox reactions that provide the basic energy source for life. It is important that such enzymes are necessary, otherwise the available chemical energy would be liberated spontaneously and rapidly by abiological processes, and there would soon be none for organisms. The abiological processes would dominate only at temperatures too high for life. In some locations where life is found deep in the crust it is possible that chemosynthetic autotrophs are the sole base of the food chain, and that the reactants they use to store energy are independent of the rest of the biosphere.

Truly, life on Earth seems to occupy everywhere that it is possible for carbon–liquid water life to exist – everywhere between the temperature range −18°C to 123°C where liquid water is available and there is a suitable energy source.

2.6 THE TREE OF LIFE

The bewildering variety of organisms on Earth has long been classified into a hierarchy with many different levels. Among plants and animals the lowest level that will concern us is the species. Two plants or two animals are said to belong to different species if a fully developed male of one species and a fully developed female of the other cannot produce fully fertile offspring under natural conditions, or if production of a fertile hybrid is extremely rare. In practice it is often difficult to put this to the test, and impossible with fossils. Therefore most organisms have been classified on the basis of their appearance and their behaviour. Species are grouped into families, and the further aggregations are called order, class, phylum, and kingdom. For example, the domestic cat (all varieties) is classified as follows: species, Catus (domestic cat); genus, Felis (wild and domestic cats); family, Felidae (all cats); order, Carnivora; class, Mammalia; phylum, Chordata; kingdom, Animalia. The phylum specifies a characteristic body plan. Chordata have backbones, and so include fish, frogs and humans, but not spiders. There are 24 phyla in the animal kingdom.

Since the 1970s it has become possible to measure the biological dissimilarity (distance) between species by comparing their RNA and DNA. This also tells us something about evolution. For example, two species A and B that differ only slightly in their nucleic acids probably diverged from a common ancestor more recently that species A and C that exhibit greater differences. With the advent of this technique some rather fundamental revisions were made to the classification of life with the outcome shown in Figure 2.13. This is commonly called the tree of life. The top level of the hierarchy is the domain. Figure 2.13 shows the three domains Bacteria, Archaea, and Eukarya, plus some of the subdivisions into kingdoms and phyla. The tip of a branch could represent a kingdom/phylum alive today, or one that has become extinct and led nowhere. Away from the tips are ancestors of the

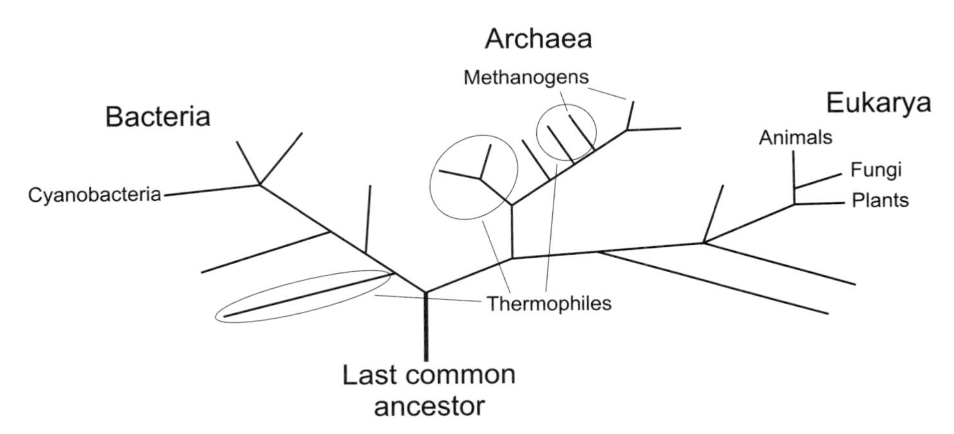

Figure 2.13 The tree of life, showing the three domains, Bacteria, Archaea, and Eukarya, into which all life on Earth is classified. For clarity, some branches have been omitted from the tree.

organisms at the tips, now extinct. At a branch point lies the common ancestor of all the organisms on the branches that diverge from that point. The distances along the branches represent evolutionary distances, as measured by differences and similarities in the nucleic acids in different organisms. Thus, animals and plants are much more closely related than either of these is to methanogens. Within each domain there are autotrophs and heterotrophs.

The Eukarya comprise all the organisms with eukaryotic cells. This domain is divided at the next level in the hierarchy into four kingdoms, and you will be familiar with members of the Animalia (animals), Plantae (plants), and Fungi kingdoms (e.g., yeast and mushrooms). Nearly all of the members of these three kingdoms are multicellular, though some fungi are unicellular. The fourth kingdom, Protoctista, consists mostly of unicellular organisms, though seaweeds are a familiar multicellular example. Protoctista comprise a very wide variety of forms so to think of it as just a quarter of the Eukarya is misleading. It might one day be split into several kingdoms.

The other two domains comprise the prokaryotes, for which the great majority of organisms are unicellular. The division between the Archaea and the Bacteria followed differences discovered in the late 1970s by the molecular biologist Carl Woese. He showed that a certain type of RNA is different in the cells of what he later named Archaea, from the RNA that performs the same functions in cells of Bacteria. The Archaea include many thermophiles, as shown in Figure 2.13, and this has led some scientists to regard them as the most similar among living organisms to the very earliest forms of carbon–liquid water life on Earth, as discussed in the next chapter. This view is supported to some extent by the early divergence of some of these types of Archaea branch from the trunk. The base of the tree trunk represents some unknown common ancestor of all life on Earth.

That we can place all known organisms on Earth on a tree reaching back to a common ancestor shows that all present life on Earth had a common origin. This view is supported by the great similarity at a fundamental biochemical level of all organisms alive on Earth today. Quite what this origin was is one of the main subjects of the next chapter.

2.7 SUMMARY

- The Earth has an iron-rich core, the outer part of which is liquid, and is the source of the magnetic field. The rest of the Earth is rich in silicates, with compositional divisions that define the mantle and the crust. The surface of the Earth is shaped by plate tectonics. The biosphere is confined to the crust and the surface, beneath an atmosphere rich in oxygen and in nitrogen.
- The basic unit of life is the cell. There are two broad types, the prokaryotic cell and the more complex eukaryotic cell.
- Cells consist largely of water and organic compounds, with proteins accounting for much of the latter. Proteins fulfil a wide range of functions – structural, transport, storage, and catalysis of chemical reactions (enzymes).

- Chemical energy is a fundamental requirement for life, particularly redox reactions. Energy is liberated from stores in the cell through respiration, aerobic or anaerobic.
- Energy stores in the cell (sugars, polysaccharides, and lipids) are constructed from food by heterotrophs, and by autotrophs either by photosynthesis or by chemosynthesis.
- Proteins are synthesised using the information contained in the base sequences of the nucleic acid DNA, mediated by mRNA and tRNA.
- DNA (rarely RNA) contains the genetic information that enables repair and reproduction. Imperfect reproduction or mutation is essential for evolution.
- Organisms occupy a wide range of habitats, including the hyperthermophiles that live at temperatures up to 123°C, many of which are anaerobes. Life on Earth seems to be everywhere that it is possible for carbon–liquid water life to exist (e.g. everywhere within the temperature range −18°C to 123°C) provided that liquid water and a suitable energy source are available. This includes hydrothermal vents deep in the oceans, and pores and crevices deep in crustal rocks.
- Organisms are divided into three domains – the Eukarya, Bacteria, and Archaea. The Eukarya are made of eukaryotic cells, and the other two of prokaryotic cells. The Archaea are thought to include organisms most similar among living species to the very earliest carbon–liquid water life on Earth.

2.8 QUESTIONS

Answers are given at the back of the book.

Question 2.1

An article in a popular science magazine contains the following statement: ‘Life on Earth requires only very few of the 100-odd chemical elements. These are hydrogen, oxygen, and carbon. The hydrogen and oxygen make water – all life needs a continuous supply of liquid water. The carbon is needed to make hydrocarbons.’ List the mistakes in this extract, citing evidence to support your conclusions.

Question 2.2

What kind(s) of autotrophs would most likely be found in the following types of environment:

(i) An oxygen-poor hot spring in a volcanic region, beneath an opaque surface-slime.
(ii) On the sea floor, an average of 10 m below the ocean surface.
(iii) At a depth of 3.5 km in a South African gold mine – the organisms were there before the mine was sunk.

Question 2.3

A biologist claims to have discovered a hyperthermophile near a hydrothermal vent. It consists of a colony of single cells, and on close examination he discovers that each cell contains a nucleus and several mitochondria. Discuss where you would place this organism in the tree of life.

Question 2.4

Place the common ancestors of the following pairs of types of organism in a time sequence stretching back from the present: Archaea–Eukarya, Animalia–Fungi, Bacteria–Archaea. Justify your sequence.

3

The evolution and origin of life on Earth

This chapter presents the second part of our survey of life on Earth, to further facilitate our search for life beyond our planet. It presents an outline of how life originated and how it has evolved to give the wide diversity we see today. All life on Earth today originated in the *last common ancestor*, as was shown in Figure 2.13, and in a modified form in Figure 3.1. What was our last common ancestor like? Scientists can't be certain of all the details, but it was surely some primitive kind of prokaryotic cell, and it certainly carried its genetic information in DNA, and relied on proteins to fulfil the wide variety of functions performed by proteins today. It was thus DNA/RNA-protein life. This kind of life is far too complex to have originated directly from non-living matter, and so there must have been considerable evolution *before* the last common ancestor. This is indicated in Figure 3.1, which extends the tree of life back to the origin of life itself. The side branches that precede the common ancestor, all of which terminate in extinction, are forms of life that differed fundamentally from our last common ancestor. The termination of these branches represents their extinction. We have no record of these other life forms, but it is thought likely that they existed.

Many of the descendents of the last common ancestor have also become extinct, again as shown by terminated branches in Figure 3.1. Extinction also lies within the branches that reach us today. These are earlier species that led to modern forms. Overall, the great majority of species have become extinct. It is this part of the history of life on our planet that we consider first – the evolution of life from the last common ancestor to the present. We will then return to the last common ancestor and work backward in time towards the true origin.

3.1 THE PROCESS OF EVOLUTION

The idea that the various lifeforms on Earth have evolved from common ancestors has a long history, and it is an idea well supported by the fossil record. Quite distinct

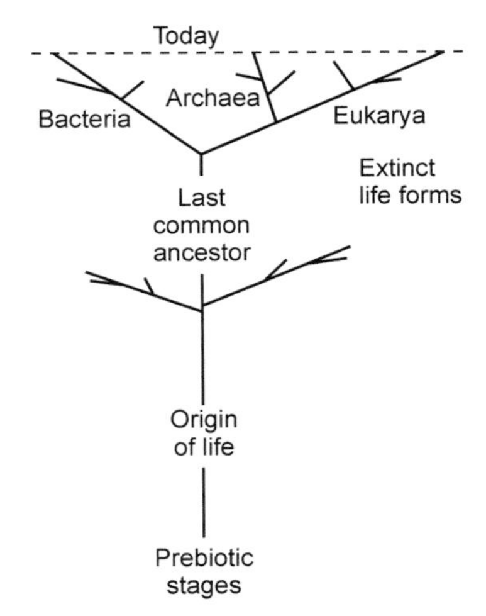

Figure 3.1 The tree of life, stretching back to the origin of life. Branches that terminate before the present represent extinction. Extinction is under-represented in this diagram.

from this is the *mechanism* by which evolution occurs. The mechanism is natural selection. Two people in particular developed this idea, revolutionary at the time. Alfred Russell Wallace was a Welsh naturalist who had collected many data in the field. In 1858 he sent an article on evolution to the English naturalist Charles Robert Darwin, describing in essence Darwin's own unpublished theory of evolution by natural selection, again based on a huge mass of observations. As a result, papers by Wallace and by Darwin were read that July to the Linnaean Society in London. Not long afterwards, in 1859, Darwin published *The Origin of Species by Means of Natural Selection*. The idea was out.

Evolution by natural selection operates through the random variations that occur between parent and offspring. Many of these variations are of little consequence, whilst others are damaging, resulting in a short lifespan, or infertility, or some other biological handicap. A few will be beneficial to the organism. Suppose that there is a variation present in one offspring that enables it to produce more progeny than average. Some or all of these progeny will inherit this variation, and so each of these will have more progeny too. In this way the variation becomes established in the population. Evolution by natural selection thus proceeds in a series of tiny steps, through a variation that results in greater numbers of progeny. In this way a huge variety of species can evolve from a common ancestor. One example of natural selection is the change in colour of a moth that has a natural predator, say, a bird. If a variation results in a colour change that provides better camouflage then this is a colour change that is likely to become established.

It is possible to see natural selection operating over a few generations, and therefore for a species with a short reproduction interval it can be seen within a human lifetime. This is notable in the Bacteria and Archaea, which have reproduction intervals of hours or days. On longer timescales, though the evidence that evolution has occurred is in the fossil record, the mechanism of evolution has to be inferred. For example, operating over many generations, in many small steps, it is possible to imagine how natural selection produced an eye from a light-sensitive patch of skin.

The discovery of natural selection did not require knowledge of the way in which variations occur in offspring. We now know that this happens in various ways, as outlined in Section 2.4.4. These include mutations in DNA, through various causes, and the variation in DNA that arises in sexual reproduction.

You might think that evolution has resulted in an ever-growing number of species. This is not so – as well as new species appearing, species also become extinct – it has been estimated that over the whole of Earth history about 99% of species have become extinct, never again to reappear. Note that extinction is the end of the line, with none of the species surviving today. One cause of extinction is the arrival of a species that is better adapted to an ecological niche than an existing species, which will therefore starve, or be otherwise eliminated. Another cause of extinction is a change in the environment (e.g., global warming), when a species can neither cope nor migrate to where the climate is still similar to that to which it is well adapted. Yet another, less dramatic cause, is the evolution of a species into other forms, without survival of the original form.

As well as a comparatively steady rate of extinction, there have been several times in Earth history when a large proportion of all species have vanished in less than a million years. These are the *mass extinctions*. Each one has provided 'ecological space' that new species can inhabit, and therefore each mass extinction is characterised not only by a huge loss in species, but in a huge number of new species. This has led to the term punctuated evolution to describe such events. Except for setbacks at each mass extinction, the diversity and complexity of life has generally increased throughout much of Earth history, though this is probably no more than a consequence of the relative simplicity of the biosphere at the time of the last common ancestor, rather than some principle of 'progress' in operation.

Evolution, punctuated or otherwise, has been in operation since our last common ancestor, and has yielded the structure of the tree of life. But what dates can we attach to the major events? We need a time line of life on Earth, and this is introduced in the next section.

3.2 LIFE ON EARTH SINCE THE LAST COMMON ANCESTOR

3.2.1 The major events and their timing

Figure 3.2 shows the time line of life on Earth, the dates having been established by radiometric dating, which will be described in Section 3.2.4. The time line starts with the origin of the Earth 4600 million years ago. We will write this as 4600 Ma, where 'a' is short for *annus*, which is Latin for year, and it is implicit that times are before the present. 4600 Ma also marks the time of formation of the Sun and the rest of the Solar System, outlined in Section 1.2. Recall that the terrestrial planets, including the

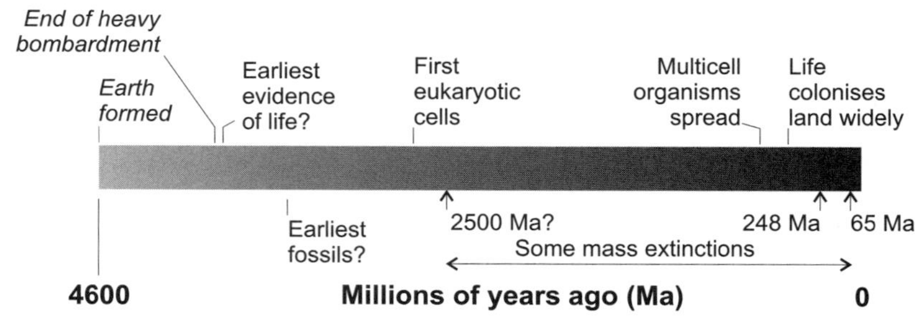

Figure 3.2 The time line of life on Earth.

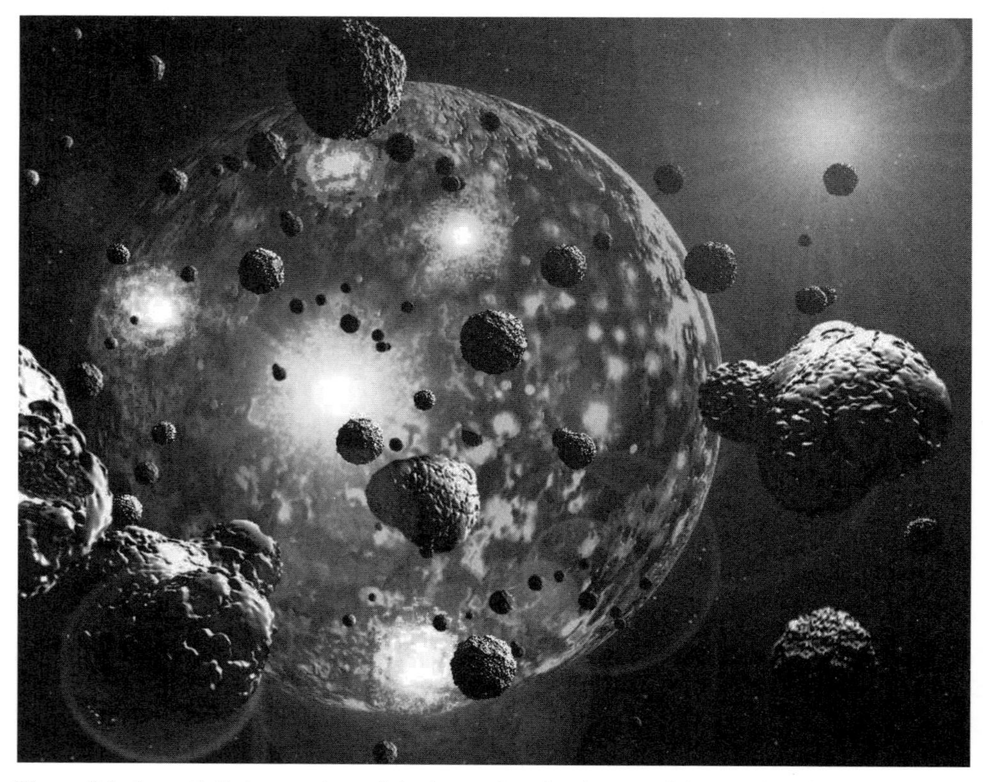

Figure 3.3 An artist's impression of the heavy bombardment of the Earth, which lasted until about 3900 Ma.

Julian Baum (Take 27 Ltd).

Earth, suffered a heavy bombardment from space by rocky and icy planetesimals that caused planet-wide mayhem (Figure 3.3). This lasted until about 3900 Ma, when it was declining fairly rapidly, presumably as the population of planetesimals decreased. The last phase of the bombardment was probably sufficiently heavy that, even if there was an earlier interval of lighter bombardment during which life originated, it is unlikely that it would have survived. Life could however have originated soon after 3900 Ma. How near to this time do we have direct evidence of life?

The earliest evidence for life

The earliest evidence dates from around 3850 Ma, the age of the oldest rocks that still survive today. Though there are no fossils in these ancient rocks, there are chemical signatures in the form of carbon isotope ratios. The common isotope of carbon is written ^{12}C to denote that, in addition to the 6 protons in the atomic nucleus that define the nucleus as carbon, there are also 6 neutrons, so that the nucleus contains a total of 12 nucleons (protons plus neutrons). A less common isotope is ^{13}C, which has 7 neutrons. In the general environment there will be a certain ratio $n(^{12}C)/n(^{13}C)$ of the numbers of ^{12}C and ^{13}C nuclei, but there is a slightly higher ratio in carbon that has been processed by photosynthesis. In small deposits of carbon in ancient rocks the ratio is enhanced compared to the value in carbonate rocks, and the best explanation is that the enhancement is due to photosynthesising organisms. The oldest rock with such a signature is from the 3850 Ma Isua Complex in the west of Greenland. There are, however, abiological ways of enhancing the isotope ratio, so the biological interpretation is controversial.

Among the oldest fossils are ones seen in the form of *stromatolites*. Figure 3.4(a) shows an example of a present-day stromatolite. It consists of layers of minerals laid down by colonies of certain oxygenic photosynthesising bacteria called cyanobacteria that flourish in shallow water in a rather restricted range of circumstances. Figure 3.4(b) shows a fossil stromatolite that seems to have been similarly created. The oldest stromatolite fossils date from 3460 Ma, from Warrawoona in Western Australia. As well as a superficial resemblance to present-day stromatolites, they also have structures that resemble fossilised bacteria, and these might well have photosynthesised, though they might not have generated oxygen. However, the fossil interpretation of these ancient stromatolites is not entirely secure – abiological formation is possible. There is also evidence for life at 3300–3500 Ma in the Baberton greenstone belt in South Africa, including what might be the remains of thermophilic prokaryotes.

However, the earliest evidence for life on which there is no real dispute, is fossil cyanobacteria from Pilbara, Australia, at 2700 Ma, and stromatolites of the same date from Fortescue, also in Australia.

Early photosynthesis

These early life forms existed on an Earth that was significantly different from the Earth today. One major difference was in the composition of the atmosphere. The Earth acquired its atmosphere (and oceans) from some combination of outgassing of the rocks that made up its bulk composition, and from the volatile-rich planetesimals that bombarded the Earth (Section 1.2.3). The volatiles delivered to the Earth's surface would have been dominated by water, CO_2, and N_2, with significant amounts of other substances such as CO, and perhaps CH_4. There would have

(a)

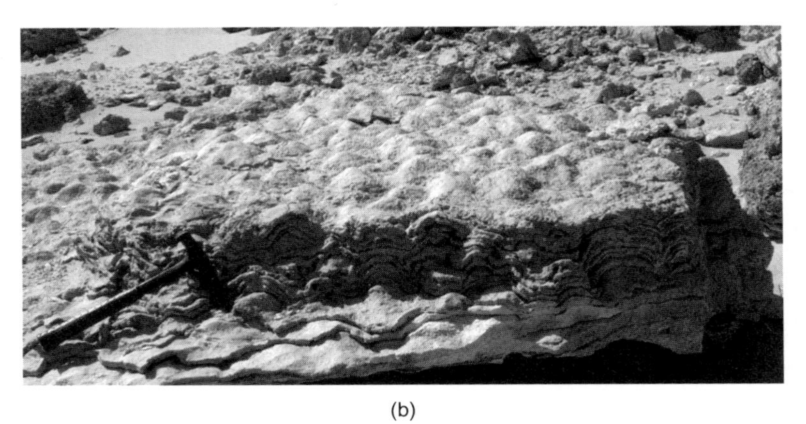

(b)

Figure 3.4 (a) Present-day stromatolites being built in the Bahamas. The divisions on the rod are 100 mm. (b) Fossil stromatolites, from Wadi Kharaza in Egypt. These are only about 15 Ma old.

(a) R. P. Reid, University of Miami. (b) Brian Rosen.

been some processing through interaction with surface rocks, and photodissociation of water by solar UV radiation would have produced some oxygen (Section 3.2.3), but overall the composition would have been little altered. This atmosphere was mildly reducing (i.e. contained molecules that tend to donate electrons (such as CO and CH_4), rather than molecules (such as O_2) that tend to accept them).

If the carbon isotope evidence has been correctly interpreted, photosynthesis had started by 3850 Ma. But was it oxygenic? This is unlikely, though it has been suggested that banded iron formations indicate otherwise. These are sedimentary rocks consisting of layers rich in iron oxides alternating with iron-poor layers, each layer typically a few millimetres thick. They formed throughout these earliest times down to about 2000 Ma. Quite how they formed is unclear, but one possibility is that atmospheric oxygen from photosynthesis, dissolved in surface waters, was oxidising iron compounds from hydrothermal vents (Section 2.5.2). Whether photosynthesis was essential for the O_2, at least at the earliest times, or whether other sources, such as the photodissociation of water, would have sufficed, is uncertain. Also, the biological evidence is that oxygenic photosynthesis evolved through a combination of two non-oxygenic forms, and it is thought unlikely that this could have happened so soon. Better evidence for oxygenic photosynthesis is at around 3500 Ma, based on the stromatolite fossils of that date, if they are the products of cyanobacteria.

Later events

It was not until around 2300 Ma that atmospheric O_2 was more than a trace – possible reasons for this are discussed in Section 3.2.3. By then the eukaryotes were well established, the first eukaryotic cells, or their now extinct ancestors, having emerged at about 2700 Ma. All earlier organisms were prokaryotic. This was one of the major developments of life on Earth, and the eukaryotes evolved rapidly.

The complex structure of the eukaryotic cell seems to have arisen through symbiosis, in which two or more prokaryotic cells united. There is evidence for this among the organelles in eukaryotic cells (Section 2.3). For example, the chloroplasts that photosynthesise in plant cells and the mitochondria that carry out aerobic respiration in plant and animal cells have DNA that show bacteria to be their closest relatives. The prokaryotes that united included anaerobic forms, but also aerobic forms, and consequently many eukaryotic cells were able to utilise oxygen for aerobic respiration. Today, almost without exception, eukaryotic cells require oxygen, either in the air or dissolved in water, whereas some prokaryotes are killed by oxygen, others can tolerate it, and only a small proportion require it.

Nearly all multicellular forms of life are made from eukaryotic cells, and their proliferation was the next great event in the time line of life on Earth. It occurred as late as about 610 Ma, when the previously very rare multicellular organisms became much less rare, larger, and more complex. Corals and creatures resembling jellyfish appeared, but also quilted creatures that have no modern counterparts. 'Shortly' afterwards, at 545 Ma, there was a huge increase in the number and variety of such organisms, and also in their range of habitats. This rather precise date defines the beginning of the Cambrian Period (545–495 Ma). Many of these new organisms had hard parts, such as shells and exoskeletons, and they have therefore been readily preserved as fossils. Figure 3.5 shows a fossil of one of the many species of trilobite, which flourished in the oceans from the early Cambrian. Though there were environmental changes occurring at the time it is not clear that they were the cause of the Cambrian 'explosion'. Further contributory factors might have been the increase in atmospheric O_2 content (Section 3.2.3), and the generation of new and diverse habitats through continental drift resulting from plate tectonics.

Up to the Cambrian, nearly all life had been confined to the oceans. At earlier times there had been isolated bacterial mats on land, even as far back as 2700 Ma, though they were sparse. It was not long into the Cambrian before life, with its enlarged potential for diversity, was able to colonise the land far more widely.

Figure 3.5 A fossil trilobite, *Ogygiocarella*, from just after the Cambrian Period in central Wales. It is 75 mm long.
Peter Sheldon.

This began around 490 Ma, at about the end of the Cambrian period, but the main colonisation was between 440 Ma and 420 Ma.

By 420 Ma all the major groups of multicellular organisms were on land and in the sea, such as invertebrates, vertebrates, and plants. Our own genus (a group of related species), *Homo*, emerged just 3 Ma ago, and our own species, *Homo sapiens*, a mere 0.1 Ma ago!

Mass extinctions

The two most dramatic mass extinctions occurred 248 Ma ago, at the end of the Permian Period, and 65 Ma ago, at the end of the Cretaceous Period (Figure 3.2). The more recent one defines the end of the Mesozoic ('middle life') Era, when 70% of marine species became extinct, plus a large proportion of land species, including, famously, the dinosaurs. The earlier mass extinction defines the end of the Palaeozoic ('old life') Era, when up to 95% of marine species became extinct plus a high proportion of land species. In each case it took 0.1–1 Ma for all the species to disappear – a mass extinction is not an overnight phenomenon. It took roughly ten times longer for the biosphere to recover. There have been three lesser mass extinctions, at about 450 Ma, 360 Ma, and 210 Ma, and probably some much earlier ones. The evidence for the earlier ones is less secure, partly because of the less complete fossil record.

3.2.2 The causes of mass extinctions

In searching for the cause of a mass extinction we must look for some global change, otherwise many species could migrate to where the environment suited them. It must also be a sudden change, otherwise many species would adapt through normal evolution. There is now considerable evidence that the mass extinction at 65 Ma was associated with a comet or asteroid impact. The remains of a large impact crater off the northern Yucatan coast of Mexico – the Chicxulub Crater – has been dated at 65.5 Ma and must have been caused by a body about 10 km in diameter travelling at about 30 km per second. The energy released by such an impact was certainly sufficient to promote a mass extinction through the change in climate it must have caused, notably through dust in the upper atmosphere that caused global cooling. This dust would have resulted from direct placement by the impact plus smoke from forest fires caused by the heat pulse from the impact. Evidence of a global distribution of impact dust comes from anomalously high concentrations of iridium around the world in sedimentary rocks 65 Ma old. Iridium is much more abundant in meteorites than in the Earth's crust. Indeed it was the discovery of this iridium-rich layer that led to the discovery of the Chicxulub Crater.

Even though an impact was a factor in the 65 Ma mass extinction, it was certainly not the only one. Some species were already in decline, and there was an increased rate of extinction before the impact. This might have resulted from the climate change that was already under way, notably global cooling and an associated lowering of sea level. Climate change has many possible causes, though a huge episode of volcanism in India at about 65 Ma might have contributed to this particular cooling by injecting dust into the atmosphere. Alternatively, a *decrease* in global volcanism might have promoted cooling due to the associated decreased output of the greenhouse gas CO_2. Thus, though the impact initiated the decline and disappearance of some species it was just the last straw for others. Many other species were unscathed, including sharks, many reptiles, and the ancestors of the modern mammals.

The 248 Ma mass extinction is less well understood. This was certainly a period of extensive volcanic eruptions in Siberia. There were also changes in the configurations of continents and ocean basins due to plate motion that altered ocean currents, lowered sea level, and removed much shallow sea shelf. There is also evidence for a modest reduction in atmospheric O_2. One plausible scenario is that the Siberian volcanism initially caused a few years of global cooling through atmospheric dust, and then, after the dust settled, caused global warming through the CO_2 increase. The GMST might have risen by 5°C. Many land species died. The oceans gradually warmed, causing huge extinctions there. The increased ocean temperatures then caused CH_4 to be released from ocean sediments. CH_4 is a powerful greenhouse gas and could have given rise to a further 5°C rise in global mean surface temperature, causing yet more extinctions.

The other three mass extinctions since the Cambrian are associated with climate change, including the onset of ice ages, though the causes of these changes are poorly understood. If, as seems likely, there have also been mass extinctions long before 248 Ma, we can only speculate on their causes. As well as the possible causes cited above, there might have been others. For example, there might have been a mass extinction when the oxygen content of the atmosphere began to increase after 2500 Ma – at that time it would have been a poison to most species, a major pollutant, and even today it is poisonous to many prokaryotic anaerobes. Another possibility is a nearby supernova. This is discussed in Chapter 7 – supernovae might have an influence roughly every few hundred million years on average.

Mass extinctions overall have probably promoted biological diversity and complexity. It is, however, possible to have too much of a good thing. If the giant planet Jupiter had been absent then comet impacts with the Earth would have been more common – Jupiter provides a measure of gravitational screening. We might then have had sterilisation of the Earth, perhaps extinguishing life forever, or holding back the emergence of multicellular organisms.

3.2.3 The effect of the biosphere on the Earth's atmosphere

The present composition of the atmosphere was given in Table 2.1. It is dominated by N_2 and O_2. This is remarkably different from what the composition is believed to have been at the time of the last common ancestor, when it was dominated by CO_2 and N_2, with significant amounts of several other gases such as CO. There was almost no O_2, though H_2O (as vapour) has always been present.

Oxygen

One way to produce O_2 is through the *photodissociation* (*photolysis*) of water vapour by solar UV radiation. This is initiated by the action of a UV photon that breaks up H_2O as follows:

$$H_2O \rightarrow H + OH \quad (3.1)$$

There are several further chemical stages, and the overall process can be summarised as:

$$2H_2O \rightarrow 2H_2 + O_2 \quad (3.2)$$

O_2 will accumulate because the H_2 with which it would recombine, escapes to space.

However, a far more copious source is oxygenic photosynthesis, in which O_2 is liberated as a by-product (Section 2.4.2), and it is this that has made O_2 a major atmospheric constituent. Yet though oxygenic photosynthesis might have been established by about 3500 Ma, and was certainly well established by cyanobacteria by 2700 Ma, it is clear from the geological record of dated minerals and isotope ratios, that around 2300 Ma the atmosphere was not oxidising, whereas after this

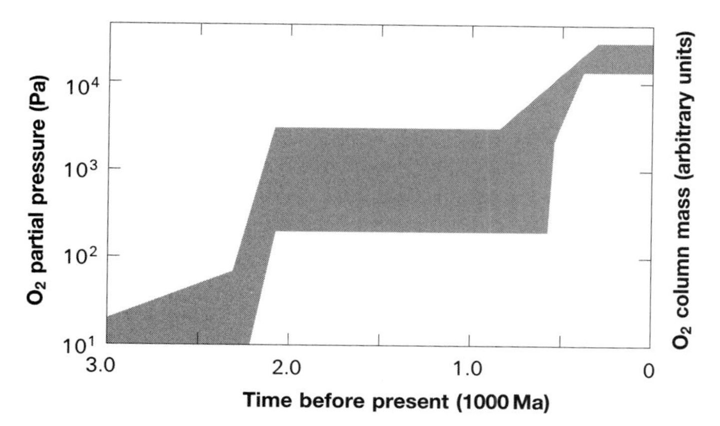

Figure 3.6 A plausible history of the O_2 content of the Earth's atmosphere. The shaded area shows the range of possible values. Note the logarithmic scale.

date it was, and that this change was due to an increase in oxygen. This must have resulted from a change in the quasi-equilibrium between the rate of generation of O_2, and its rate of removal. The rate of generation of O_2 through photosynthesis seems to have been roughly constant through 2300 Ma, in which case the rate of removal must have declined. O_2 is removed through the oxidation of organic compounds that occur in respiration (Section 2.4.2) and in the decay of dead organisms. It is also removed by the oxidation of suboxidised rocks and gases. One suggestion is that the anaerobic decomposition of photosynthetically produced organic matter led to crustal rocks becoming more oxidised, thus reducing the rate at which they took up atmospheric O_2. Anaerobic decomposition generates O_2 and CH_4, and if the CH_4 escapes to space, or at least the H_2 that it contains, then increased oxidation of the rocks would result. However, the cause of the rise of atmospheric O_2 is far from fully understood.

Adjustments in the quasi-equilibrium throughout Earth history have resulted in the atmospheric O_2 content varying for much of the time – Figure 3.6 shows one interpretation of the limited data. In the absence of oxygenic photosynthesis the atmosphere would always have contained much less than 1% O_2. The high abundance of O_2 accounts for the very low abundance of reducing substances such as CO and CH_4. Note that CH_4 is so readily oxidised that its abundance would be even lower than its present number fraction of 1.6×10^{-6} if it were not produced copiously in the biosphere, by methanogens (Section 2.4.2).

The O_2 content of the oceans has also varied. The top few hundred metres is in good contact with the atmosphere and has followed the atmospheric content closely. Deeper down there has always been less O_2, and aerobic respiration must have become possible only after the atmospheric content had risen significantly. There are estimates for the deep ocean that are as late as 1000 Ma.

Ozone (O_3)

In Section 2.1.3 it was mentioned that O_3 shields life from the damaging effects of solar UV radiation. This it does by absorbing much of this radiation. O_3 is derived from O_2 by the action of solar UV photons, and so its history follows that of O_2. However, rather little O_2 is needed to provide sufficient O_3 for shielding. Therefore, there has been sufficient O_2 in the atmosphere to shield life from solar UV from the earliest days of the biosphere.

Carbon dioxide

The biosphere has been responsible for removing much of the CO_2. Most of it has been locked up in rocks, particularly in carbonate minerals such as calcium carbonate ($CaCO_3$), which are the dominant constituents of the various types of limestone (which include chalks). Limestones are sedimentary rocks that on prolonged heating and pressurisation become metamorphic rocks such as marble. Most carbonate rocks form on ocean floors, incorporating the CO_2 dissolved in seawater. The CO_2 is replaced from the atmosphere, which is consequently depleted in CO_2. This removal would proceed at a very slow rate in the absence of a biosphere, and so the regeneration of CO_2 through volcanic emissions and the weathering of rocks would sustain a good deal more atmospheric CO_2 than we see today. The biosphere raises the rate of removal through the formation of skeletal parts of sea creatures. The biosphere also removes CO_2 through the formation of organic compounds. This is only partly reversed by respiration and decay because of the burial of a proportion of the organic carbon to form various deposits, including fossil fuels.

Though the atmosphere now contains only a trace of CO_2 it is important because it makes a major contribution to the greenhouse effect, water vapour being the other major contributor (Section 2.1.3). Without its overall greenhouse effect the Earth's GMST would be about 33°C lower than it presently is. The trace of CO_2 also enables photosynthesis to take place, and it is thus essential for the continued existence of green plants, algae, certain prokaryotes, and for all life that depends on them. CO_2 also helps solve another problem, as follows.

At around the earliest estimated time for the last common ancestor, shortly before 3850 Ma, the Sun had only about 75% of its present luminosity. This presents us with the *faint Sun problem*, because with its present atmosphere the Earth might then have had a frozen surface, a state from which it would have been hard to recover. But there is geological evidence that liquid water was widespread at that time and ever since. One solution to the problem is through the dynamics of the *carbonate–silicate cycle*. Suppose that the surface of the Earth cools. This reduces the rate at which silicate rocks are weathered by the action of CO_2 and H_2O to form carbonates. This in turn reduces the rate at which CO_2 is removed from the atmosphere. CO_2 is emitted by volcanic activity, so with the rate of reduction smaller, the atmospheric content of CO_2 rises. This increases the greenhouse effect, and so the Earth's surface warms up. This might have prevented the Earth from freezing when the Sun was faint. In the longer term, as the Sun's luminosity increased, the rate of weathering rose, and so the CO_2 content fell, thus preventing a large rise in temperature. This was additional to the removal of CO_2 by the biosphere.

An alternative solution to the faint Sun problem is CH_4. This is such a powerful greenhouse gas that a trace could have prevented freezing. More is required than is present today, but so little O_2 would then have been present that sufficient CH_4 could have been sustained.

Nitrogen

Whereas the biosphere has decreased the atmospheric abundance of CO_2 it has had the opposite effect on the most abundant atmospheric constituent, N_2. This is removed by electrical discharges, notably lightning, which forms nitrogen oxides. Rain then removes these oxides from the atmosphere. In the absence of a biosphere only the slow weathering of crustal rocks would return the N_2 to the atmosphere. In the biosphere some prokaryotes release N_2 from organic matter, and though others also remove N_2, the net effect is an increased rate of release, thus sustaining N_2 at a level higher than would be the case on a sterile Earth.

The present atmosphere of the Earth is therefore very much a product of its biosphere.

3.2.4 Radiometric dating

The dates of the various events that we have been describing have been determined by *radiometric dating*. Consider the example of rubidium–strontium dating. Imagine that a rock solidifies and that it contains a mineral that includes the element rubidium. A small proportion of the rubidium nuclei will be the unstable isotope ^{87}Rb. This radioactively decays to form the stable strontium isotope ^{87}Sr, as shown in Figure 3.7. Assume that there was no ^{87}Sr in the rock when it solidified, and that neither the ^{87}Sr that appears nor any ^{87}Rb escapes from the rock. The ratio $n(^{87}Sr)/n(^{87}Rb)$ of the numbers of nuclei will thus increase gradually. If, at some time t, we measure $n(^{87}Sr)/n(^{87}Rb)$ then we can tell how long ago the rock solidified, provided that we know the rate at which ^{87}Rb decays into ^{87}Sr. This rate can be expressed as the half-life – the time taken for half of the nuclei to decay. For this decay the half-life is 48 800 Ma, and so, with $n(^{87}Sr)/n(^{87}Rb) = 0$ at solidification ($t = 0$), $n(^{87}Sr)/n(^{87}Rb) = 1.0$ at $t = 48\,800$ Ma, 3.0 at $2 \times 48\,800$ Ma, and so on. The method is complicated by the possibility that both isotopes were present at solidification, though the actual value of $n(^{87}Sr)/n(^{87}Rb)$ at $t = 0$ can be inferred from the relative abundances of other isotopes in the rock, and an allowance made.

A solidification age is important because it tells us the time at which the rock formed. Rocks formed by solidification are called igneous rocks. However, fossils would not be preserved in rocks that were molten, so in order to obtain dates for the development of the biosphere, use is made of igneous rocks that sandwich the sort of rocks in which fossils are preserved, namely sedimentary rocks and a few metamorphic rocks. The ages of the igneous rocks bracket the age of the interposed sedimentary rock. With sedimentary rocks dated in this way we can also put dates

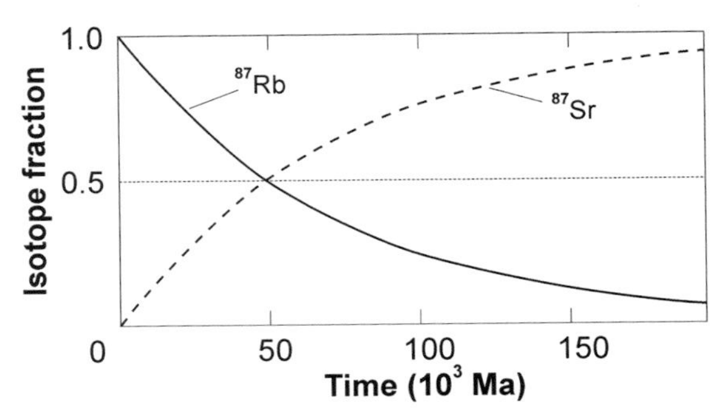

Figure 3.7 The decay of ^{87}Rb to form ^{87}Sr, which enables radiometric dating to be performed.

on other events, such as the increase of oxygen in the atmosphere – sediments formed in oxygen-rich oceans differ from those in oxygen-poor oceans.

With such a long half-life the Rb–Sr system can provide dates reaching back to the origin of the Solar System. Other pairs of isotopes also have long half-lives and between them have been used to date a wide variety of events. These include the early impact history of the Earth in the time before the 3800 Ma age of the oldest surviving terrestrial rocks (only a few rare exceptions are older). This history has been deduced from our nearest neighbour, the Moon, determined from the solidification ages of impact-melted lunar samples – much of the lunar surface is older than 3800 Ma. The 4600 Ma age of the Earth, and of the Solar System as a whole, has been obtained from the radiometric ages of components in the most primitive meteorites.

For more recent events, in the past 0.1 Ma or so, isotopes with shorter half-lives have to be used, so that measurable changes have occurred. The isotope ^{14}C has a half-life of only 5730 years, and is the basis of carbon-14 (^{14}C) dating (or radiocarbon dating). Atmospheric CO_2 contains a small proportion of ^{14}C, produced through the action of cosmic radiation on ^{14}N, the common isotope of N. Living organisms exchange carbon with the atmosphere through photosynthesis, eating, and respiration, and acquire a ratio of ^{14}C to carbon as a whole that is not very different from the atmospheric ratio. At death this exchange slows right down, and so the ratio $n(^{14}\mathrm{C})/n(\mathrm{C})$ declines. Thus, the death date of organic matter can be determined from the ratio in the dead matter compared to the ratio in the living biosphere. The limit on precision is due to various uncertainties (e.g. in the atmospheric ratio in the past). Thus, at 20 000 years the uncertainty is about 3000 years, with accuracy worsening further back, so that 70 000 years is about the limit for radiocarbon dating.

3.3 THE ORIGIN OF LIFE ON EARTH

As stated at the beginning of this Chapter, the last common ancestor was surely some primitive kind of prokaryotic cell. It certainly carried its genetic information in

DNA, and relied on proteins to fulfil the wide variety of functions performed by proteins today. But this could not have been the first life form on Earth that had sprung directly from relatively small organic molecules – this is far too great a step. A major difficulty is the 'chicken and egg' problem. Proteins require the genetic information in DNA for their synthesis from small molecules, yet proteins as enzymes are needed to promote this synthesis (Section 2.4.3). Furthermore, proteins, as enzymes, are needed to copy DNA during reproduction (Section 2.4.4). Thus, we can't have DNA without proteins and we can't have proteins without DNA.

3.3.1 RNA world

One way out of this dilemma was suggested in the late 1960s, independently by several scientists. It was proposed that the precursor of life's last common ancestor used RNA rather than DNA as the repository of genetic information. Crucially, the RNA in this theory replicates without the aid of proteins and catalyses all the chemical reactions necessary for the precursor to survive and reproduce, including the synthesis of proteins. The world in which organisms rely on RNA in this way is known as *RNA world*. The basis for this proposal was that:

- the nucleotides in RNA are more readily synthesised than those in DNA;
- it is easy to see how DNA could evolve from RNA and then, being more stable, take over as the repository of genetic information (it has since been discovered that when retroviruses invade eukaryotic cells they can make DNA from their RNA); and
- it was difficult to see how proteins could replicate in the absence of nucleic acids (though it has recently been discovered that some proteins can self-replicate).

The concept of RNA world did, however, face a major difficulty in the late 1960s, namely that RNA was not known to act as a catalyst for any biochemical reactions in the modern biosphere. Then, in 1983, the US biochemist Thomas R. Cech and the Canadian biologist Sidney Altman independently discovered enzymes made of RNA – *ribozymes*. Though these ribozymes could do little more than cut and join pre-existing RNA they gave weight to the idea that some ancient forms of RNA could have done a lot more. Moreover, though some stages in the RNA-catalysed reproduction of RNA remain elusive, other stages have been demonstrated in the laboratory. Since 1983 it has also been established that in the ribosomes (Section 2.4.3) it is probably the RNA and not the protein that catalyses protein synthesis. There is, however, a huge gulf between what we know RNA can do, and what it would be required to do in RNA world.

Assuming that RNA world existed, it would have been a huge step backwards from the last common ancestor towards the origin of life. But it by no means takes us the whole way, because in order to get to RNA we need a supply of its components (i.e. bases, the sugar ribose, and phosphates). It is difficult to make these, particularly some of the bases and ribose itself. Many experiments have been performed in which gas mixtures that model the Earth's mildly-reducing (perhaps neutral) atmosphere at the time, are exposed to simulated solar radiation, lightning, and other energy sources. A great range of organic compounds has been produced, including small yields of amino acids and other small organic molecules of relevance to life. Unfortunately, other essential molecules have been produced only in negligible proportions, including those troublesome RNA bases and ribose. Hydrothermal systems at temperatures not much above 100°C could have supplied some missing components, but not in large enough quantities, and probably not all of them. One possible solution is that relevant biomolecules arrived from space.

Infall from space in the form of volatile-rich meteorites and comet debris must have delivered significant quantities of what are called prebiotic molecules that presumably aided the emergence of carbon–liquid water life on Earth. Meteorites, which are derived from asteroids, come in a range of types, and in one of these, the carbonaceous chondrites, we see a wide range of organic compounds, including amino acids. Observations of comets indicate that they too contain organic compounds. However, the bases and ribose have not been seen, so they might have been absent in the past.

But even if we had all the components there are the further huge problems of assembling them into nucleotides, and assembling the nucleotides into an RNA strand with the order of 10^5 bases required as a basis for life. An additional difficulty would have been the likely abundance of molecules related to the RNA nucleotides. Their incorporation into RNA would have destroyed its functionality. A possible way ahead is outlined in Section 3.3.3.

3.3.2 The origin of cells

As well as the biochemicals of life, we also need to provide a protected environment – a cell. Quite how the first true cell appeared, like much else, is unknown, but droplets called *coacervates* might provide a clue. When a solution of suitable organic polymers is shaken, droplets of higher concentration are formed, bounded by membranes. Small molecules such as amino acids can diffuse through the membrane and join to form short strings that cannot escape. This is a mechanism for providing a protected environment for producing long polymers. However, it is a long way from a living cell. Like much else in origin of life studies, it gives us a plausible hint, but no more.

3.3.3 The role of minerals

The difficulties outlined in the previous sections have led to the hypothesis that the surfaces of minerals, and tiny compartments in minerals, played an essential role in the early development of life. Tiny compartments can shelter simple molecules, and surfaces can promote the combination of these molecules, ultimately to form long biopolymers. It is also the case that a surface can acquire a concentration of molecules that is much higher than that in the surrounding environment, thus facilitating the build-up of larger molecules from smaller ones. Clays are central to

Figure 3.8 Clay surfaces with intricate structures of possible biological significance. The scale bars are as follows: (a) 20 mm, (b) 10 mm, (c) 2 mm, and (d) 1 mm.
F. R. Entensohn and Raymond Bayan.

the mineral hypothesis because of their intricate surfaces and interior structures (Figure 3.8). Clays form through the recrystallisation of silicates and other minerals dissolved in H_2O, and consist of minerals modified by the inclusion of H_2O molecules and the hydroxide (OH) fragment. There is plenty of evidence that the Earth had a large amount of liquid H_2O at its surface 3800 Ma ago (e.g. from radiometrically dated metamorphic rocks at Isua in Greenland), and there are observational and theoretical reasons to suppose that liquid H_2O was present far earlier. It seems certain that there was liquid H_2O before any RNA world, and that therefore clays were also present at that time.

There is evidence for the potential importance of clays to the origin of life. When a water-based solution of amino acids is evaporated from a vessel containing clays, the surfaces of the clays build short chains that resemble proteins of the sort found in cells today. The similarity with evaporation from a shallow pond or tidal pool with a muddy bottom on the early Earth must be noted. Clay surfaces have also been found to promote the assembly of RNA from its constituents. In these examples some selectivity of the clays is apparent – many types of complex molecules *could* have been synthesised, but biomolecules are prominent. This selectivity is characteristic of other minerals too. It is also possible that the enclosure of biomolecules within a cell was first accomplished on or within a mineral, though lipid-like vesicles might have been present first in a watery environment and then colonised by biomolecules.

In recent years attention has turned to the role of minerals deep under the Earth's surface. On the ocean floors, at hydrothermal vents (Section 2.5.2), simple minerals such as the iron oxide magnetite (Fe_3O_4) and the iron sulphide pyrite (FeS_2) might be able to accomplish much. Magnetite could have catalysed the formation of NH_3 from the H_2 and N_2 that these vents emit – biological reactions involving N_2 require it in the form of NH_3. More profoundly, groups of minerals, particularly iron and nickel sulphides, could have acted as templates, catalysts, and as the energy sources that produced the first carbon-based biomolecules, and enabled them to form quasi-living systems. The minerals might have acted as reactants too, giving up some of their atoms to various biomolecules. Biomolecules can exist inside minerals at temperatures greater than outside, and so the high temperatures near such vents are not a problem.

Graham Cairns-Smith of the University of Glasgow, UK, has taken the role of clays to the limit, by proposing since the 1970s that clays, not carbon compounds, provided the first genetic material. It was *life*, though not as we know it. He has come to this view because though clays (and other) minerals seem able to promote the formation of carbon-based biomolecules, and would thus have made it far easier for carbon-based life to emerge, they are insufficiently selective, with many non-biological outcomes. Therefore, there seems to be no opportunity for evolution to act with such minerals, and consequently it is unlikely that specific molecules like RNA and certain proteins emerged.

The proposed solution is a *clay-based* genetic material that can produce offspring imperfectly, thus enabling evolution by natural selection. A mineral can fulfil this function in a variety of ways. Figure 3.9 shows just one possibility, in which the genetic information is encoded as an aperiodic sequence of stacked layers. The sheet could adsorb organic molecules from the surroundings, and might catalyse

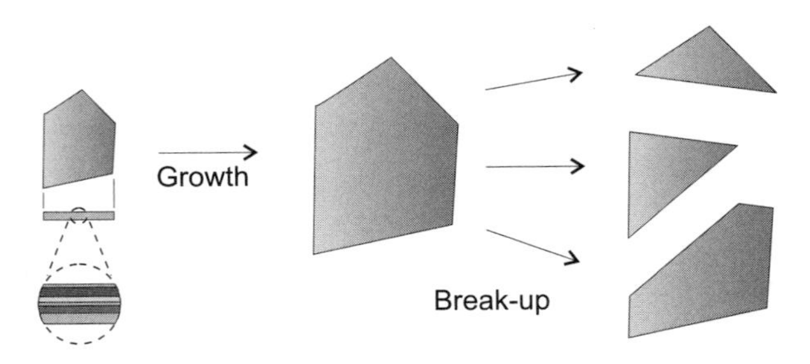

Figure 3.9 A mineral-based genetic system, in which imperfect reproduction occurs when the sheet grows sideways and break up.

reactions among them. This activity would depend on the particular stacking sequence. The sheet could become embedded in a jelly that would help control the interaction of the sheet with its environment. It would be rather like a very primitive cell. Reproduction is when a sheet grows sideways and breaks into pieces. Each layer is likely to have imperfections within it, or there might be stacking faults, and so the pieces (offspring) might not all be identical to the parent. Through evolution by natural selection the cells could become ever more complex in a way that promoted their survival and made their (imperfect) replication more efficient.

New structures would emerge and disappear many times. Then, at some point a molecular structure would emerge which, in addition to whatever useful function it served (such as structural), happened itself to be a genetic material, in that it could promote the production of copies of itself, perhaps still using components provided through the action of the original clay genetic material. It could, through imperfect copying and natural selection, make powerful structures with the clay becoming obsolete. This might have happened several times, until RNA or DNA appeared. The jelly would also have evolved, and would have become enclosed in a membrane – a true cell would have appeared.

The first crystal-based genetic material could have taken many other forms than that in Figure 3.9. Indeed there are now too many possibilities! Even if life did originate in this general way, we certainly do not know any details. We do, however, know that crystal replication can go wrong very easily. In the example in Figure 3.9 we need sideways growth that on the whole preserves the layers, otherwise the cell structure of the offspring is very likely to be unviable. The general difficulty of preserving a reasonable degree of fidelity goes some way towards answering the question 'why is life not originating in this way all the time?'. Also, selection pressures that favour complexity might be lower in today's different environment. Nevertheless, if there is anything in minerals as the precursor of carbon-based life, then minerals must still be assembling structures capable of evolution towards life. Perhaps it is because any complex quasi-biotic molecules are now gobbled up by existing organisms that life no longer springs from clay.

3.3.4 Chirality in biomolecules

One feature of life on Earth is associated with chirality. Figure 3.10 shows the amino acid alanine and the sugar ribose. In each case two forms are shown, labelled L and D, that differ only in that they are mirror reflections of each other, rather as your left hand is a mirror reflection of your right hand. This means that the molecule cannot be superimposed on its mirror image. In this case the molecule is said to be *chiral.* A striking feature of life on Earth is that in all DNA and RNA the sugars occur only in the D form, and very nearly all amino acids in proteins occur only in the L form. The other forms are of no biological use at all. For example, a ribosome constructed to assemble L amino acids will not attach D forms – the external shape of the D form is wrong, rather like a left-handed glove will not fit a right hand. Why should some biomolecules have to be D and others L? Minerals offer one possible solution.

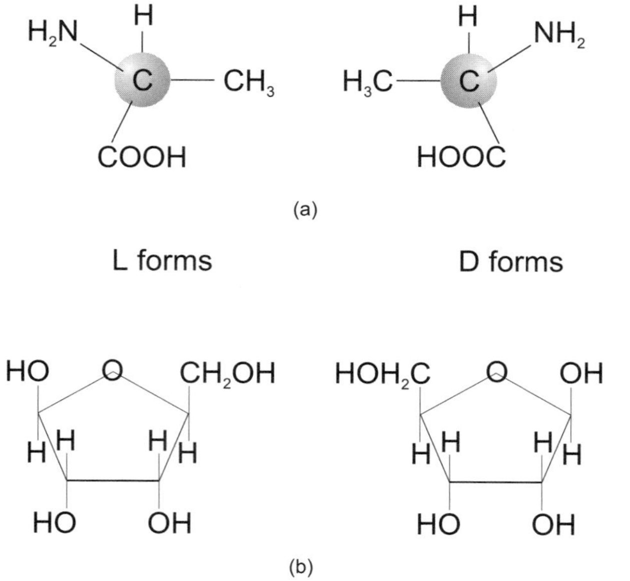

(a)

L forms D forms

(b)

Figure 3.10 (a) Alanine in L and D forms. The sphere around the central carbon atom is to help indicate the 3-D bond directions. (b) Ribose in L and D forms. The carbon atoms at the apices of the hexagonal ring of bonds are not shown.

Some minerals are chiral, in that they have faces that are mirror reflections of each other, for example, the faces of a crystal of calcite ($CaCO_3$). It has been found that if calcite is exposed to amino acids, one type of face acquires an excess of the L form on its surface, and the other type acquires the D form, particularly if the surfaces are finely terraced. The two types of face are equally abundant, so we need to assume that it was by chance that the L form appeared first and through (self) replication quickly dominated the biosphere. Another example is provided by certain clays, where, due to geometrical space limitations, a surface can acquire the property of either being able to adsorb only the L form of a chiral molecule, or only the D form. Again it would seem to be chance which chiral form came to dominate our biosphere.

There are other explanations of chiral biomolecules. One is that circularly polarised light bathed the cloud from which the Solar System formed. In this case a slight excess of one hand of molecule over the other could have been produced in the prebiotic molecules. Circularly polarised light can be crudely modelled as waves spiralling through space like a corkscrew, twisting either clockwise or anticlockwise, and the favoured molecular hand depends on which sense the light happened to have during the molecule's formation in space. The light in star-forming regions has been observed to be circularly polarised, and some of the amino acids in carbonaceous meteorites have a slight excess of one hand over the other. However, this explanation, like that involving minerals, is no more than plausible. The origin of chirality in biomolecules is yet another mystery.

3.3.5 Where did life originate?

It is perhaps natural to think that life must have originated in the oceans, near the surface within reach of sunlight and perhaps in shallow pools or tidal flats where small component molecules could have been concentrated by evaporation, perhaps on clay surfaces. There are however four reasons for us to look deeper.

First, the heavy bombardment must have frustrated the origin of life at the surface many times. If life originated beneath the surface it would have been protected from this and other environmental hazards, giving several hundred million years of opportunity that were denied to life on the surface.

Second, photosynthesis was not operating at the origin of life – it is a very complex process that must be some distance along in evolution. It is possible that a primitive system existed in which solar photons were used as a source of energy without acquisition of carbon from CO_2, but it is more likely that a non-solar source powered the earliest life. There are biologically potent energy sources well away from the surface. These are based on redox reactions between materials coming from deep in the Earth and the environment into which they emerge, such as at hydrothermal vents (Section 2.5.2). Therefore, though much (perhaps all) life on Earth now relies ultimately on photosynthesis, this was surely not the case at the beginning. We thus need not confine ourselves to the surface of the Earth in looking for life's origin.

Third, much of the biosphere today, perhaps most of it by mass, is well below the surface of the solid Earth, in the crust (Section 2.5.2). We need not assume that this deep biosphere spread from the surface – it could have been the other way.

Fourth, the tree of life (Figure 2.13) shows that among the earliest forms of life, still with us today, is a large group of Archaea that are thermophiles, flourishing in the sort of temperatures found at hydrothermal vents and deep in crustal rocks. Though these Archaea must have evolved from the (unknown) last common ancestor, and though later colonisation by species that originated in cooler environments cannot be ruled out, they do indicate that life might have originated at high temperatures.

For all these reasons the view that life originated deep in the Earth and spread to the surface has gained considerable support among biologists, regardless of the extent of mineral involvement. By thus removing the origin of life from where the Sun keeps a planetary surface suitably warm, the number of potential habitats in the Solar System increases, as you will see in Chapter 4.

One other possible source of life on Earth is extraterrestrial. In modern times this idea dates back to the early 20th century, to the Swedish physical chemist Svante August Arrhenius, and the British physicist Lord Kelvin. They each proposed the idea of panspermia in which dormant cells travel through interstellar space, seeding planets. In Lord Kelvin's version the dormant cells are carried inside meteoroids, and are thus protected from radiation. Panspermia doesn't solve the problem of how life arose – it merely transfers the origin elsewhere – though it does seem to be a viable idea, at least within the Solar System. The possibility that life could have emerged on our planetary neighbours and that we have about thirty meteorites that seem to have come from Mars, raises the intriguing possibility that we are all martians! More on this in Chapter 5.

3.3.6 When did life originate?

In Section 3.2.1, 3850 Ma was given as the earliest time for which there is any evidence at all for life on Earth, and this would have been fully-fledged RNA/DNA-protein life. The origin of life would clearly predate this, but we really have no idea by how much. It is possible that life originated much earlier, perhaps a few times, only to be made extinct by large impacts during the heavy bombardment. What kind of life any such extinct form might have evolved into today is of course completely unknown, but given the random nature of evolution by natural selection it is certain it would have differed considerably from the life around us. Alternatively, life might have been protected deep in the crust, never being completely extinguished. But if life was not protected in this way, then the present habitable phase began towards the end of the heavy bombardment, 3900 Ma ago. Therefore, if fully-fledged RNA/DNA-protein life was established by 3850 Ma, then there is only 50 Ma for a staggeringly huge amount of evolution. On the other hand, if our last common ancestor lived not long before the latest really firm evidence for life, at 2700 Ma (Section 3.2.1), then there is a lot more time.

As soon as a self-replicating carbon-based biomolecular system emerged, it began to use up the environmental resources on which it relied. Evolution by natural selection would then have led to greater efficiency and greater complexity, until life in the form of the last common ancestor arrived. Subsequently, there would have been a negligible chance of any other self-replicating, evolving system emerging – life as we know it would have been a powerful 'predator', occupying all possible niches.

3.3.7 Conclusions

The origin of life on Earth is one of the greatest of all unsolved scientific mysteries. A huge amount of progress has been made in recent decades, but there are still great gaps in our knowledge. Fortunately, for the rest of the book we don't need to understand the details of the origin of life on Earth, just that it happened early in Earth history, and that it might have arisen very soon after the Earth became habitable.

We have far better knowledge of the essential characteristics of life on Earth, and we must remember these in searching for life elsewhere, to optimise our search strategies. It makes sense to concentrate on carbon–liquid water life, and on the conditions it requires. However, our strategies must not be based too closely on the particular version of carbon–liquid water life that we have on Earth, namely life based on RNA, DNA, and a particular set of proteins, or we might miss variants. Furthermore, in some types of search we might be able to find life based on very different chemistry (e.g., no carbon and/or no water). We therefore need to decide what it is we are really looking for out there – this is where we start the next chapter.

3.4 SUMMARY

- The last common ancestor of all life on Earth today certainly predates 2700 Ma, probably predates 3460 Ma, and might even predate 3850 Ma. Since then, evolution by natural selection, and punctuated evolution at mass extinctions, has produced the great variety in present-day organisms.
- For most of Earth history the biosphere was dominated by unicellular organisms with multicellular organisms flourishing only in the last few hundred million years.
- The biosphere, through oxygenic photosynthesis, is responsible for sustaining the high O_2 content of the atmosphere, and the chemical imbalance between O_2 and CH_4. It is also responsible for lowering the CO_2 content by accelerating the formation of carbonate rocks and creating organic carbon deposits such as fossil fuels. The biosphere maintains an enhanced atmospheric content of N_2 through prokaryotes that liberate N_2 from organic matter.
- There is evidence for liquid water at the Earth's surface throughout Earth history, in spite of the lower luminosity of the Sun early on. One solution to this faint Sun problem is through the carbonate–silicate cycle, which led to higher CO_2 content and therefore an enhanced greenhouse effect. CH_4 might also have contributed to an enhanced greenhouse effect at that time.
- The 'chicken and egg' problem, that there can be no proteins without DNA and no DNA without proteins, has led to the idea of RNA world. In this world, RNA acted as the repository of genetic information, and as the catalyst in the synthesis of proteins and RNA. The discovery in 1983 of ribozymes – enzymes made of RNA – has boosted this idea, though there are many remaining difficulties.
- Though the RNA nucleotides are easier to synthesise than those of DNA, it is still sufficiently difficult that scientists have looked to minerals for help. The roles suggested for minerals include promotion of the growth of biopolymers at clay surfaces, mineral reactions as energy sources, and clay-based genetic material that led to the development of RNA/DNA.
- That some chiral biomolecules occur in organisms only in the D form and others only in the L form is unexplained. Possible explanations include the influence of minerals with chiral surfaces in the formation of the first biomolecules, and the influence of circularly polarised light on the precursor molecules in the solar nebula.
- The origin of life on Earth is very poorly understood, though there is growing support for the idea that it originated beneath the surface of the Earth, either deep in the oceans at hydrothermal vents or in pores and crevices deep in the Earth's crust.

- If there were RNA/DNA-protein organisms 3850 Ma ago, and if the heavy bombardment sterilised the Earth until around 3900 Ma, then life emerged promptly, no longer than about 50 Ma after the Earth became habitable. There might be a lot more time available than this.

3.5 QUESTIONS

Answers are given at the back of the book.

Question 3.1

On a 24-hour clock, with the Earth's origin at 00:00, and the present at 24:00, give the times of the following dates: the end of the heavy bombardment; the appearance of prokaryotes; the appearance of eukaryotes; the rise of atmospheric oxygen; the spread of multicellular organisms; the Permian mass extinction; the appearance of *Homo sapiens*; and the start of the Common Era (1 AD).

Question 3.2

Outline what might happen to the composition of the Earth's atmosphere if the biosphere died. Give reasons.

Question 3.3

If the last common ancestor existed at about 3000 Ma rather than 3850 Ma, what enormous advantage does this give us for tracing the steps that led to the last common ancestor.

Question 3.4

Suppose that extraterrestrial RNA is discovered in a meteorite.

(a) Discuss to what extent this could boost our understanding of how life originated on Earth, stating an important proviso involving chirality.
(b) State why it would not necessarily boost our understanding of the origin of life in the Solar System.

4

Where to look for life elsewhere in the Solar System

We now move beyond the Earth and embark on the search for extraterrestrial life, a broad and rapidly growing subject. In this chapter we will discuss what sort of life we are searching for and where in the Solar System we might find it. In subsequent chapters we consider Mars and Europa in more detail. We then consider potential habitats beyond the Solar System.

4.1 WHAT SORT OF LIFE ARE WE SEARCHING FOR?

What is life? It is not easy to come up with a general definition, and some people have resorted to 'I can't define it but I'll know it when I see it'! One general definition widely used is that life is characterised by imperfect replication that permits evolution. Imperfect replication alone is insufficient. Many entities replicate more or less imperfectly, such as mineral crystals, but evolution, particularly by natural selection, allows the sort of progression from biological precursors to multicelled creatures that has occurred on Earth. Another general definition, fairly widely used, is that life directs its own actions.

Such general definitions provide little guidance to the best places to search for extraterrestrial life. Considerably more guidance is provided by the restriction that we are looking for life that has the same general basis as life on Earth (i.e. complex carbon compounds and liquid water). This is not as parochial as it might seem. Carbon, among all the chemical elements, has by far the greatest potential for forming the huge variety of complex compounds that are necessary to support the processes of life and to store the information needed to produce each new generation. The element with the next greatest potential is silicon, particularly when it combines with oxygen, which is why silicates are the basis of such a wide variety of rocky materials. However, even though clays, which are hydrated silicate minerals, might have played a crucial role in the emergence of life on Earth (Section 3.3.3), there is no evidence that silicon could form the basis of a fully developed form of life. We therefore set silicon aside, though note that complex compounds involving silicon are stable at much higher temperatures than are complex carbon compounds, are therefore it just might be the case that silicon-based life exists somewhere at temperatures too high for carbon-based life.

A less radical alternative is to substitute liquid water with, for example, liquid ammonia. This would enable life to operate at lower temperatures. But this is conjecture, and there is no evidence as yet that such an alternative is possible.

It therefore makes sense, at least initially, to focus our searches on potential habitats for carbon–liquid water life. The one exception is the search for extra-terrestrial intelligence (SETI) – in this case it is an *activity* that we are searching for, namely communication across interstellar space, and the chemical basis is of secondary concern. SETI is the subject of Chapter 13.

4.1.1 Potential habitats for carbon–liquid water life

Potential habitats need to have significant quantities of water and carbon, plus sufficient quantities of other elements found in complex carbon compounds, such as nitrogen. There also needs to be a suitable energy source, such as solar radiation or redox pairs of chemical compounds (Section 2.4.2).

The water needs to be in the liquid phase. Figure 2.10 showed the ranges of temperature and pressure over which pure water exists in stable form as a liquid, solid, and gas. That Figure is reproduced here as Figure 4.1. It is truncated at about the maximum pressures at which life has been found on Earth (Section 2.5.2), though there is no known reason why life cannot exist at even higher pressures. The lower pressure limit for pure water to be liquid is 610 Pa, though this could be sustained in a cell even if the pressure in the surrounding environment was lower. Thus, life could exist at any of the pressures in Figure 4.1, though it is difficult to see how life could have emerged if pressures everywhere had never exceeded 610 Pa.

Temperature is a more severe constraint. On Earth, life is found over the temperature range −18°C to 123°C (Section 2.5.2). Below 0°C, water in cells is kept liquid by having certain proteins and other substances dissolved in it, thus lowering the freezing point. The low-temperature limit might be even lower, if, for example, the water in the cells were mixed with ammonia or rich in salts. The boiling point is raised above 100°C by high pressure, though dissolved substances can also contribute. At even higher pressures pure water can be liquid at temperatures greater than 123°C, but above about 160°C the complex biochemical carbon compounds break down. Therefore, for carbon–liquid water life, potential habitats will probably have temperatures below about 160°C, down to several tens of degrees below 0°C (Figure 4.1).

Potential habitats could exist at or below the surface of a planet. First, consider the surface. This brings us to the concept of the habitable zone around a star.

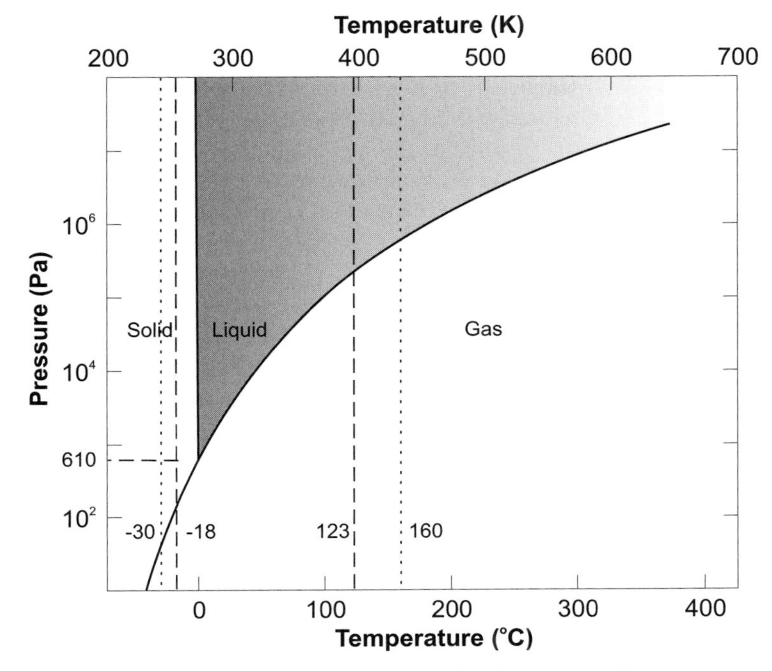

Figure 4.1 The phase diagram of pure water, showing the ranges of temperature over which life has been found on Earth (−18°C to 123°C) and over which carbon–liquid water life might be possible (approximately −30°C to 160°C).

4.2 THE HABITABLE ZONE (HZ)

The HZ around a star is *defined* as that range of distances from the star within which the stellar radiation would maintain water in liquid form over at least a substantial fraction of the surface of a rocky planet. If the planet is closer to the star then all the surface water evaporates. If the planet is further from the star then the surface water everywhere is permanently frozen.

Even if it is in the HZ, the planet must be sufficiently well endowed with volatiles to form an atmosphere. The planet must also have sufficient mass (m), for three reasons. First, the atmosphere could then provide enough pressure for water to be liquid. Atmospheric pressure at the surface, the force per unit area of surface, is given approximately by:

$$p_s = \left[\frac{m_c G m}{R_s^2}\right] \tag{4.1}$$

where m_c is the column mass of the atmosphere (Section 1.1.3), R_s is the surface radius, and G is the universal constant of gravitation. Thus, for a given column mass, the greater the mass of the planet the greater the surface pressure. Second, with sufficient mass the planet is likely to be geologically active enough to keep a large proportion of its volatiles from being incorporated in crustal reservoirs. Mars, at about one-tenth the Earth's mass, is below the limit, and has lost atmosphere to the crust, as you will see in Chapter 5. Mars now has only a thin atmosphere.

Third, with sufficient mass, little of the atmosphere will have leaked away into space. If a molecule is moving outwards fast enough from the top of the atmosphere it will escape. The escape speed is given by:

$$v_{esc} = \left[\frac{2Gm}{R_a}\right]^{1/2} \tag{4.2}$$

where R_a is the distance from the centre of the planet to the top of the atmosphere. Equation (4.2) shows (reasonably) that the greater the mass the greater the escape speed, and the less readily the planet loses its atmosphere. Table 4.1 lists v_{esc} for some representative bodies in the Solar System. We need to compare v_{esc} with molecular speeds. Molecules in an atmosphere have a great range of speeds, even at a given temperature. In the midst of this range lies the most probable speed, v_{mps}. This depends on the mass of the molecule and on the local temperature. The higher the mass and the lower the temperature the smaller is v_{mps} and the less likely these molecules are to escape. Table 4.1 lists $6v_{mps}$ for some common molecules at the present-day mean temperatures high in the atmospheres from where escape to space is possible – these temperatures are quite different from the surface temperatures. For the Moon, which has no atmosphere, the mean surface temperature has been used, though it could be argued that the peak temperatures are more appropriate. The factor 6 emerges from detailed studies showing that if $6v_{mps} \sim v_{esc}$ throughout (most of) the 4600 Ma history of the Solar System, then this is the approximate threshold for significant loss by thermal escape (see planetary science books in Resources). Table 4.1 shows that Venus, Earth, Mars, and Jupiter should not lose H_2O, H_2, and CO_2 by thermal escape.

It is important to note that thermal escape is not the only way that a planetary body can lose volatiles to space. Even if a planet is sufficiently massive to avoid thermal escape it might still lose copious quantities in other ways. In the case of Mars there has been non-thermal escape, where molecular speeds are boosted by chemical reactions and by the impact of a wind of charged particles from the Sun – the solar

Table 4.1. The escape speed v_{esc} of some planetary bodies and $6v_{mps}$ for some common molecules for these bodies, where v_{mps} is the most probable speed.

Speed (km s^{-1})	Venus	Earth	Moon	Mars	Jupiter
v_{esc}	10.4	11.2	2.4	5.0	59.5
$6v_{mps}$ H_2	8.2	17.6	8.9	8.6	16.3
$6v_{mps}$ H_2O	2.7	5.8	3.0	2.9	5.4
$6v_{mps}$ N_2	2.2	4.7	2.4	2.3	4.4
$6v_{mps}$ CO_2	1.7	3.8	1.9	1.9	3.5

wind. There have also been losses through impact erosion by large bodies. These have been important for Mars because its escape speed is low.

For a planet in the HZ to have sufficient mass to have liquid water at its surface it thus needs to be rather more than the mass of Mars, one-tenth that of the Earth. The term '*Earth-mass planets*' will denote planetary bodies from a few times the mass of Mars to a few times that of the Earth. Suppose, therefore, that we have a sufficiently massive planet, sufficiently well endowed with volatiles. The actual distances of the HZ boundaries are determined from a model of the planet's atmosphere. These are hard to construct and always contain approximations, such as modelling vertical structures only, or the models hedge around certain poorly understood factors, notably the effects of clouds. This has led to a considerable range of boundary distances. Also, the boundary criteria can be specified in different ways, which adds to the range of boundary values.

HZ boundaries

For the inner boundary, the criterion adopted here is the distance from the star at which the planet loses all the liquid water from its surface rapidly via what is called a *runaway greenhouse effect*. Water is a powerful greenhouse gas (Section 2.1.3). It has a tendency to cause a runaway effect because an increase in surface temperature increases the quantity of water vapour in the atmosphere, which further increases the surface temperature, which results in a further increase in water vapour, and so on. The surface temperature might stabilise short of complete evaporation, and the planet is then, by definition, in the HZ. But if the stellar radiation is sufficiently intense the evaporation will become complete. Not only will the surface then be devoid of liquid water, the surface temperatures will be at least 374°C (647 K), well above the 160°C limit for complex carbon compounds, so carbon–liquid water life would be impossible.

At the outer boundary of the HZ the low temperatures ensure that there is little mass of water in the atmosphere. CO_2, among the common volatiles is then the controlling substance, and its greenhouse effect is crucial. The criterion for this boundary adopted here is the maximum distance from the star at which a cloud-free CO_2 atmosphere could maintain a mean surface temperature of 0°C. This maximum occurs at a CO_2 pressure of 8×10^5 Pa. At greater pressures the surface is cooler because of the stellar radiation scattered back to space by the atmosphere, and at lower pressures the CO_2 greenhouse effect is weaker, so the surface is again cooler. These are commonly used boundary criteria, and though there are others leading to a narrower HZ, there are yet others for which the HZ is even broader.

At the inner boundary the water cloud in the model is fixed at the present terrestrial value. In reality, the cloud cover would increase and this would increase the scattering of solar radiation back to space. The surface would then be cooler and so the boundary would move inwards. At our outer boundary the formation of CO_2 clouds would also increase the scattering of solar radiation back to space, but this would probably be more than offset by an enhancement of the greenhouse effect, and so the outer boundary would move outwards. The greenhouse effect is enhanced through the scattering of the infrared radiation emitted by the planet's surface by the tiny CO_2 ice crystals that would constitute the clouds. The outer boundary could also be moved outwards by traces of powerful greenhouse gases, such as CH_4.

The HZ boundaries vary with the age of the star. A star is born when it is stabilised by the nuclear fusion of hydrogen to helium (Section 1.1.2). It has then started the main sequence phase of its lifetime, and its HZ is denoted by HZ(0), termed the zero age value. By about this time, or soon after, any terrestrial planets will also have been born. The star's effective temperature (Section 1.1.2) will then gradually rise, shifting its spectrum to shorter wavelengths, and its luminosity will gradually increase. The increase in luminosity tends to cause an increase in surface temperature of any planets, but the increase in effective temperature, because of the associated shift in the stellar spectrum to shorter wavelengths, tends to cause a decrease in surface temperature. However, the increase in luminosity has the larger effect, and so the inner boundary of the HZ shifts outwards. The outer boundary will also shift outwards unless it is pinned at the HZ(0) value. This would happen if any frozen planet could never thaw in spite of the changes in the star. In this case a planet initially outside HZ(0), experiences a *cold start*, and would remain frozen. Cold starts would not be a problem if, for example, the carbonate–silicate cycle (Section 3.2.3), were sufficiently effective.

In spite of its name, the HZ does not encompass all the locations in a planetary system where potential habitats can be found. Other locations will be discussed in Section 4.3. Nor does it follow that if a sufficiently massive planet lies in the HZ, and is sufficiently well endowed with volatiles, it will be inhabited. One reason among many is that it has not been in the HZ long enough. You will encounter other reasons later.

4.2.1 The HZ in the Solar System

Figure 4.2(a) shows HZ(0) for the Solar System, and, 4600 Ma later, the present day habitable zone HZ(now), according to our adopted criteria. Figure 4.2(b) shows the increase in the luminosity and the change in effective temperature of the Sun throughout its 11 000 Ma main sequence lifetime. The orbits of the terrestrial planets have been added to Figure 4.2(a), and you can see that Mercury is well interior to HZ(0) and HZ(now), so we would expect it to have lost any water it might have had, particularly because of its small mass. Mercury indeed is very dry, and has barely any atmosphere at all. Moreover, as Mercury rotates, its surface temperature varies from far in excess of 160°C in daytime (up to about 430°C) to as low as about −180°C at night. Only in isolated pockets near the poles might temperatures be maintained in a range suitable for liquid water, but this would require an atmospheric pressure far in excess of that found. It would not be a good use of resources to search for life on the surface of Mercury.

Venus is at the inner boundary of HZ(0), but the evolution of the Sun would soon have carried the boundary beyond Venus. However, clouds could have placed the inner boundary of HZ(0) closer to the Sun than shown in Figure 4.2(a), and so Venus might have been in the HZ for the first billion years or so of its history. It is

the radius of the body. Thus, the smaller the body the greater the loss-rate per unit source volume, and the cooler the interior becomes. This simple model neglects possible differences in heat transfer mechanisms from the interior to the surface, but is the basic reason why a rocky satellite a few hundred kilometres across has a cooler interior today than a 10 000-km planet with the same composition.

With life deep in rocks a possibility, how many more potential habitats can we find in the Solar System? Mercury, which has a frigid globally averaged surface temperature of about $-100°C$, is just about large enough to have an interior that is still sufficiently warm in its outer few tens of kilometres for water to be liquid. Very near the surface, within a few tens of metres, the huge diurnal swings in surface temperature are too great for liquid water to persist. However, there is no evidence for water in any form on Mercury, except perhaps near the poles. We therefore set Mercury aside until we know more about its volatile endowment. Venus's surface, almost everywhere close to 462°C at all times, is too hot with temperatures even greater in the interior, so we set Venus aside too. Mars is a case where we know that water is abundant, and where temperatures in the crust could sustain it as a liquid. Mars will be explored in Chapter 5.

Among the planets, that leaves the four giant planets and Pluto. In Jupiter the atmospheric temperature increases with depth below the topmost layer of cloud and passes through a range of altitudes where the temperature and pressure allow water to be liquid and complex carbon compounds to exist. Water and carbon compounds are present at these altitudes and there is sufficient solar radiation to provide an energy source. All the conditions for life are there, and this has led to the fanciful speculation of life in the jovian atmosphere. Unfortunately, the atmosphere is highly convective, causing downward mixing to where the temperatures are very high, so it is difficult to see how life could have emerged in such an unstable environment. Similar remarks apply to Saturn. In the cases of Uranus and Neptune, much of their interiors are probably liquid water, and though most of it is probably far too hot for life, it is possible that in its upper reaches it is sufficiently cool, though convection currents might again have frustrated the emergence of life. Overall, the giant planets are very unlikely to have life.

With Pluto we come to a body considerably smaller than the other planets, and indeed smaller than several satellites (Figure 1.7). Such small bodies would initially have had warm interiors from primordial heating, but the great majority must have cooled in less than 100 Ma, and are therefore unlikely ever to have seen life. At first sight it seems that only the very largest, Ganymede, Titan, and Callisto, could have remained warm enough for sufficiently long for life to emerge, and just might possibly be warm enough today to sustain subsurface life. But this ignores the supplementation of internal heat sources by tidal heating. This can be regarded as an external source of heat, and it can make a difference to subsurface habitability.

4.3.2 Tidal heating

Tidal heating arises from the gravitational distortion of a body by one or more other bodies. Figure 4.4 shows the distortion of a satellite caused by the giant planet it

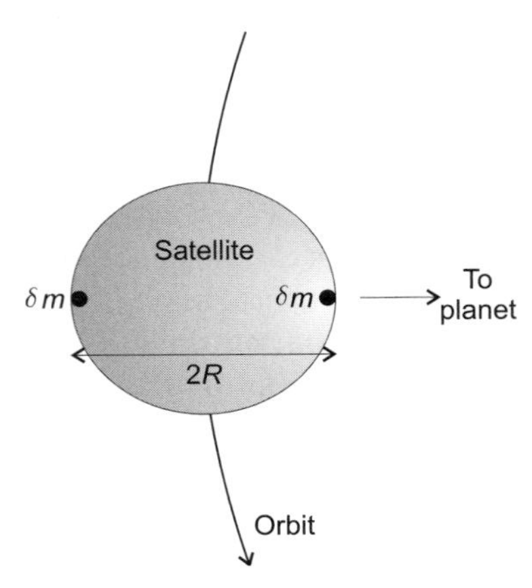

Figure 4.4 The gravitational distortion of a satellite caused by the giant planet it orbits (exaggerated).

orbits. To see why this occurs, consider the gravitational force exerted on a satellite with a mass m by a giant planet with a mass M a distance d away. This force has a magnitude (size) given by (see physics books in Resources):

$$F = G\frac{Mm}{d^2} \tag{4.3}$$

The force is directed towards M. Strictly, this equation requires either that the bodies are extremely small compared to their separation, or that they have spherically symmetrical mass distributions with d the distance between their centres. To sufficient accuracy the equation applies to a giant planet and any one of its satellites. It also applies to any small portion of a satellite. If we consider a portion with a mass δm nearest the giant, and a portion with a similar mass furthest from the giant (Figure 4.4), then clearly the magnitude of the force on the near side is greater than that on the far side. The difference in the magnitude of the forces on the extremities is:

$$\Delta F = \frac{GM\delta m}{(d-R)^2} - \frac{GM\delta m}{(d+R)^2} \tag{4.4}$$

where R is the radius of the satellite. With d very much greater than R this equation simplifies to the approximation:

$$\Delta F = \left[\frac{4GM\delta m}{d^3}\right]R \tag{4.5}$$

This differential force tends to stretch the satellite along the direction to the planet.

In addition, there is a differential gravitational force perpendicular to the direction to the planet, half that of ΔF in Equation (4.5), and arising from the slight difference in the *direction* to the planet from the leading and trailing extremities of the satellite in its orbit. This force tends to squeeze the satellite. Differential gravitational forces exist throughout the satellite, all contributing to the tidal distortion. The overall tidal force, as in Equation (4.5), is proportional to $1/d^3$. Therefore, even though the Sun is far more massive than the giant planet, it is so much further off that its tidal force on the satellite is negligible.

How does tidal distortion lead to heating? The satellite needs to be in a non-circular orbit so that its distance from the giant planet varies. When it is closer to the planet the tidal force is greater, and so the satellite is more distorted than when it is further from the planet. Thus, as it orbits the planet, the whole satellite flexes, and this causes heating of the interior. If the satellite were highly fluid then the tidal distortion would always lie close to the line from the satellite to the planet, as in Figure 4.5(a). Additional tidal heating arises as follows. Consider the common case when the orbital period of the satellite equals its rotation period, and the two rotations are in the same direction. This is *synchronous rotation*, and were the orbit circular the satellite would keep the same face towards the planet. This is not quite the case when the orbit is non-circular, as Figure 4.5(a) shows, where P is a particular point on the satellite's surface. The satellite rotates at a uniform rate, but its orbital speed varies, being greatest when the satellite is closest to the planet and least when it is furthest. As seen from the planet, P oscillates around the line from the satellite to the planet. Thus, the distortions swing to and fro within the body of the satellite, and this flexing further heats the interior. If the satellite is much more rigid, the tidal distortion is carried considerably off-line, as in Figure 4.5(b), but both components of tidal heating still occur. In non-synchronous rotation the details differ again, but both components are still present.

The average rate of tidal heating depends on the composition of the satellite, but is proportional to other parameters approximately as follows:

$$W_{\mathrm{tidal}} \propto (Mm)^2 R \frac{e}{a^6} \tag{4.6}$$

where e is the eccentricity of the satellite's orbit and a is its semimajor axis. The derivation of this expression can be found in advanced texts on the Solar System or planetary science (see Resources), but even on qualitative grounds it is to be expected that W_{tidal} increases as e increases, and decreases as a increases.

4.3.3 Tidally heated bodies

The most dramatic examples of tidal heating are among the Galilean satellites of Jupiter – Io, Europa, Ganymede, and Callisto. These are planet-sized worlds, ranging in size from somewhat smaller than the Moon (Europa) to somewhat larger than Mercury (Ganymede, the largest satellite in the Solar System). Interior models are shown in Figure 4.6. Io and Europa are predominantly rocky bodies. Io contains little or no water, but Europa is topped with a shell of water about 100 km thick. Ganymede has a very thick shell of water around a rocky centre. Callisto is poorly differentiated, consisting of a mixture of rock and water throughout, and therefore its primordial heating did not include any heat of differentiation. The question is, to what extent is the water in Europa, Ganymede, and Callisto liquid?

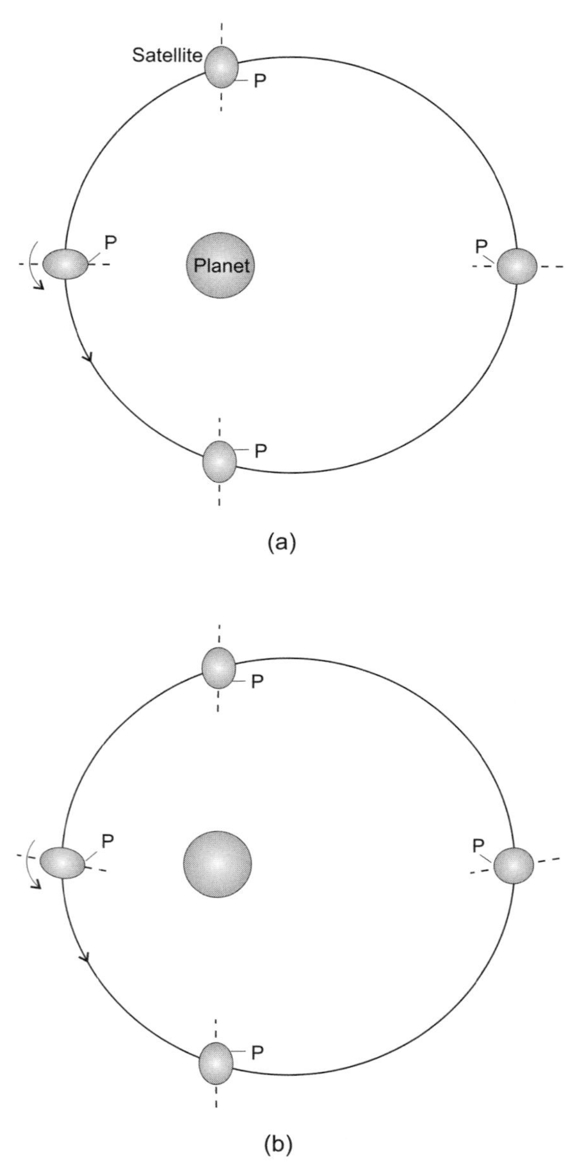

Figure 4.5 A satellite in a non-circular orbit around a planet (not to scale), showing its varying tidal distortion. (a) Highly fluid satellite (b) More rigid satellite.

It is certainly ice at the surface, but what about deeper down? Tidal heating would be required to melt the ice, and that depends on the orbits.

The orbits are shown to scale in Figure 4.7, and though it is not apparent on the scale of this figure, all of the orbits are slightly elliptical. Therefore, all four satellites

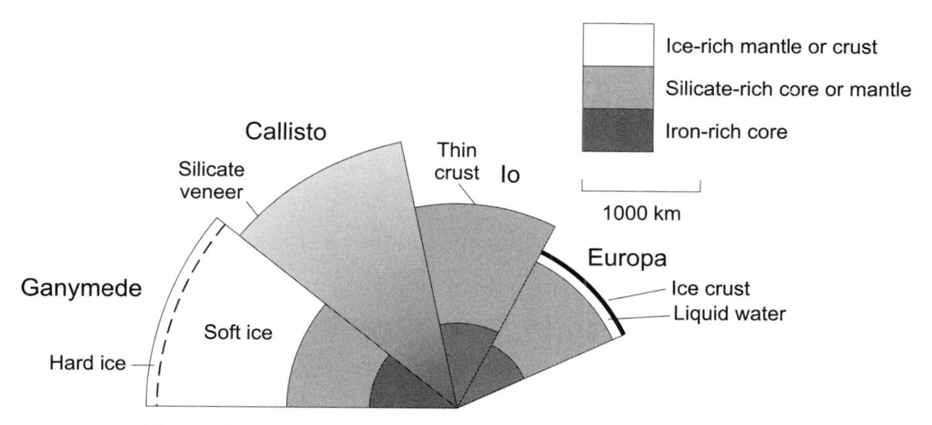

Figure 4.6 Internal models of the Galilean satellites of Jupiter.

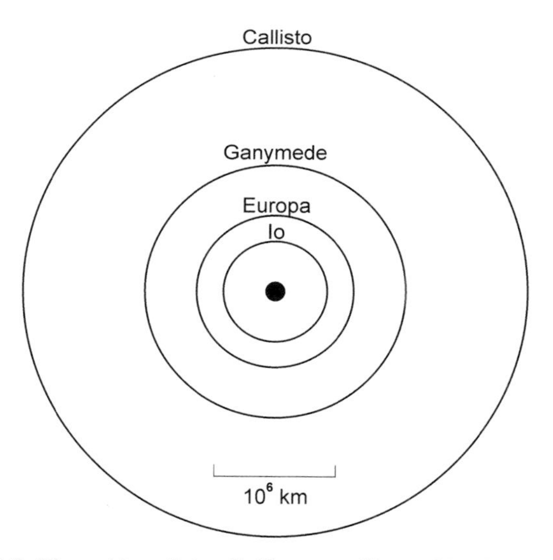

Figure 4.7 The orbits of the Galilean satellites of Jupiter (to scale).

are tidally heated. Moreover, the orbital periods of the inner three are in the ratios 1 : 2 : 4. Consequently, each time Ganymede goes around Jupiter once, Europa goes around twice, and Io goes around four times. These are examples of *mean motion resonances*. As a result, the satellites periodically tug each other in a cumulative way so that the orbital eccentricities are higher than they would otherwise be, and the tidal heating is correspondingly greater (Equation 4.6). Table 4.2 gives the present values of the orbital eccentricities, and other data.

Io is the most tidally heated of the four. This is because it is in the smallest orbit. It is so strongly heated by tides and by the decay of long-lived radioisotopes that it is the most volcanically active body in the Solar System (Plate 10 and Figure 4.8).

Table 4.2. Properties of the Galilean satellites of Jupiter, and their orbits.

	Mass of satellite (10^{21} kg)	Radius of satellite (km)	Semimajor axis (10^3 km)	Orbital eccentricity
Io	89.3	1821	422	0.004
Europa	48.7	1565	671	0.010
Ganymede	149.0	2635	1070	0.001
Callisto	107.5	2405	1885	0.007

Figure 4.8 Volcanic activity on Io. Loki, one of many active volcanoes, is easily visible against the backdrop of space.
NASA/JPL.

If it had a suitable atmosphere then it could have a habitable surface, but its atmosphere is extremely tenuous, and even if at shallow depths there are places at present suitable for liquid water, these are unlikely to be suitable for long, as volcanic activity sweeps across this world.

Europa is less tidally heated than Io – its greater distance from Jupiter and its

Table 4.3. Properties of various satellites and their orbits.

	Mass of satellite (10^{21} kg)	Radius of satellite (km)	Semimajor axis (10^3 km)	Orbital eccentricity
Enceladus (Saturn)	0.065	250	238	0.004
Triton (Neptune)	21.4	1350	354	<0.0005
Charon (Pluto)	1.6	600	19.4	~ 0

smaller size more than offset its greater present orbital eccentricity (Equation 4.6). As a consequence it is not obviously volcanically active, and indeed it is ice-covered. However, there is evidence that there is (or recently has been) an extensive ocean of liquid water under the icy crust, and it might have a modest level of volcanism on the rocky ocean floor that could support life, rather like the hydrothermal vents in Earth's oceans (Section 2.5.2). Europa is the subject of Chapter 6.

Ganymede and Callisto are only weakly tidally heated today, though the orbital eccentricity of Ganymede can be episodically larger than the present value. Both of these satellites have magnetic fields that require an electrically conducting fluid in their interiors, and this could be liquid water. In bodies of this size it is certainly possible that heat from long-lived radioisotopes could sustain liquid water at depths below about 100 km. However, among the Galilean satellites, Europa is the best prospect for habitability. We will therefore leave Ganymede and Callisto here, as long shots.

Other bodies that might be sufficiently tidally heated to have liquid water include Enceladus, a 250 km radius satellite of Saturn (Table 4.3, Figure 4.9). In spite of its small size, some of its icy surface is so free of impact craters that it must have been resurfaced relatively recently, in which case tidal heating is presumably responsible, and this could melt ice at some depth. The eccentricity of the orbit is currently the same as that of Io, though it might have been higher in the past, and it is in any case only 238 000 km from Saturn.

Beyond Saturn, the low temperature of the solar nebula during planetary formation could have led to NH_3, CH_4, and other icy materials being present in considerable quantities. In this case, the lowering of the freezing point of water by the admixture of such materials increases the chance of finding subsurface liquid with a high water content. Neptune's large satellite Triton might be one such case. Today tidal heating is negligible (Table 4.3), but in the past the orbital eccentricity might have been much larger. This is because it is believed that Triton is a captured embryo. The process of capture would have resulted in an eccentric orbit that it could have taken a billion years to circularise. It is just possible that water-rich liquid is still there today, sustained by rather feeble radiogenic heat. The evidence that Triton was captured is its retrograde orbit – it goes around Neptune the opposite way to all the other large satellites in the Solar System, and opposite also to the direction the planet orbits the Sun.

The other bodies where tidal heating might be or might have been important are Pluto and its satellite Charon. The orbit of Charon is almost circular (Table 4.3), and

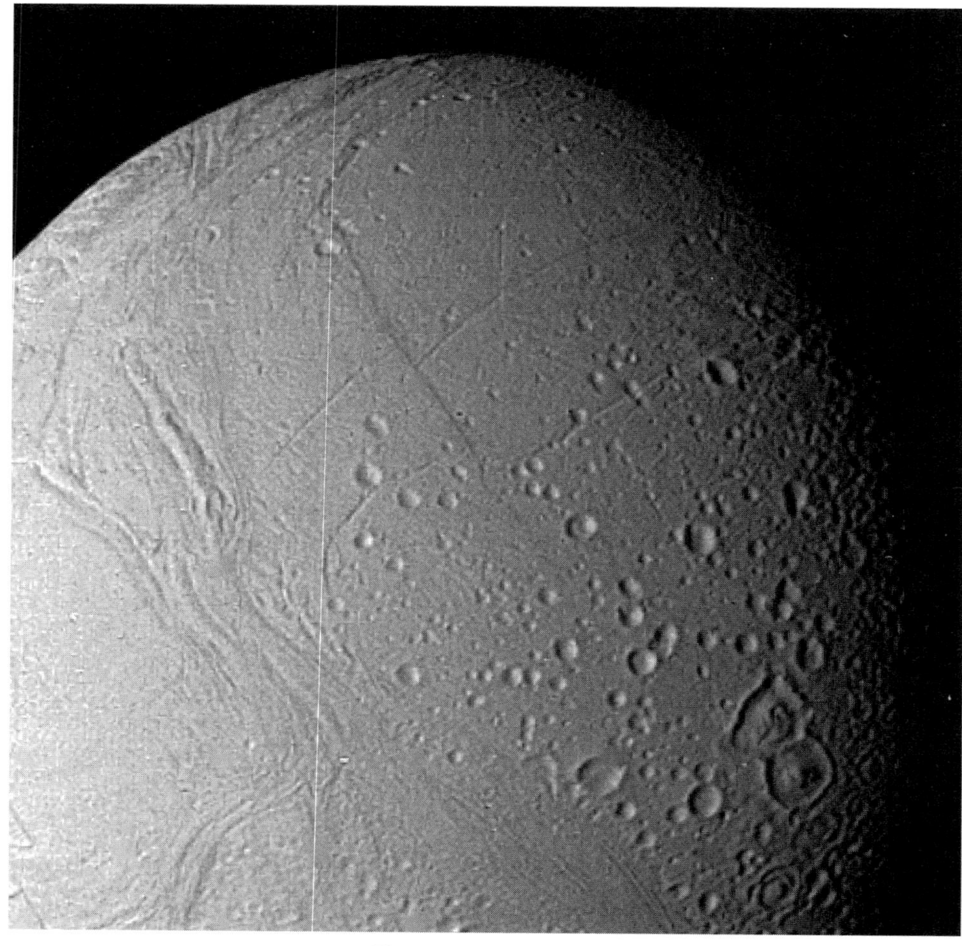

Figure 4.9 Enceladus.
NASA/JPL.

they are each in synchronous rotation with respect to each other. This would seem to rule out tidal heating. However, the orbit of Charon is almost perpendicular to the orbit of Pluto. It turns out that the resulting distortions of Charon can enhance the tidal heating, and it is just possible that in Charon today there is a layer of water-rich liquid. Pluto is now unlikely to have any such liquid, though for its first several hundred million years it is possible that enough heat was left over from any giant impact on Pluto that might have formed Charon. In addition to this, tidal heating before the orbits settled into their present configuration, could have sustained a water-rich liquid long enough for life to emerge. These possibilities are somewhat remote, and so we set Pluto and Charon aside until we have more data on these distant worlds.

Table 5.2. The atmosphere of Mars, with the Earth for comparison.

	Major gases (number fractions)		Column mass ($kg\,m^{-2}$)	Acceleration of gravity ($m\,s^{-2}$)	Mean surface pressure (Pa)	Mean surface temperature (°C)
Earth	Total		10 300	9.78	101 000	15
	N_2	0.78				
	O_2	0.21				
	Ar	0.0093				
	H_2O	0.01				
	CO_2	0.000 345				
Mars	Total		150	3.72	560	−55
	CO_2	0.95				
	N_2	0.027				
	Ar	0.016				
	O_2	0.0013				
	CO	0.0007				
	H_2O	0.0003				

follows:

$$p_s = m_c \frac{Gm}{R_s^2} = m_c g_s \tag{5.1}$$

where m is the mass of the planet and R_s is its surface radius. In this equation $g_s = Gm/R_s^2$ is the force per unit mass, or the value of the acceleration of gravity at the surface. Table 5.2 shows that the mean surface pressure on Mars is less than the 610 Pa triple point of water (Section 4.1.1). Therefore, only in very low-lying areas, where the column mass and consequently the pressure will be greater than the mean, could water be stable as a liquid, if the temperature exceeded 0°C.

The mean surface temperature is very low, −55°C in equatorial regions and −119°C at the poles, which might suggest that even with greater pressure it would be too cold for liquid water. But these are average temperatures, and the diurnal temperature variations on Mars are huge so that it can reach about 10°C in tropical regions in daytime. Figure 5.4 shows the approximate ranges of surface pressure and temperature on Mars, on a phase diagram of water. It is possible for water to be stable in low-lying tropical regions during a few hours around midday. The large diurnal temperature variations are due to the infrared transparency of the atmosphere (infrared radiation from the surface escaping freely to space). This transparency also results in only a weak greenhouse effect, which raises the mean surface temperature by about 8°C. On Earth the greenhouse effect is larger (about 33°C), mainly because there is a lot more water vapour in the atmosphere.

The large diurnal and seasonal changes in temperature across the martian surface lead to strong winds of 10 $m\,s^{-1}$ or more. These winds, even in such a thin atmosphere, are sufficient to raise the widespread micron-sized dust particles into the atmosphere to form yellow dust clouds. These range in size from local dust storms to dust clouds that can envelope almost the whole planet for several months.

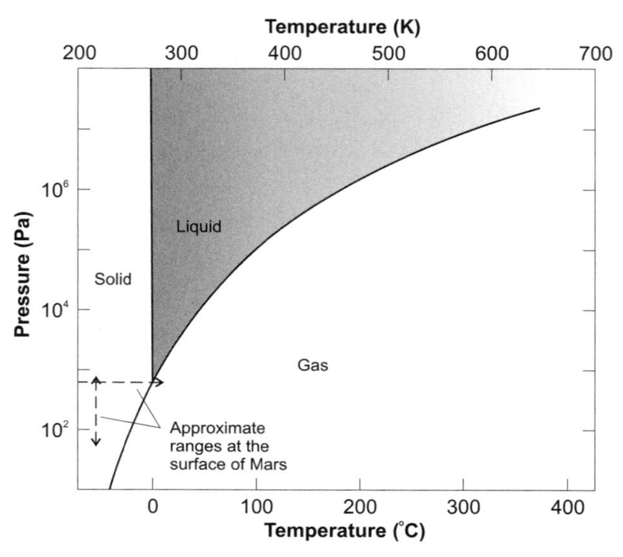

Figure 5.4 The approximate pressure and temperature ranges that occur on the surface of Mars, on a phase diagram of water. The low pressures at the highest altitude peaks have been excluded.

5.1.4 The martian surface from space

Mars has long been observed from space, and by the end of the 19th century it was well known to have a surface exhibiting dark areas on a light red background and white polar caps, overlain by a thin atmosphere that supported white clouds and dust storms. Plate 5 shows one of the best images of Mars taken from Earth, in which most of these features can be seen. Until the exploration of Mars by spacecraft it was thought by many scientists that the dark areas on the surface were some form of vegetation. These areas certainly exhibit seasonal changes in their shape and contrast. Then, in July 1965 the NASA spacecraft Mariner 4 flew by Mars and revealed very little difference between the dark and light areas. Both were impact cratered, like the Moon. Moreover, Mariner 4 discovered that the atmospheric pressure, 400–610 Pa, was too low to allow liquid water on the surface. Therefore, the belief that the dark areas were vegetation died almost overnight.

We now know, from the various spacecraft since Mariner 4 that have flown by, orbited, or landed on Mars, that the light areas are covered in light-reddish dust, and the dark areas in dark-reddish dust. The light dust is probably derived from the dark dust by a variety of physical and chemical processes. The dust is everywhere dominated by basaltic silicates (Section 2.1.2), but the light dust additionally contains clay minerals (Section 3.3.3). The ubiquitous red tint is due to iron-rich minerals in the dust. The separation between the two types of dust probably depends on prevailing winds interacting with small-scale topography. Seasonal changes are then the result of seasonal changes in the winds.

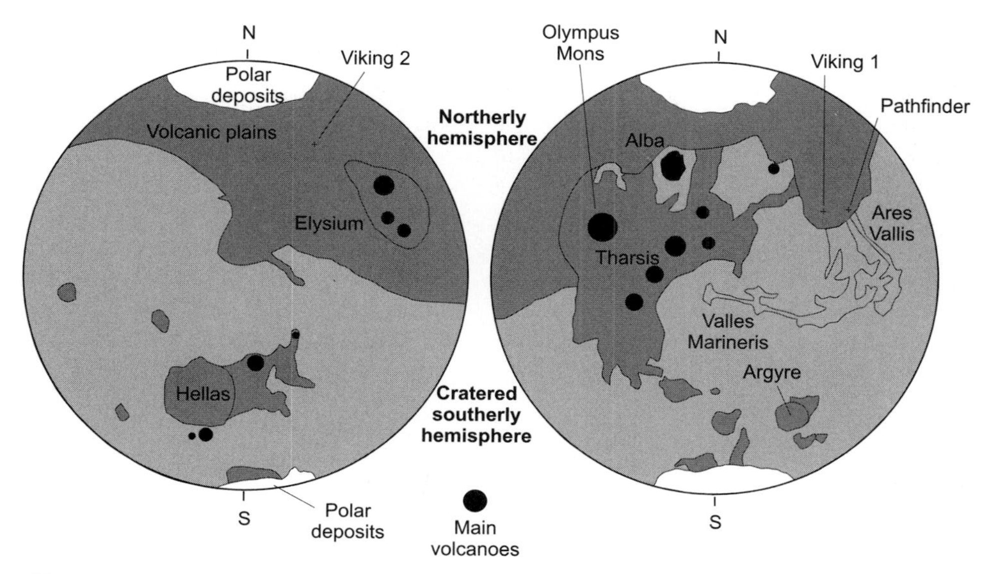

Figure 5.5 The two hemispheres of Mars, with some major topographic features marked.

Seasonal changes also occur in the polar caps, which advance in the autumn and retreat in the spring. These changes are due to the deposition and sublimation of CO_2 ice. The northern cap loses all its CO_2 in the summer so that a permanent cap largely of water ice mixed with dust is exposed. The retreat of the southern cap halts with CO_2 still in place, possibly mixed with water ice, and presumably overlying water ice. This is in spite of the more extreme seasonal changes in the south (Section 5.1.1). The reason for this is not yet known, though one suggestion is that it might be the greater proportion of dust in the northern cap, darkening it and promoting summer warming.

Mars as a whole divides into two hemispheres, with a ragged dividing line inclined at about 30° to the equator (Figure 5.5). The southerly hemisphere is at an average altitude about 5 km higher than the northerly hemisphere (altitudes are measured using the atmospheric pressure at the local surface). This correlates with the martian crust being generally thicker in the south, though this crustal boundary does not coincide with the altitude boundary. At present the reason for this discrepancy is unknown.

The two hemispheres display striking differences in their surface features. The southerly hemisphere is dominated by impact craters and impact basins (Figure 5.6), whereas the northerly hemisphere is dominated by vast plains, domes, volcanoes, and rift valleys (Figure 5.7). The scarcity of impact craters in the northerly hemisphere shows that it is the younger hemisphere – the older a surface the more time it will have had to accumulate impacts. Lava flows are one way to remove impact craters and thus rejuvenate a surface, and there is widespread evidence of lava flows in the northerly hemisphere. The thinner crust in the north would give magma

Figure 5.6 Features of the southerly hemisphere of Mars, impact craters, and, lower left, the 900-km diameter Argyre impact basin.
NASA/JPL.

more ready access to the surface than in the south, so there is a consistent picture here.

Though there are volcanic features in the northerly hemisphere, and though this is the younger hemisphere, this does not mean that the surface there is *young*, nor that volcanism is active today. Much of the northerly hemisphere's surface could be billions of years old. Overall, Mars today is far less geologically active than the

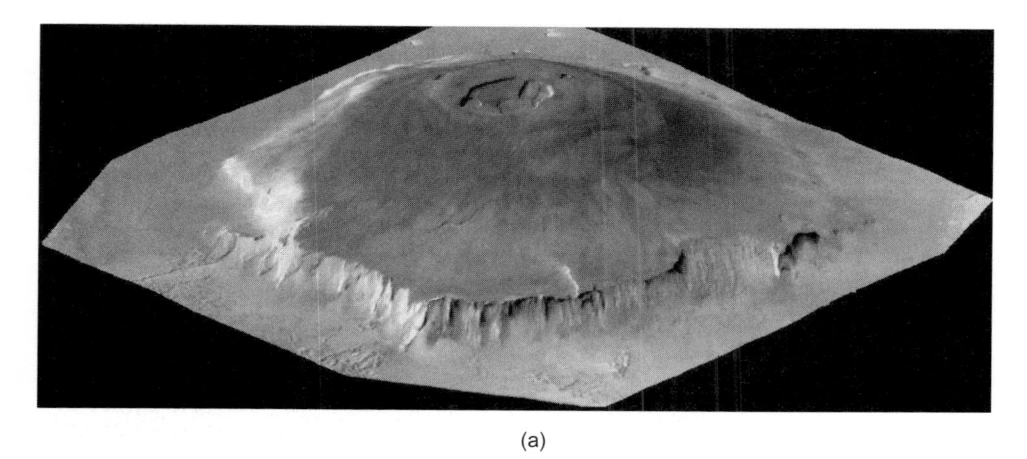

(a)

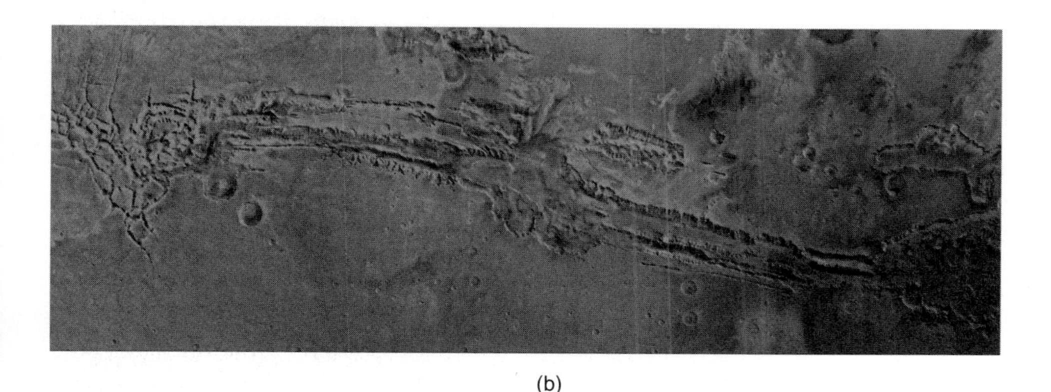

(b)

Figure 5.7 Features of the northerly hemisphere of Mars: (a) Olympus Mons, a volcano 550 km across reaching 25 km above its surroundings (vertical scale exaggerated); (b) Valles Marineris, a rift valley about 2000 km long.
NASA/JPL.

Earth, in accord with its thicker lithosphere. There are, however, two regions that have substantially younger surfaces, the Tharsis and Elysium regions (Figure 5.5), huge bulges that are home to most of the martian volcanoes. Several of the Tharsis volcanoes have almost certainly been active in the last 1000 Ma, and the least cratered areas of the lava flows from Olympus Mons (Figure 5.7(a)) might be as young as 30 Ma, indicating that this volcano is probably dormant rather than extinct. Some modest level of magma activity at no great depth cannot be ruled out today, as indicated by some of the widespread liquid-water features, to which we now turn. Such features are of obvious importance to Mars as a potential habitat.

5.1.5 Features that indicate the presence of liquid water

There are several types of surface feature on Mars that suggest the involvement of liquid water. One type is shown in Figure 5.8, two examples from a small proportion of impact craters that have peculiar ejecta blankets. These are probably the result of impact melting of near-surface ice and its subsequent flow.

Other types of suggestive surface feature are shown in Figure 5.9, a variety of channels, all of which could be the result of flowing liquid water. Figure 5.9(a) shows an example of an outflow channel. These are of order 1000 km

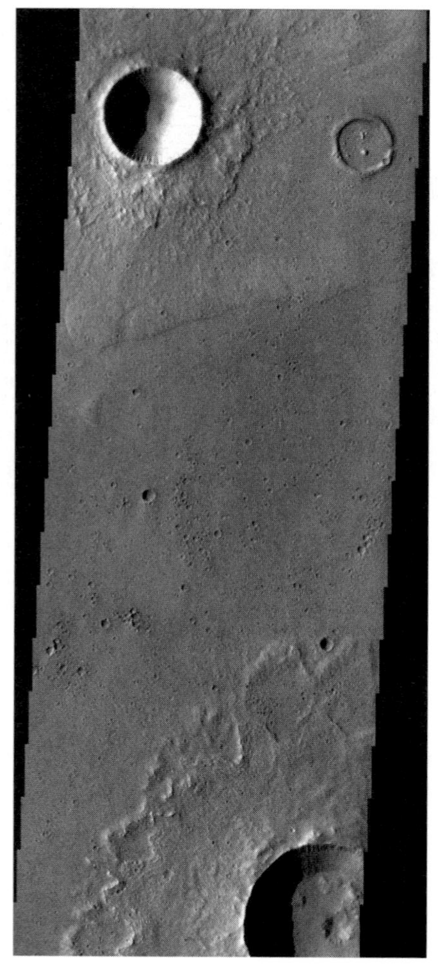

Figure 5.8 Martian craters a few kilometres across, with ejecta blankets suggesting a surface flow of water plus entrained rocky materials.
NASA/JPL/Arizona State University.

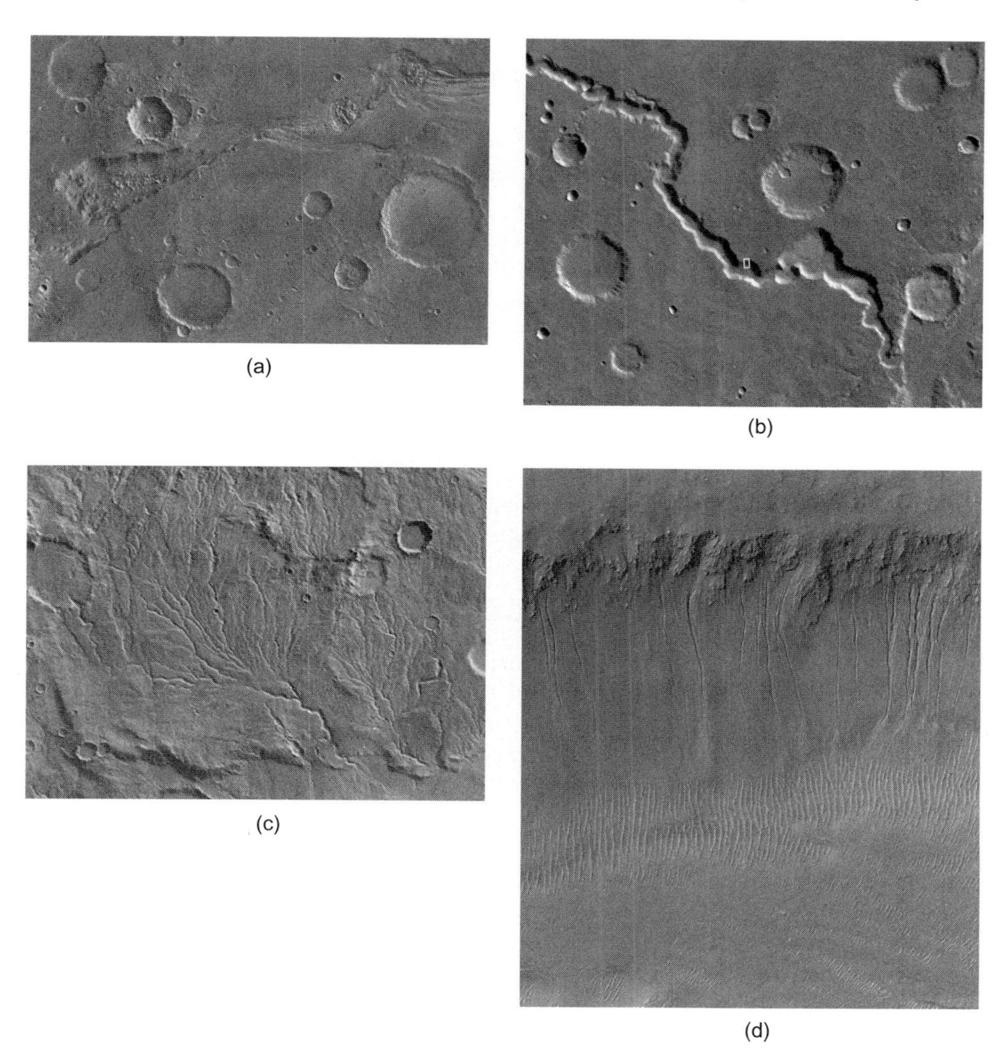

Figure 5.9 Martian channels. (a) The outflow channel at the head of Simud Vallis. The width of the frame is about 300 km. (b) Part of the fretted channel Nirgal Vallis. Frame width about 160 km. (c) A valley network. Frame width 130 km. (d) Gullies in the south-facing wall of Nirgal Vallis, in the small box in (b). Frame width 2.3 km.

(a), (b), and (c) NASA/JPL. (d) NASA/JPL/Malin Space Science Systems.

long, 100 km wide, and a few km deep, with teardrop islands that have tails downstream. They seem to be the result of the sudden outflow of subsurface water – lava flows would have generated different morphologies. Most of the outflow channels originate in aptly named chaotic terrain, where water seems to have been released from an aquifer previously sealed beneath a permafrost layer. The release could have been caused by an impact, by fault movement, by melting due to the subsurface movement of magma, or by the pressure of water in the aquifer exceeding a critical value.

Other outflow channels originate in canyons. Groundwater seepage into these canyons could have created lakes with ice-covered surfaces. Any weakening of a canyon wall or rising water levels would then lead to an overflow. Evidence of the former shorelines has been found in some of these canyons. Yet other outflow channels are associated with fissures from which lava has flowed, the channels indicating that water was also released. A series of such fissures extending for over 1000 km across the lava plains of Cerberus, south of the Elysium region (Figure 5.5), seems to have been active fairly recently, perhaps 10 Ma ago, and could become active again.

Most of the outflow channels flow into low-lying plains in the northerly hemisphere and into the Hellas Basin in the southerly hemisphere (Figure 5.5). Lakes must have formed on these plains, and there are features consistent with this possibility, such as possible shorelines and extensive deposits of layered sediments. It has been conjectured that oceans the size of the Indian Ocean have occupied much of the northerly hemisphere, the water now stored in huge aquifers made of porous sediments and lavas, but this remains controversial.

Fretted channels (Figure 5.9(b)) are narrower and more sinuous than outflow channels, and have short, stubby tributaries. They stretch hundreds of kilometres into the southerly uplands from parts of the boundary between the northerly and southerly hemispheres.

Valley networks (Figure 5.9(c)) are particularly interesting. Each consists of channels with typical widths of 1–10 km, typical depths of 100–200 m, and having well developed tributary systems. Some of them seem to connect with depressions that might be ancient lake beds. Most of the networks are in the southerly hemisphere, where they are widespread, particularly at high altitudes and low latitudes. The creation of valley networks was thus concentrated at some earlier period in martian history.

Though it is widely accepted that the valley networks were created by the flow of liquid water, there remains the question of whether it was surface run-off following precipitation, or the release of groundwater near the head of each network. The channel morphologies, compared to terrestrial examples of known cause, indicate that all, or nearly all, were created by groundwater, with at most a small minority created by run-off, though precipitation might have contributed to the groundwater supply. Precipitation of liquid water is certainly not possible on Mars today, because the atmospheric pressures and temperatures are generally too low. Therefore, if there was widespread precipitation in the past the atmosphere must have been warmer and more massive. This might also have been necessary in order for liquid water to have survived for the time it took to flow along hundreds of kilometres of narrow channels, though liquid water protected by surface ice might be able to survive long enough even at the present time.

A minority view is that the fretted channels and valley networks were not caused by liquid water alone, but by general subsidence, mobilised by groundwater at the base of the debris, in which case it would have been rocky rubble and dust that comprised most of the flow. It is however difficult to see how such a process could have produced such long narrow channels.

Surface water today, or in the recent past

Gullies were discovered in 2000 by the orbiting spacecraft Mars Global Surveyor. They are small V-shaped features found on the edges of channels, canyons, and craters. Figure 5.9(d) shows one example. The gullies and their immediate surroundings are so fresh-looking that these flows must have occurred within the past few million years, and perhaps very recently, with continuing activity possible. A possible source is the melting in spring of snow at the base of thin winter deposits. This liquid water could form and flow because it would be protected by the overlying snow. Later in spring the rest of the snow disappears to reveal the gullies. Another possibility is that the gullies, or at least some of them, are caused by the release of liquid water from subsurface deposits of ice somewhere in the region, perhaps by heat from magma. Alternatively, the variations in the axial inclination of Mars (Section 5.1.1) might be the main cause. When the inclination is greater than about 45° the poles get sufficiently warm in summer that water vapour from subsurface ice could enter the atmosphere to produce a significant rise in atmospheric pressures and temperatures. Near-surface ice deposits could then melt and flow over the surface, to produce the gullies. The inclination last exceeded 45° a million or so years ago. A quite different possibility is that the gullies are the result of CO_2, rapidly released as a gas in spring, with no water involved at all. Curiously, there are no old, degraded gullies – they seem to be solely a relatively recent phenomenon.

Confirmation that near-surface water is widely available came in 2002, when the orbiting spacecraft, Mars Odyssey, returned data from its gamma-ray spectrometer and two neutron spectrometers. Neutrons are ejected from atoms in the outer metre or so of the martian surface by cosmic rays – very fast charged particles from space. Some of these neutrons reach the spacecraft directly, others indirectly, and yet others collide with atoms resulting in emission of gamma rays, some of which also reach the spacecraft. A gamma ray photon has an energy characteristic of the atomic nucleus it came from, and so we can identify the elemental composition of the atoms. The best interpretation of the data from the three spectrometers is that there is a hydrogen-rich layer at increasingly shallow depths as latitude increases. It lies below about 1 m at mid-latitudes and below about 0.3 m at polar latitudes. These depths correlate with those below which water ice should be stable, and so it seems likely that the hydrogen is in the water molecule, in ice, which would account for about half the volume of the material at these depths at mid-latitudes and up to 90% at high latitudes. The spectrometers also indicate that at middle to equatorial latitudes, where ice would be too deep for cosmic rays to reach, there is still some hydrogen. If so, this is likely to be chemically bound to minerals, such as clays, as water molecules, or as OH. Near-infrared observations by the Hubble Space Telescope (HST) also indicate that minerals incorporating water or OH are present on at least some parts of the surface.

There thus seems to be plenty of water near the martian surface, probably as ice most of the time. What have we learned at the surface itself?

5.1.6 The martian surface from landers

So far (2003) five spacecraft have landed on Mars. In 1971, as part of the Soviet Mars 2 and Mars 3 missions, two robotic rovers reached the surface, but the Mars 2 rover crash-landed and the Mars 3 rover unexpectedly ceased transmission after 20 seconds. In 1976, two NASA orbiters reached Mars, Viking 1 and Viking 2, and each of these successfully placed a lander on the surface. In 1997, the NASA spacecraft Pathfinder also landed successfully. Of the three successful landers, only Vikings 1 and 2 directly targeted the search for life, as will be described in Section 5.3.2.

The Viking Landers are located in well-separated areas in lowland plains in the northerly hemisphere (Figure 5.5). The view from both sites is broadly similar, a gently undulating, dusty landscape strewn with boulders in the 0.01–1 m size range, most of which were probably ejected from nearby impact craters. Bedrock might be visible here and there. The limited photometry that the Lander cameras could perform on the boulders is consistent with basalt. At the Viking 2 Lander site some of the surface has the same crusty appearance as duricrust on Earth, where it is the result of dust grains becoming bound together by materials precipitated in evaporating pools of water. In the case of Mars today, the water could have been delivered by upward percolation, or from the melting of permafrost. Plate 15 shows a view from Viking 2 Lander, where the horizon is about 3 km away. Note the water frost in the shade of the boulders.

At both sites the dust was sampled and analysed, and the outcome is consistent with orbital spectrometry on bright dust. The relative abundances of chemical elements and isotopes, and sparse data from other observations, can be matched by clays rich in magnesium and iron. On Earth such clays result from the action of water on iron- and magnesium-rich basalts, and on both planets such basalts are to be expected in volcanic lavas. The high iron content of the dust, through oxidation (by an unknown process) has produced hematite (Fe_2O_3), and this gives Mars its red tint. The dust at the sites is also rich in sulphur, perhaps a result of sulphates left by water evaporation. The action of water is also expected to produce carbonates, and though the Landers had no means of detecting these, the infrared spectrometer on the Mars Global Surveyor did detect them, fairly widespread over the surface.

Pathfinder landed near the mouth of the outflow channel Ares Vallis (Figure 5.5). The view to the west is shown in Figure 5.10. The two peaks on the horizon, Twin Peaks, are 30–35 m tall. Though the last outflow from Ares Vallis might have been as long as 2000 Ma ago, evidence of a copious outflow is still abundant, in topographical features such as ridges, troughs, and a distant streamlined island, and also in the boulders – the assortment of types, their size distribution, the roundness of some, and the way that many are stacked. A few boulders, and the more southerly of the Twin Peaks, might be layered, suggesting sedimentary deposition. The surface dust is much the same as at the Viking Lander sites. Bright dust was found to be more common than dark dust, the latter predominantly being found where winds

Figure 5.10 A view of Mars from Pathfinder, which landed near the mouth of the outflow channel Ares Vallis. The horizon is about 1 km away.
NASA/JPL.

had swept the bright, finer dust away. Airborne dust contained maghemite, an iron oxide that on Earth is precipitated from water rich in iron compounds.

The Pathfinder mission included a briefcase-sized rover, called Sojourner, which made several trips up to 12 m from Pathfinder, and measured the relative abundances of the elements at the surfaces of six boulders – the Viking Landers could only carry out such analyses on the dust. These analyses were supplemented by photometry in more wavelength bands than the Viking Landers could perform, covering the range 0.44–1.0 μm. Some of the boulders do not have a basaltic composition, but have a lower proportion of iron and magnesium somewhat like parts of the continental crust on Earth (Section 2.1.2). As on Earth this could be the result of partial melting of basaltic rocks. Overall, the composition of martian boulders and dust is consistent with volcanically produced basalts, modified in some cases by subsequent partial melting and by the action of water.

The skies at the three sites were usually milky-red, caused by dust in the atmosphere. At times of low dust content, and particularly towards the zenith, the sky was a very dark blue.

5.2 MARS IN THE PAST

Previous sections have indicated that conditions on Mars today seem to differ from those in the past, and in particular that it might once have been warmer and wetter. Whatever the present potential of Mars as a habitat, this raises the possibility that it was more habitable in the past.

5.2.1 The three epochs of martian history

It has already been noted that, on the basis of the density of impact craters, the southerly hemisphere is older than the northerly hemisphere. Impact crater densities are the basis of the division of the history of Mars into three epochs. The Noachian Epoch is the earliest and is defined by surfaces that have crater densities near to saturation, that is, a new crater, on average, will remove a pre-existing crater (e.g. Figure 5.6, lower right). The Hesperian Epoch followed the Noachian. Surfaces from this epoch have moderate crater densities. The Amazonian Epoch followed the Hesperian, with surfaces that are lightly cratered (e.g. Figure 5.7). The Amazonian Epoch continues today. Much of the southerly hemisphere is Noachian. Most of the rest of this hemisphere, and some of the northerly hemisphere is Hesperian.

The absolute ages for the boundaries between these epochs are poorly known. We have to rely on lunar data, where surfaces with different crater densities have been dated absolutely using radiometric methods. These dates are difficult to apply to Mars, because of Mars's different gravity, its proximity to the asteroid belt, the different nature of the martian surface, and the higher degradation rate of craters. Different models lead to different ages. The Noachian Epoch clearly embraces the heavy bombardment that pervaded the Solar System in its youth, and that was in steep decline on the Moon 3900 Ma ago. Even so, the Noachian–Hesperian boundary on Mars could have an age anywhere in the range 3500–3800 Ma. The other two epochs lie in the far lighter bombardment that has occurred since the end of the heavy bombardment. The Hesperian–Amazonian boundary could have an age anywhere between 1800 Ma and 3500 Ma (Figure 5.11).

5.2.2 Atmospheric change on Mars

Valley networks (Section 5.1.5) are concentrated in the Noachian areas of the southerly hemisphere, with a few in the Hesperian. This indicates that their formation was concentrated very early in martian history. The atmospheric pressures and temperatures might then have been high enough for liquid water to be stable over much of the martian surface. It certainly indicates that the conditions for their formation had gone soon after the start of the Hesperian Epoch. In particular there was no longer an ample supply of liquid water to keep the groundwater supplies charged. This could well have been associated with a thinning and cooling of the atmosphere. Further evidence for this is that Noachian craters are far more eroded than even the oldest Hesperian craters.

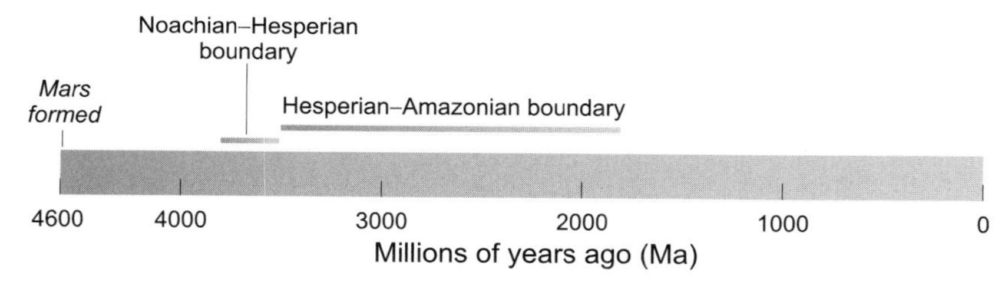

Figure 5.11 The three martian epochs, showing the uncertain boundaries between them.

What sort of atmosphere might Mars have had in the Noachian? An important factor is that the Sun started its main sequence lifetime with only about 70% of its present luminosity, and that even by the end of the Noachian it had only reached about 75% (Figure 4.2(b)). At Mars's distance from the Sun this required a powerful greenhouse effect to attain the minimum mean surface temperature of about 0°C for liquid water to be widespread. It could have been achieved in a dry atmosphere with the greenhouse effect of a minimum of about 10^5 Pa of CO_2, which is about 180 times the present atmospheric quantity. It is essential that the CO_2 clouds that would surely have formed did not promote cooling, and it seems they would not have done. It has been shown (Section 5.2) that the tiny CO_2 ice crystals in high-altitude clouds would have contributed to the greenhouse effect by scattering infrared radiation back to Mars's surface, at the very least offsetting the cooling from the solar radiation that the clouds would have reflected back to space. The atmosphere would have contained some water vapour, also a greenhouse gas, so the CO_2 requirement is reduced, by up to a factor of about two. Any other greenhouse gases that might have been present, such as NH_3 or CH_4, further reduce the CO_2 requirement.

If liquid water was indeed widespread at the surface in the Noachian then it would have promoted removal of CO_2 by chemical weathering of rocks to form carbonate rocks on the short timescale of 10 Ma. Moreover, meteors and larger bodies were plunging into the martian atmosphere and ejecting gases into space. The low gravity of Mars facilitated such erosion, which would have been significant during the heavy bombardment of the Noachian. To maintain the atmosphere, volcanism and impact cratering would have had to break up the carbonates and return the CO_2 to the atmosphere at a sufficient rate. Plausible models of these competing processes show that the atmosphere could have been kept in place throughout the Noachian, and therefore we can account for the denser, warmer conditions that might have prevailed at that time.

The bombardment that characterises the Noachian must also have increased the water content of the atmosphere. The impactors themselves are likely to have been water-rich, and the heat of impacts from impactors 100 km across or larger would have liberated this water and also water beneath the surface of Mars. It is even possible that between large impacts Mars was too cool for liquid water at its surface, in which case the Noachian atmosphere would only episodically have been massive enough to sustain liquid water.

The atmospheric loss that is presumed to have been ongoing since the end of the Noachian is thought to have resulted largely from the decline in geological activity expected for a small planet as it cools, in spite of the decline in impact erosion at the end of the heavy bombardment. The recycling of CO_2 from carbonates then proceeded at a very slow rate, and so the atmospheric quantity fell. Much of the initial complement of CO_2 was thus lost to carbonates. Further losses have been a slow persistent leakage to space, facilitated by Mars's low gravity. This leakage requires a molecule near the top of the atmosphere to reach the escape speed (Equation 4.2), and this it can acquire in various ways, including impacts by the charged particles that constitute the solar wind. If Mars once had a substantial magnetic field that it then lost, solar wind erosion would have since increased.

The detection of carbonates at the martian surface by Mars Global Surveyor (Section 5.1.6) supports the view that Mars has indeed lost some of its atmospheric CO_2 to surface deposits. These carbonates are widespread. Additionally, there's probably a lot more CO_2 in the polar caps alone than in the present atmosphere, and deep deposits of carbonates in the polar regions cannot be ruled out. It is beyond reasonable doubt that Mars has lost most of its CO_2 to its surface and to space. This is also the case for significant proportions of its initial endowment of other volatiles, including water and N_2.

Were there once oceans or large lakes on Mars? Perhaps not. Such large bodies of water would have led today to basins where carbonates are concentrated, for example in limestone, with this concentration being preserved at the surface. But the carbonates detected by Mars Global Surveyor are not concentrated into basins, indicating that liquid water was not widespread at the surface. This could mean that any warmer, wetter period in the Noachian was a rather modest enhancement of present conditions.

5.3 THE SEARCH FOR LIFE ON MARS

The warmer, wetter Mars that might have existed in the Noachian would have been habitable, and based on the history of life on Earth, for a long enough period for unicellular life to have emerged. Did it emerge? Are there fossils to be found? Even today there is water not far below the surface, and beneath a kilometre or so it should be liquid. On Earth there are organisms that live at such depths in the crust (Section 2.5.2). Is the same true for Mars today?

5.3.1 Before the space age

The belief by many scientists that the dark areas of Mars were probably vegetation stretch back to the 1860s. Indeed, by the end of the 19th century some astronomers believed that life on Mars had advanced far beyond vegetation, and that a technologically intelligent species was present. The most famous proponent of this view was the US astronomer Percival Lowell. From 1894, at his observatory at Flagstaff, Arizona, he drew thousands of pictures of Mars showing numerous fine lines

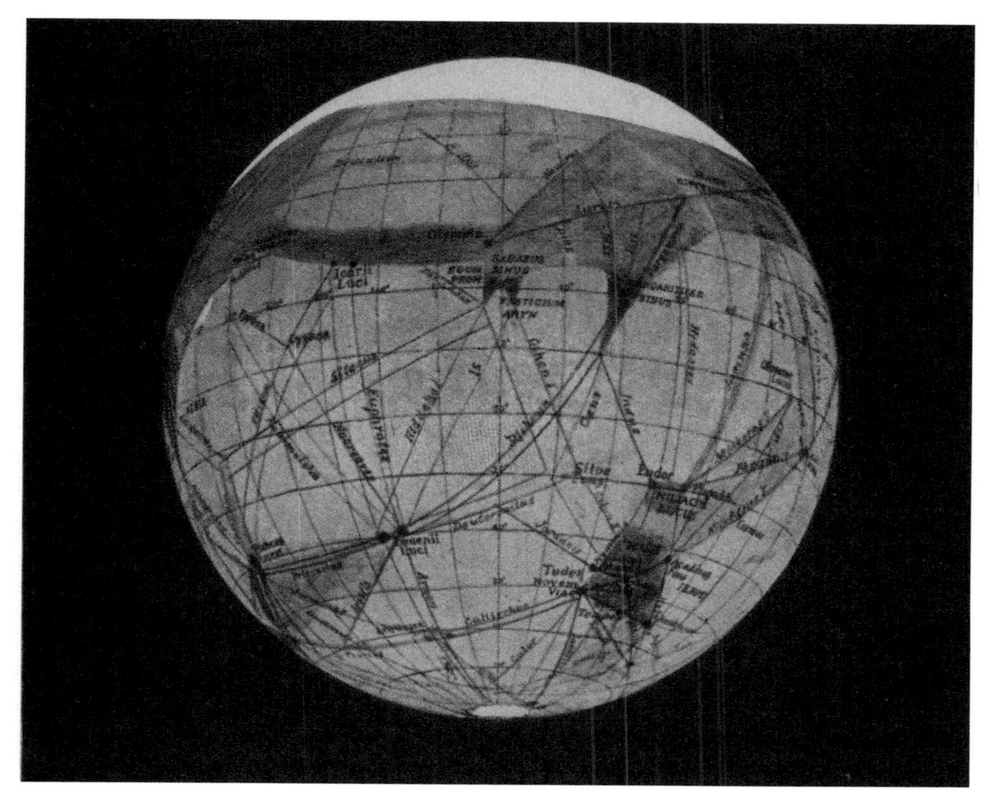

Figure 5.12 A drawing of Mars by Percival Lowell, showing many canals. The hemisphere shown is roughly that shown in Plate 5, except that in this drawing south is at the top, and the northern hemisphere is tilted more towards us.
Percival Lowell.

(Figure 5.12) that he interpreted as irrigated strips of land following the tracks of canals eking out the meagre water supply. Earlier, the Italian astronomer Giovanni Virginio Schiaparelli, at the Brera Observatory, Milan, had observed Mars during the favourable opposition of 1877, and he drew linear markings that he termed canali, Italian for channels, with no implication that they were artefacts. The majority of astronomers at that time also believed that the linear features were natural, and many observers could not see them at all! At the favourable opposition of 1909 most observers failed to confirm the existence of canals, and they became regarded by many as artefacts of the human brain, struggling at the limits of what could be seen. This view was confirmed much later by spacecraft, which showed that the canals, and associated dark strips, do not exist.

The evidence that the dark areas were vegetation was more convincing, and was bolstered by Earth-based measurements in 1964 of an atmospheric pressure of 2500 Pa, sufficient for water to be liquid over the range 0°C to 20°C. But then, in

1965 Mariner 4 showed that the dark areas could not be vegetation, thus bearing out the few dissenting voices that believed the markings to be due to mineral differences (Section 5.1.4).

5.3.2 The Viking Landers

So far (2003) only the Viking 1 and 2 Landers have specifically targeted the search for life on Mars. They landed in 1976 at well-separated areas on lowland plains in the northerly hemisphere (Figure 5.5). A view from the Lander 2 camera is shown in Plate 15, and the view from the other in Figure 5.13. Neither camera, in their several years of operation, saw any macroscopic life forms, nor evidence of their activity, but surface samples were also collected at each site, and the Lander 1 scoop that did this is visible in Figure 5.13. The samples were drawn into the Lander and subjected to four investigations. Three of these were specifically biological, based on the assumption that martian life would, like that on Earth, be based on complex carbon compounds and liquid water. The fourth was a more general chemical analysis.

In the Gas Exchange experiment a sample was placed in a chamber and moistened with an aqueous solution of nutrients that became known as 'chicken soup'! If any martian organisms digested the nutrients some change should have occurred in the simulated martian atmosphere in the chamber. Terrestrial organisms would cause changes in the amounts of O_2, CO_2, or CH_4 as they metabolised. The atmosphere was analysed with a gas chromatograph. In the Labelled Release experiment the nutrients that were fed to the martian sample had been labelled by substituting some of their carbon atoms ^{12}C with the isotope ^{14}C. This is radioactive and therefore any respiration by martian organisms that resulted in the emission of, say, CO_2 or CH_4 would release ^{14}C into the atmosphere in the chamber. Therefore, the gases were analysed for ^{14}C. The Pyrolitic Release experiment avoided the possible problem that martian organisms were unused to aqueous solutions, and that their biochemistry might be sufficiently different for the nutrients to be unusable, even poisonous. The surface samples were kept dry, and the martian atmosphere in the chamber was modified by adding CO_2 and CO that incorporated the radioactive isotope ^{14}C. Each sample was exposed to simulated sunlight, and after a while it was heated to 750°C. Any gases evolved from the sample *other* than CO_2 and CO were tested for ^{14}C radioactivity. This would reveal any carbon atoms from CO_2 and CO that had been incorporated into biological materials, such as occurs in terrestrial photosynthesis.

All three biology experiments initially gave positive results, consistent with the existence of living organisms. Unfortunately, further investigations led to the almost unanimous conclusion that no life had been detected. The Gas Exchange experiment evolved huge quantities of oxygen, but it was shown that this could be due to the action of water on inorganic compounds that had been dry for millions of years. The Labelled Release experiment detected lots of ^{14}C, but it was soon realised that this could come from CO_2 produced by the reaction of nutrients with oxygen-rich substances. Subsequent work has shown that these might include ions like O_2^- attached to mineral surfaces. Such ions can be produced under the intense solar UV radiation

Figure 5.13 Viking 1 Lander on Mars, showing the 80 mm wide trench left by the scoop that collected a surface sample.
NASA/JPL.

that penetrates the ozone-free atmosphere, and the ions can then survive in the cold, dry conditions. Moreover, a second feed of nutrients gave no further release – there was nothing growing. The Pyrolitic Release experiment gave a small positive signal, but also gave a signal, albeit even smaller, after a sample had been heated to 175°C for three hours. In the likelihood that martian organisms could not survive such heating, an alternative explanation was sought for the initial signal, and it is plausible that it was due to a small amount of contamination of the sample by NH_3 from the Viking Lander descent engines.

The surface samples were also analysed by a gas-chromatograph mass-spectrometer (GCMS). This instrument showed that any organic compounds in the sample were below the limits of detection, at most a few tens of parts per billion. This is below the organic content of Antarctic soil on Earth, and also well below that in primitive meteorites (called carbonaceous chondrites). In such meteorites the organic compounds are of abiological origin. The martian dust really should contain organic compounds from the infall of such meteorites. That it does not must be due to the strong oxidising power of the martian surface, the oxygen-rich ions converting the carbon in the organic compounds to CO_2. Impacts and dust storms would turn over the dust to result in full oxidation to perhaps as deep as a metre, much deeper than the Viking Lander sampling scoops penetrated. Surface oxidation would also destroy organic compounds from dead organisms and the waste products of living organisms.

At the very best, life at the Viking Lander sites is so sparse that it is below the detection limits of the experiments. There are barren places on Earth where this is the case, yet that are marginally inhabited. Martian life might be more abundant elsewhere, though this seems unlikely given the global redistribution of dust by martian winds. A widespread surface biosphere is also contradicted by the atmospheric composition. There is no sign that it has been driven far from chemical equilibrium in the way that the Earth's atmosphere has been by the terrestrial biosphere (Section 3.2.3). The most likely place to find living organisms on Mars today is deep under the surface, where the interior heat keeps water liquid. It might be difficult to find such life on Mars in the foreseeable future, though if it was sufficiently widespread it could be releasing short-lived gases such as CH_4 that we might be able to detect as they seep through the dust before being destroyed in the atmosphere. We return to the possibility of life at depth in Section 5.3.4.

What about martian life in the distant Noachian? Is there any evidence for that? None has yet been obtained by spacecraft or by Earth-based observations of Mars, but some scientists believe that there is positive evidence in martian meteorites.

5.3.3 Martian meteorites and fossils

Meteorites are small pieces of rock that reach us from other bodies. In space they are called meteoroids, as they streak through the atmosphere we call them meteors, and on the ground they become meteorites. Of the many meteorites known to science, there is a small group (currently over 30) that very probably came from Mars. That they *are* a group is indicated by the similarity in the Oxygen isotope ratios among them. A martian origin is suggested because these ratios are unlike those found on Earth, the Moon, and the main meteorite families. That leaves Venus and Mars as plausible sources, but Venus has to be ruled out because its massive atmosphere would make impact ejection of material from its surface to space highly unlikely, whereas from Mars it is easy. A martian origin is put beyond reasonable doubt by small bubbles of gas trapped in glassy material in some of the group. The relative proportions of the inert gases argon, krypton, and xenon in these bubbles closely match those in the martian atmosphere – no other group of meteorites gives such a

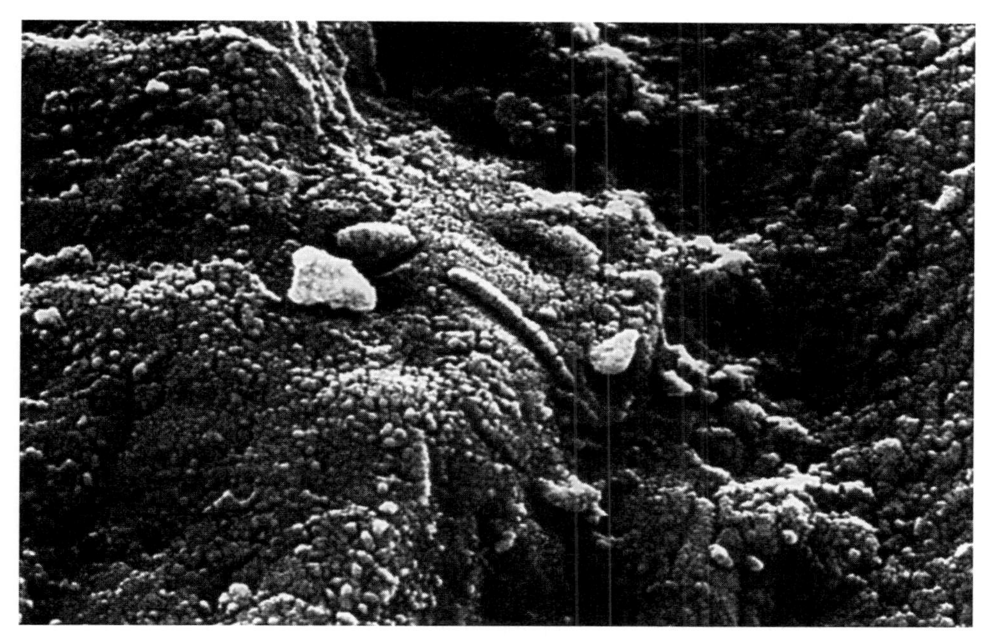

Figure 5.14 An elongated, segmented structure that typifies the larger of such structures (~0.1 μm long) found in the martian meteorite ALH84001. A martian fossil? Probably not. NASA/Stanford University.

close match. Other chemical and isotopic similarities between the group and Mars have further strengthened the case.

The martian meteorites are broadly basaltic in composition, and seem to have crystallised from magma at or below the martian surface. The radiometric solidification ages range from 4500 Ma to 180 Ma. Since solidification they have been modified by liquid water. One of these meteorites, the 1.9 kg ALH84001 has aroused enormous interest since it was claimed in 1996 that it contains evidence of past life on Mars. ALH84001 was discovered on the surface of Antarctica. Its name shows that it was discovered in the Allen Hills (ALH) region, in 1984 (84), and that it was the first meteorite of that season's harvest to be catalogued (001), though it was not recognised as martian until 1993. At 4500 Ma it is by far the oldest of the martian meteorites. This is the solidification age. Further dating has been performed by various techniques, which show that it was fractured between 4000 and 3800 Ma, probably by an impact. Carbonate globules formed in the cracks between 3800 and 1300 Ma. It was ejected from Mars by another impact at about 16 Ma, and it arrived on Earth about 13 000 years ago.

The claim that ALH84001 contains evidence of life rests on several observations: the tiny (50–200 μm) carbonate globules; far smaller grains of magnetite contained within the globules; traces of organic material called polycyclic aromatic hydrocarbons (PAHs); and elongated, segmented structures typically 0.02–0.1 μm in length (Figure 5.14). The claim is that these structures are fossil bacteria, or bacterial filaments, and that the carbonates, magnetite, and PAHs are of bacterial origin. The claim is based on Earth-analogues that are clearly biological. However, there are also abiological and terrestrial contamination explanations for all these things, and though much work has been done on ALH84001, it is not clear whether it bears any evidence at all for life on Mars.

Even if we have not yet discovered martian fossils, it is certainly possible that they await discovery in martian meteorites. It is also possible, in the distant past if not today, that martian meteorites carried living organisms from Mars to Earth, with interesting implications for cross-planetary biological contamination, or seeding! But regardless of what awaits us in martian fossils, many astrobiologists think there is an excellent chance of finding them on Mars itself, particularly in the southerly hemisphere where much Noachian landscape is preserved. For present-day life perhaps the northerly hemisphere, is a better prospect. What are the plans for future exploration?

5.3.4 Prospects for the future

In January 2004, NASA's Mars Exploration Rovers A and B (now called Spirit and Opportunity) are scheduled to land in two separate areas and gather data on rocks and minerals intended to help advance our understanding of the history of water on Mars and of martian climate change. They will not look specifically for life, but there will be cameras that could see any macroscopic life (unlikely), and microscopes that could see any smaller life forms or fossils. The landers will be able to travel up to 100 m per day, and they will be active for at least 90 martian days (sols). Rover A will land in Gusev Crater, a 150 km diameter impact crater near the boundary between the northerly and southerly hemispheres, and Rover B will land on Meridiani Planum, a plain near the equator in the southerly hemisphere. At both sites there is evidence that liquid water was once present.

Beagle 2 (Figure 5.15) is being carried to Mars aboard ESA's Mars Express. It is scheduled to land on Christmas Day 2003 in the northerly hemisphere in the sedimentary basin Isidis Planitia, and carry out a range of investigations, several of which are aimed at detecting life. A robotic arm will extend towards a suitable-looking rock. It will grind away the weathered surface, and then use a microscope to examine the interior for bacterial fossils and for mineral structures indicative of past life. It will carry out other *in situ* studies of the rock. The arm will then extract a sample of the rock and take it back to the body of the lander where it will be analysed with a mass spectrometer to determine its carbon isotope ratios. Organisms on Earth preferentially incorporate ^{12}C, and thus contain an enhanced $^{12}C/^{13}C$ ratio compared to their environment (Section 3.2.1). If such enhancements are seen in the martian rock this would indicate the possible presence of life now or in the past.

The robotic arm also carries a 'mole' that will burrow beneath any large rock to a depth of at least a metre, obtain a sample, which will then be carried back to the lander where it will be subject to the same sort of investigations as the rock sample. The mass spectrometer will also look for CH_4. Any life in the crust could well include

Figure 5.15 The ESA lander, Beagle 2, on the martian surface, showing the robotic arm extended (artist's impression).

methanogens (Section 2.4.2), in which case small quantities of CH_4 could diffuse upwards and last long enough to be detected near the surface before being destroyed. Beagle 2 also has a camera.

If there are any living organisms on Mars today they are probably deep in the crust. In Section 5.1.5 we saw evidence for water at shallow depths across much of the martian surface, though for most of the time it must be frozen. However, temperatures in all planetary interiors increase with depth, and so too does the pressure, therefore water will be liquid below some depth. In Mars this depth is uncertain, because of our poor knowledge of internal heat sources and the thermal conductivity of the crust. At present, estimates of the depth at which 0°C is attained vary from 1–11 km in the tropics, with a best estimate of 2.3 km. At the poles the best estimate is 6.5 km. There is as yet no firm plan to burrow to these sort of depths in Mars, but the forthcoming landers, and more distant sample return missions, just might solve the question of whether Mars is, or once was, inhabited.

5.4 SUMMARY

- Mars is about half Earth's radius and one-tenth its mass, and today is a cool world under a thin atmosphere which consists almost entirely of CO_2.

- The northerly hemisphere is the younger and carries widespread evidence of volcanism, that might persist today in restricted regions. The southerly hemisphere is heavily cratered and thus is older, with much of it belonging to the earliest, Noachian Epoch of martian history, that ended 3500–3800 Ma ago.
- Features that indicate the action of liquid water include outflow channels that might have produced transient lakes or small oceans, valley networks that are particularly common in the oldest terrain, and the relatively recently formed gullies. This evidence for water near the surface of Mars has been borne out by the detection of hydrogen in at least the outer metre of the surface, probably incorporated in the water molecules in ice, and in clay minerals. Near-infrared observations from the HST support this possibility.
- The prevalence of valley networks in the oldest terrain, and a sharp decline in erosion rate at the end of the Noachian, indicate a denser warmer atmosphere at that time. This could have been a 10^5 Pa CO_2 atmosphere with water vapour. Most of this has since been lost, by incorporation into carbonates and water-rich deposits, by the impacts of the heavy bombardment, and subsequently by other losses to space. The small size of Mars is the root cause of these losses, through the decline of geological activity that would liberate volatiles from surface and subsurface deposits, and through the low escape speed. The absence of a strong magnetic field would assist loss by the impact of solar wind particles.
- The possibly more clement conditions in the Noachian could have made Mars habitable for long enough for life to evolve. Therefore, even if Mars is uninhabited today, it might have been inhabited in the distant past – there could be fossils.
- Only the Viking Landers in the 1970s looked specifically for life. Beagle 2 is about to do so again.
- Living organisms have not yet been discovered on Mars though they might be present beneath the surface, perhaps over a kilometre deep where water could be sustained as a liquid by internal heat.

5.5 QUESTIONS

Answers are given at the back of the book.

Question 5.1

Discuss why it is reasonable to suppose that water in the crust of Mars should have been liquid at generally shallower depths during the Hesperian Epoch than it is thought to be today. What are the implications for life on Mars?

Question 5.2

If Mars had been several times its actual mass, it is likely that its surface would be inhabitable today. State the reasons for this.

Question 5.3

(a) Suppose that no progress towards the origin of life on Earth was possible until after the end of the heavy bombardment. Suppose also that on Mars, where the heavy bombardment was probably lighter because of Mars's lower mass, such progress started at 4400 Ma. Discuss the chances of there being martian fossils of eukaryotic cells and of multicellular eukaryotes.

(b) How could the early origin of life on Mars implicit in part (a) have helped to give rise to life on Earth?

6

Life on Europa?

Europa was identified in Chapter 4 as a place where life might be present in an ocean beneath its icy crust. In this Chapter the evidence that it has such an ocean is presented, and then its potential as a habitat is explored.

6.1 EUROPA

Europa is one of the four Galilean satellites of Jupiter (Section 4.3.3). Table 6.1 compares some basic data on Europa with the Earth. Though Europa is considerably smaller, it is still a planetary body in size, and would be regarded as a planet if it were in its own orbit around the Sun rather than in orbit around Jupiter. The mean solar distance of Europa is the same as that of Jupiter, and in Table 6.1 it is the semimajor axes of the orbit of Jupiter that is given. Jupiter is so far from the Sun that solar radiation at the surface of Europa is feeble, and so its surface might be expected to be cold. So it is. Europa is covered in frozen water, as shown in Plate 11, with a surface temperature of about −140°C at the equator with the Sun high in the sky. But though solar radiation is the dominant energy input at the *surface*, the interior of Europa is warmed by the decay of long-lived radioactive isotopes and by tidal heating (Section 4.3.3).

Table 6.1. Basic data on Europa, with the Earth for comparison.

	Average solar distance (AU)	Equatorial radius (km)	Mass (mass of Earth)
Europa	5.203	1561	0.00804
Earth	1.000	6378	1

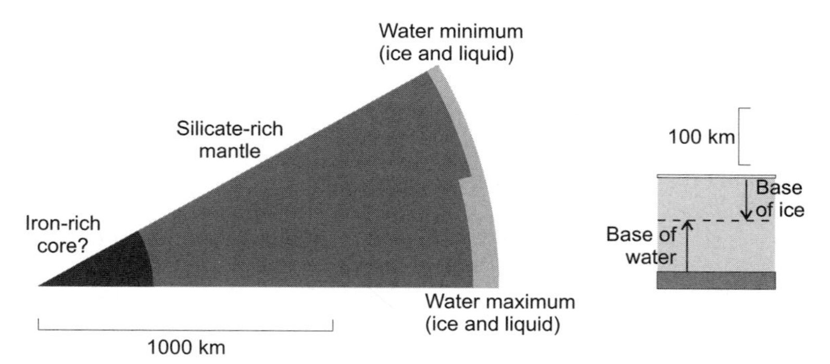

Figure 6.1 An interior model of Europa, showing possible ranges in the thickness of the shell of water and the ice carapace.

Gravitational data from spacecraft trajectories, notably the flybys of the Galileo Orbiter from 1995 to 2001, have provided the mass of Europa allowing its mean density to be established and also the density increase with depth. On the basis of these data, plus the water-ice detected spectroscopically at the surface, the interior model in Figure 6.1 has been established. Water, as ice or liquid, accounts for about the outer 80–170 km. On the basis of some estimates of the tidal and radioisotope heating, the water-ice gives way to liquid water at a depth of about 10 km, but other estimates range from a few km to considerably more than 10 km. The thicker estimates raise the question of whether there is much liquid water present at all.

6.2 IS THERE AN OCEAN ON EUROPA?

Voyagers 1 and 2 in 1979, and later the Galileo Orbiter, have revealed many features of the icy surface of Europa that are evidence of a widespread ocean under the ice, if not today then in the geologically recent past. One is the relatively high albedo of the surface (i.e. the proportion of sunlight scattered back to space). For Europa this is 70%, significantly higher than the 45% for Ganymede and the 20% for Callisto. Ice darkens very slowly with exposure to space, and a value of 70% indicates resurfacing more recently than in the case of the other two satellites, though Callisto might have a silicate veneer. Europan resurfacing could have been through liquid water escaping through cracks in the icy shell, perhaps produced by tidal stresses. Further indication of a liquid is provided by the significant magnetic field of Europa. This requires an electrically conducting liquid, and though one possibility is a partly molten iron-rich core, another is a salty ocean, the ions from the salts providing the conductivity. The salts would include the carbonates and sulphates that have been detected in surface spectra. These spectra also show that the saltiest areas of the surface are those that appear freshest.

Another indication of resurfacing is the scarcity of impact craters, apparent in Plate 11. The most prominent is Pwyll ('Pooh-eel'), visible in the lower right of Plate

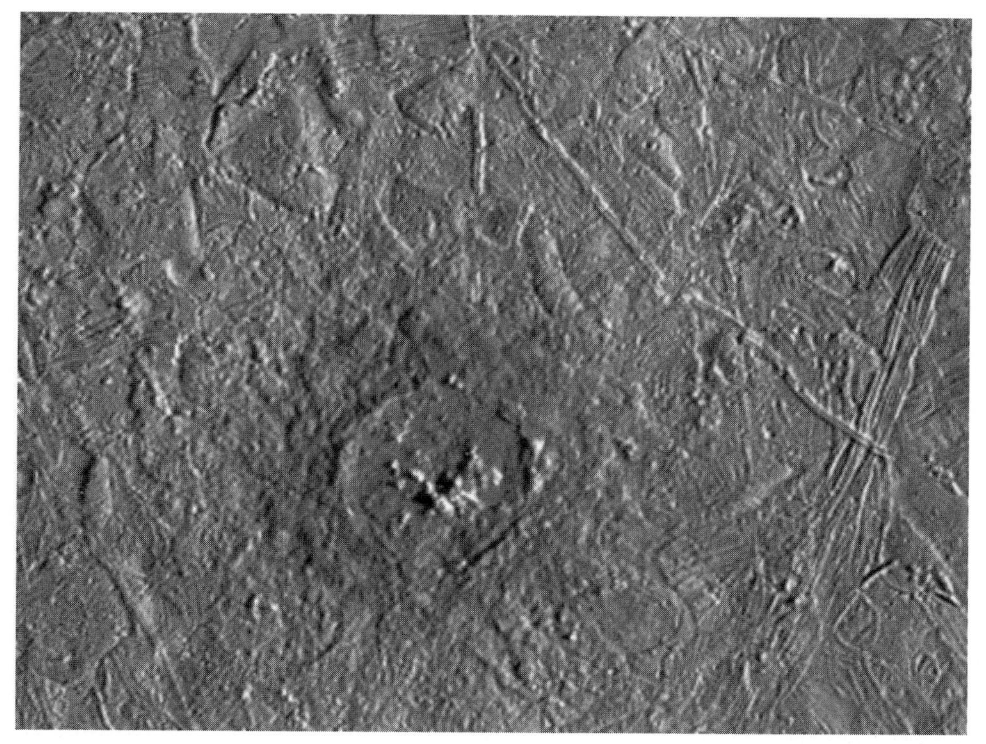

Figure 6.2 A close-up of the impact crater Pwyll ('pooh-eel'). This frame is 120 km across.
NASA/Arizona State University.

11 and in more detail in Figure 6.2. This 26-km-diameter impact crater is relatively young, as shown by the rays of bright ice that spread out from it. The probability of an impact of sufficient size places an upper limit of about 20 Ma on the age of Pwyll. It is of low relief, with a rim only about 200 m above the surrounding terrain, and a floor hardly lower than this terrain. This indicates an impact into thin ice underlain by a weak medium, such as an ocean or slushy ice. Thin ice, over at least some of Europa's recent history, would also contribute to the general smoothness of the surface, as topography 'relaxed' away, leaving a surface today over which the altitude range rarely exceeds a few hundred metres. However, some studies of the depths of Europa's impact craters versus their diameter have been interpreted to mean that the ice is at least 19 km thick.

Figure 6.3 shows abundant evidence of resurfacing. Light terrain has been cut by processes that have produced darker bands up to a few tens of kilometres in width. The bands must be younger than the terrain they cross – the terrain must have been there first. Close examination suggests that the bands are the result of extension and cracking of the icy crust, for example through tidal stresses, allowing fresh material to rise up. In the upper right of this figure the bands have been encroached upon by

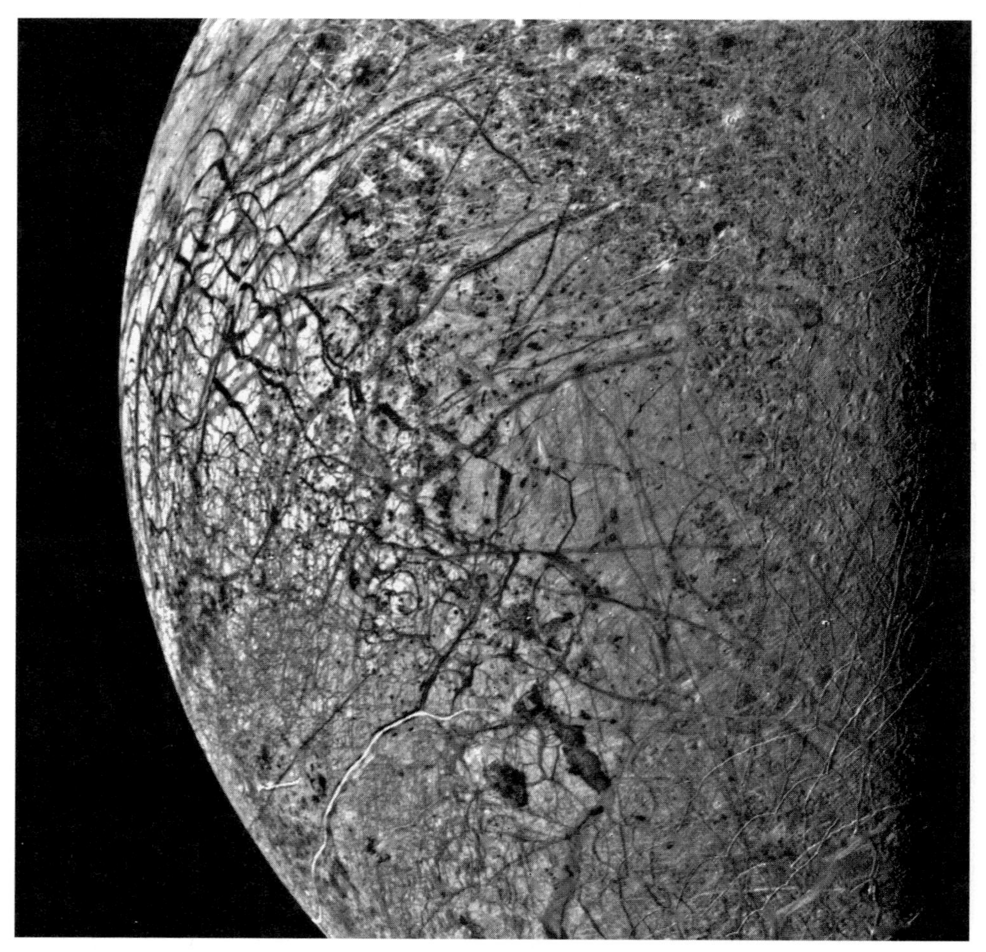

Figure 6.3 Evidence for resurfacing of Europa, where dark bands have resulted from extension and cracking of the icy crust, allowing fresh material to rise up.
NASA/JPL.

even younger material with a mottled appearance. At the extreme right the low Sun angle has revealed curved ridges (Plate 12). Some of these cut across the bands, so the ridges are also younger than the bands. That the material constituting the bands is fresh but dark is *not* at variance with the presumed darkening that has afflicted the ancient surfaces of Ganymede and Callisto. This latter darkening has occurred over a much longer period than the timescale for brightness changes at the surface of Europa, where the whole surface is many orders of magnitude younger that the surfaces of Ganymede and Callisto.

The light terrain between the bands, and the bands themselves, display many low ridges, forming what is called 'ball-of-string' terrain (Figure 6.4), each ridge pre-

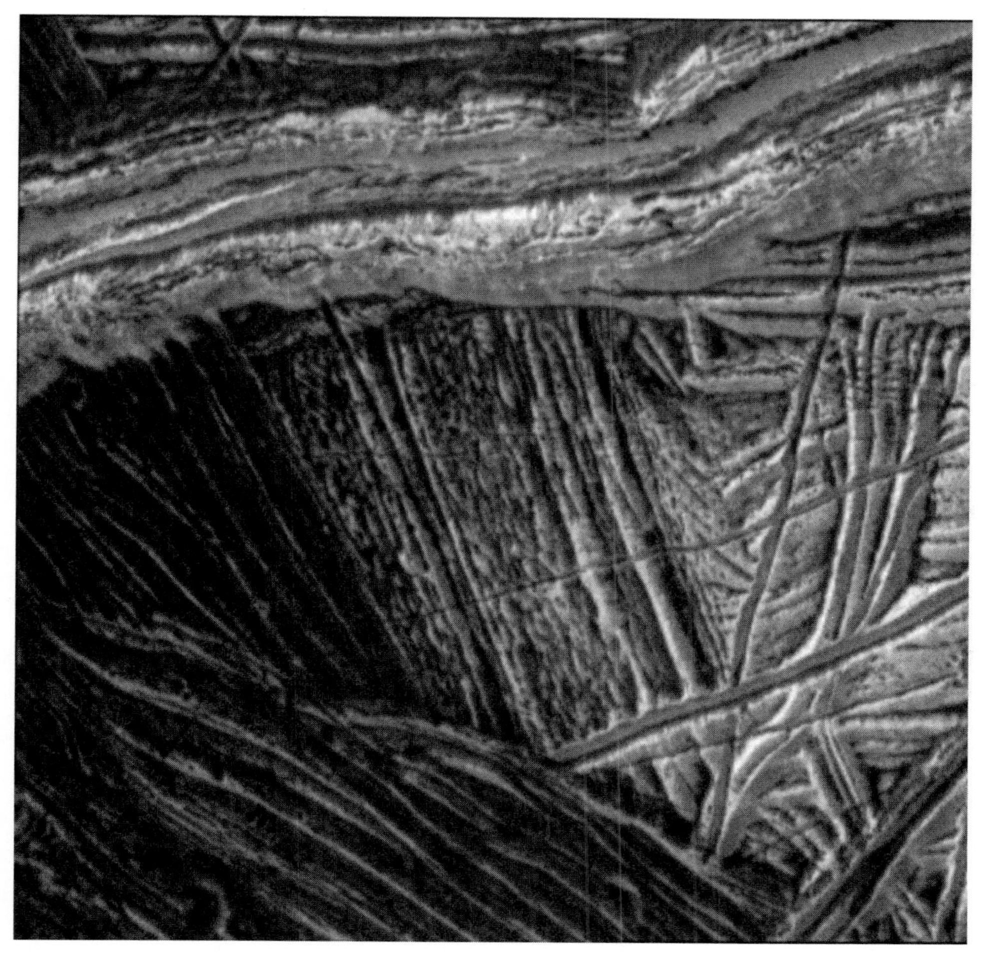

Figure 6.4 'Ball-of-string' terrain on Europa, with ridges presumed to be the result of cryovolcanism at cracks. Frame width 20 km.
NASA/JPL.

sumably the result of cryovolcanic eruption (icy lavas) along a crack. In places this terrain is disrupted by domes, caused presumably by upwelling material made buoyant by its relative warmth. Elsewhere (Figure 6.5) there seems to have been melting that created rafts of ball-of-string terrain that drifted in a medium that solidified to form a low-lying hummocky matrix. Rafts plus matrix is often called 'chaos'. This is evidence of a local ocean or a highly plastic surface layer. There are fewer secondary craters and rays from Pwyll on the matrix than on the rafts, showing that the matrix was plastic enough to remove such features. Chaos could be where there is a net destruction of surface, thus balancing the intrusion of the new surface in the bands.

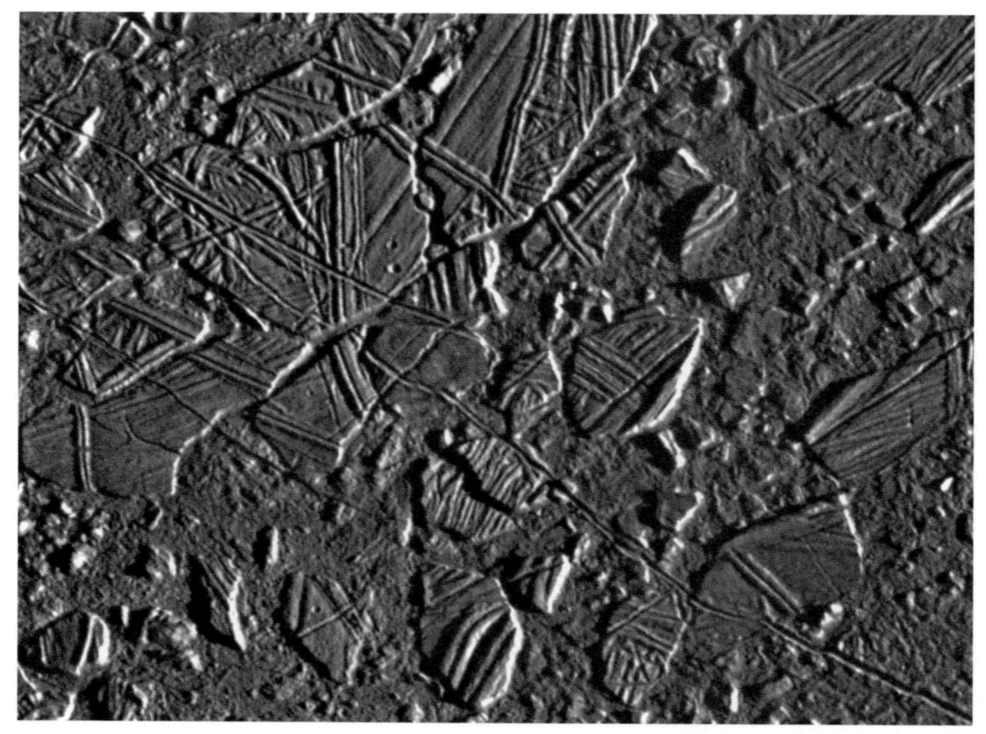

Figure 6.5 Rafts of 'ball-of-string' in a hummocky matrix, constituting 'chaos'. Frame width 42 km.

NASA/Arizona State University.

The rafts provide a further indication of the ice thickness. Assume that the matrix was liquid, and that the rafts were floating in equilibrium in it, rather like icebergs in the Arctic Ocean. Then, from the 100 m elevation of the top of the rafts above the matrix, a raft thickness in the range 0.5–2.2 km is obtained, depending on the difference in density between the rafts and the matrix, which depends on their salt contents. If the matrix was not fully melted, but was a skin overlying liquid, it would have supported the rafts to some extent, and consequently the rafts could be thinner than 0.5 km. On the other hand, if the event that heated the matrix also melted the base of the rafts then they could originally have been thicker than 2.2 km. Therefore, just before the rafts were created, the ice sheet of which they became fragments could have had a thickness anywhere between a few hundred metres and several kilometres, perhaps tens of kilometres.

Resurfacing could be ongoing, or it could be concentrated into certain epochs. For example, greater tidal heating would arise from greater orbital eccentricity – changes in eccentricity are caused by the intricate gravitational interactions between the Galilean satellites (Section 4.3.3). The ice thickness could correspondingly fluctuate by a large amount. It is therefore possible that Europa's ocean today lies much deeper than several kilometres, and even that there is at present no widespread liquid water at all. But even in this case it could be as recent as about a million years ago that liquid water lay at no great depth. It is also possible that there never is a totally liquid layer, but a slurry of liquid and ice. In spite of these uncertainties, most planetary scientists believe the data to show that there probably is a widespread ocean on Europa today, beneath a carapace of ice with a thickness somewhere between a few kilometres and a few tens of kilometres. This ocean could have been present throughout much of Europa's history. Could there be life in any such ocean?

6.3 THE POTENTIAL OF EUROPA AS A HABITAT

6.3.1 Current knowledge

Europa probably has all the ingredients necessary for carbon–liquid water life. We know that carbon is present from the spectroscopic detection of carbonates at the surface, and in any case the ice-covered satellites of the outer planets are expected to contain carbon from the numerous carbon compounds that must have been available in the regions where they formed. We know that water is present, and we will assume that from some unknown depth down to the rocky mantle it has long been liquid.

Life also needs a photosynthetic or chemosynthetic energy source (Section 2.4.2). The attenuation of sunlight by ice would restrict photosynthesis to depths less than the order of 10 m. Temperatures at such shallow depths, even with the most optimistic of heat outflow rates, would not exceed about −20°C. Therefore, any communities based on photosynthesising autotrophs would at best be dormant, awaiting rare moments of local heating, such as via tidal cracks in the ice that enabled water from below to bathe them. It is unlikely that any such communities exist. A more likely prospect is primary energy obtained through chemosynthesis at hydrothermal vents on the ocean floor. Whether these vents exist, and their abundance and power, depends on where within Europa the tidal heating is concentrated. This depends in turn on unknown factors such as the strength and other properties of Europa's ice and rocky interior. Concentration of heating in the rock is best, because if the rock is merely warmed, the circulating water would quickly deplete it of useful chemicals, including those that could supply energy. If the heating were stronger, so that there was partial melting at certain locations, vigorous circulation including hydrothermal vents would resupply and regenerate compounds. Life could have originated and be sustained at such vents.

On Earth, hydrothermal vents are known to support colonies of hyperthermophiles and other creatures (Section 2.5.2). The energy sources are redox reactions such as the one utilised by methanogenic bacteria on Earth with the aid of catalysis by enzymes (Section 2.4.2):

$$2CO_2(\text{aqueous}) + 6H_2(\text{aqueous}) \rightarrow \text{`}CH_2O(\text{solid})\text{'} + CH_4(\text{aqueous}) + 3H_2O(\text{liquid}) \quad (6.1)$$

This clearly relies on CO_2 and H_2 being present in the hot water gushing from a vent. On Earth the CO_2 is supplied partly from oxidation of crust and mantle carbon, and partly from the break-up of carbonates that have been taken into the mantle at subduction zones. The H_2 probably comes mainly from the oxidation by water of iron in new crustal rocks, though some could be produced biologically.

Production of CO_2 on Europa might require oxygen or oxygen-rich species from the break-up of H_2O at the surface. Hydrogen escapes to space, but only some of the O_2 escapes. The rest is dissolved in the ice and some must find its way into any ocean. Gaseous O_2 has been detected at the surface. H_2O break-up is not only caused by UV photons, but also by charged particles (radiolytic break-up). At the surface, the particles are supplied by the solar wind and cosmic rays. Cosmic rays also penetrate beneath the surface, causing further radiolytic break-up of H_2O, supplementing the local action of radioactive isotopes in the water, notably ^{40}K. Therefore, even though the subduction of carbonates into the rocky mantle probably does not occur, CO_2 would still be produced from oxygen-rich species in the water percolating into the rocks near hydrothermal vents. There is also likely to be some H_2, for example, produced from new crustal rocks, as on Earth. Therefore, if H_2 and CO_2 are present in hydrothermal vents on Europa, Reaction (6.1) could be a chemosynthetic energy source. Another possible chemosynthetic source, not involving CO_2, is a redox reaction between $2Fe(OH)_3$ (oxidant) and H_2 (reductant) in the vented water:

$$2Fe(OH)_3(aq) + H_2(aq) \rightarrow 2FeO(solid) + 4H_2O(liq) \qquad (6.2)$$

The organism gets energy from the reduction of Fe^{3+} to Fe^{2+}.

If the Europan ocean contains little oxygen then the vented water might contain CH_4 rather that CO_2 and H_2. A variant of Reaction (6.2) then occurs, in which H_2 is replaced by CH_4, though the organism still gets energy from the reduction of Fe^{3+} to Fe^{2+}. In another variant H_2 is replaced by H_2S.

Thus, there are plausible schemes for life to exist at Europan vents. Vents come and go, so it would be essential for organisms to be able to migrate from one vent to another. For there to be life on Europa today we thus require a minimum of one body of liquid water large enough for there to have been at least one vent, not necessarily the same vent, stretching into the past for long enough for life to have developed. If this condition is met, there is a reasonable possibility that life is present on Europa. How might we find it?

6.3.2 The future exploration of Europa

Further space missions are needed to show whether Europa really has a widespread ocean, and whether it is inhabited. The development by NASA of a Europa Orbiter was reinitiated in October 2002, for possible launch late in this decade. It would establish whether there is an ocean, and if so, it would map it. It would also advance our understanding of the processes that have formed the surface. It will not do much to further the search for life, but it would identify landing sites for later missions that will explore the ocean, to see if it is inhabited. In order to explore the ocean a lander

Figure 6.6 A future mission to explore Europa's ocean (artist's impression).
JPL/Caltech/NASA.

will need to carry a probe that will somehow be able to bore through at least a few hundred metres of ice, and probably many kilometres. It will then search for life, particularly in the region of any hydrothermal vents (Figure 6.6). Its findings will be transmitted to the lander, and the lander will then send them to the Earth, where scientists and the general public alike will doubtless be waiting eagerly. This will not happen until about 2020 at the earliest.

Before then, some indication of what we might find in any Europan oceans could come from Lake Vostok, which lies beneath nearly 4 km of ice near the Earth's South Pole. It is one of the largest of the Antarctic's subglacial lakes, about 10 000 km^2 in area and up to 500 m deep. In 1998 a Russian–French–US team drilled to within 150 m of the lake, stopping short to avoid contaminating any aquatic life (which would probably be unicellular). The ice cover is nearly half a million years old, and so it could have preserved life forms significantly different from those elsewhere on the planet. Such life forms would also indicate that life could survive under conditions that on Earth are the closest to those we might find on Europa. Future exploration of the lake will require the development of ultra-clean technologies, that could pave the way for the technologies we will need to explore Europa.

6.4 SUMMARY

- Gravitational data from spacecraft have established that Europa is a rocky world, covered with a layer of water about 80–170 km thick. The water is ice at the surface, but because of a combination of radiogenic and tidal heating it could be liquid below some unknown depth.
- The surface of Europa displays considerable evidence for widespread subsurface liquid water or a slushy mixture of liquid and ice. This has been present in the last few million years, and could be present today. It could also have been present throughout much of Europa's history.
- If the tidal heating is concentrated in the rocky interior then Europa might have hydrothermal vents. These could support colonies of creatures with chemosynthesis as the basic source of energy.
- It will probably not be earlier than 2020 before any oceans on Europa will be explored to see if life is present. Before then, exploration of Lake Vostok will aid the development of technologies for the exploration of Europa.

6.5 QUESTIONS

Answers are given at the back of the book.

Question 6.1

(a) Why are the orbital eccentricities of the Galilean satellites of Jupiter subject to variation?
(b) Explain what would probably happen to any widespread ocean of liquid water on Europa if its orbital eccentricity were reduced close to zero and remained close to zero for a long time.
(c) If the orbital eccentricity of Ganymede were to increase considerably, explain why this might promote the origin of life there.

Question 6.2

What manifestations of life should a mission like that in Figure 6.6 be able to recognise?

Question 6.3

Even if there were no oceans on Europa, discuss whether life could still be present, and where it might be located.

7

The fate of life in the Solar System

Just as there was a beginning to the Solar System's habitability, probably when the heavy bombardment subsided some 3900 Ma ago, so there will be an end, not just to the habitability of the Earth, but to the Solar System as a whole. This end might be brought about by a variety of cataclysms originating outside the Solar System. One possibility is the demise of a massive star, more than 8 solar masses, when it is no longer able to support itself by thermonuclear fusion. Its collapse will be seen as what is called a Type II supernova, a huge explosion (Figure 7.1). As the name implies, there are also Type I supernovae. There are three subtypes, one of which (Ia) occurs in binary systems where the two stars are so close together that they exchange mass, resulting in the explosion of the one that receives material. In the other subtypes (Ib and Ic) the explosive end point of a massive star has been modified by earlier copious mass loss, in some cases in a close binary system.

The particle flux and electromagnetic radiation from a supernova near the Solar System, particularly the gamma rays, would be damaging to surface life, though radiation-resistant bacteria and subsurface life might be little affected. The gamma rays might be beamed along the rotation axis of the star, and if the Solar System lay in the beam, damage could be sustained even at a range of a few thousand light years. Estimates of how frequently the Solar System has suffered a damaging supernova vary from about once per 100 Ma to about once per 1000 Ma on average. If the lower figure is correct then supernovae must have been a factor in at least some of the mass extinctions on Earth (Section 3.2.2), though it appears that the latter figure might be nearer the truth. Supernovae that would *sterilise* the Solar System must be less frequent – there could not have been one in the last 3900 Ma, and it is unlikely that one will occur before the end of the Sun's main sequence lifetime in about 6000 Ma.

Another cataclysm from beyond the Solar System would be the approach of a star close enough to gravitationally disrupt the planetary orbits. An approach within a few tens of AU would suffice. However, the stars in the solar neighbourhood are on

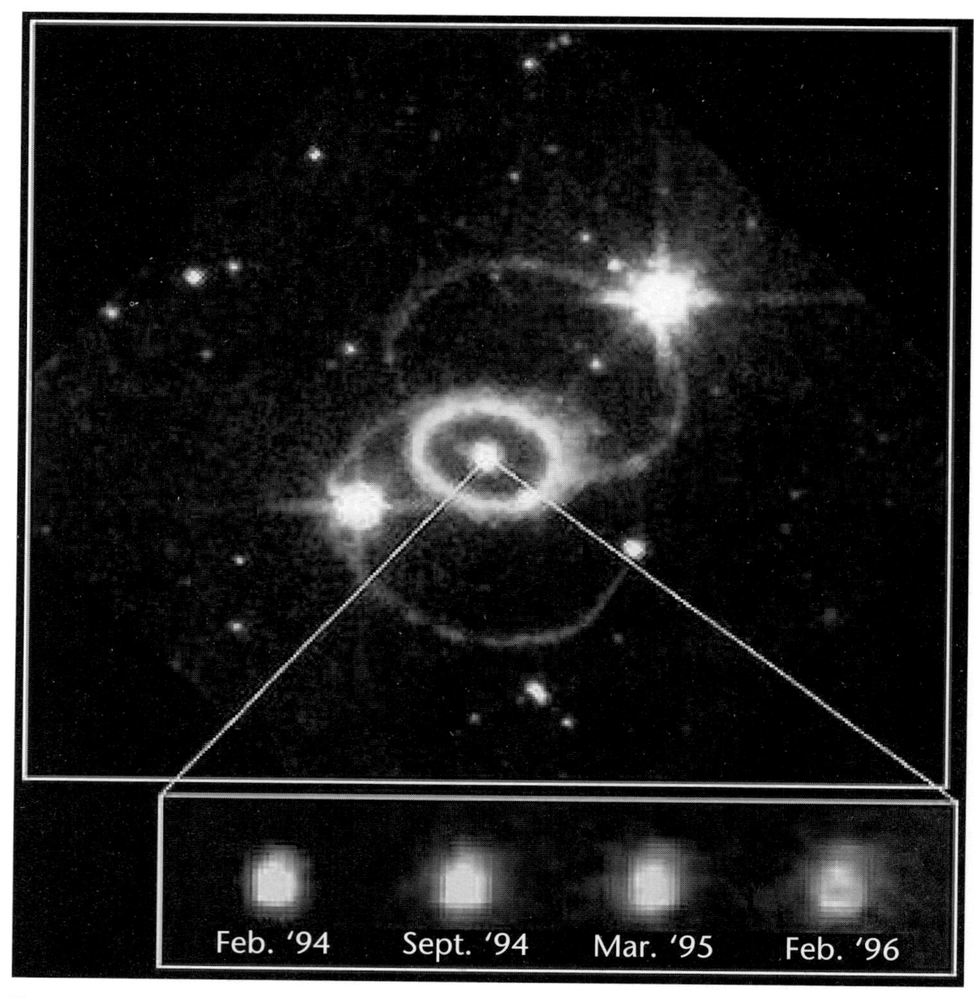

Figure 7.1 The Type II supernova (1987A) that exploded in the Large Magellanic Cloud in 1987. The fireball can be seen growing at the heart of ring structures created by the explosion earlier.

Chung Shing Jason Pun (NASA/GSFC), Robert P. Kirshner (Harvard–Smithsonian Centre for Astrophysics), and NASA.

average several thousand times further apart than this, and they move with relative velocities of only a few $km\,s^{-1}$. It can then be shown that the probability of such an approach is only about once in 10^6 Ma. A far bigger target is the 10^5 AU diameter Oort cloud of comets (Section 1.1.1). A star or dense interstellar cloud passing through the Oort cloud could scatter many comets into the planetary region. However, though the resulting bombardment could cause mass extinctions (and perhaps has done so in the past), it is unlikely to cause surface sterilisation.

Such cataclysms are hard to predict and at present there is no sign of one. By contrast, if life in the Solar System survives long enough, it will certainly be brought to an end by the Sun's evolution. Moreover, whereas there could be recovery from an external cataclysm, there is no recovery from what the Sun will do.

7.1 THE EVOLUTION OF THE SUN

7.1.1 The main sequence phase and the transition to the giant phase

It was seen in Section 1.1.2 that the Sun is presently nearly half way through the 11 000 Ma main sequence phase of its lifetime. In this phase the solar luminosity is sustained by the thermonuclear fusion of hydrogen into helium. The temperature increases with depth into the Sun, and it is only in a central core that it is high enough for the fusion to be important, at about 1.5×10^7 K. This is a relatively stable phase of the Sun's life, in that short-term changes in luminosity and effective temperature are small. There are however long-term trends (Section 4.2), and these become more dramatic after the main sequence phase. These external changes will be detailed in Section 7.2. First, consider the events inside the Sun that drive these external changes, starting with the main sequence phase.

Deep in the solar interior there are two balances that maintain the Sun in quasi-equilibrium. First, the increase in pressure with depth – the pressure gradient – tends to expand the Sun, and this balances the inward force of gravity. Consequently, the Sun, to a very good approximation, is nowhere undergoing expansion or contraction. The pressure gradient is sustained by the energy released by the thermonuclear fusion in the core. Second, the rate of release of energy in the core equals the luminosity, so there is no build-up or depletion of energy in the interior. These balances are not quite exact. For example, there is a long-term increase in the Sun's luminosity. This is caused by the gradual accumulation of helium in the core of the Sun. Each helium nucleus replaces four hydrogen nuclei. This causes a drop in pressure in the core, creating an imbalance between the force of gravity and the core pressure. The core thus shrinks slightly, converting gravitational energy into thermal energy. As a result the core temperature rises, which, along with the increase in density, causes a rise in pressure until there is again a balance of forces. At the higher temperature and density the rate of thermonuclear fusion is greater, which sustains the new temperature and the pressure gradient. Also, with this greater rate of release of energy in the core the Sun becomes more luminous. This goes on throughout the main sequence phase.

The end of the main sequence phase will occur when the Sun's core runs out of hydrogen, and though there will still be a thin layer of hydrogen at the surface of the core hot enough to undergo fusion, its small volume will release energy insufficiently rapidly to sustain the balance of forces. The core, now made almost entirely of helium, will shrink considerably. The gravitational energy released will raise the temperature not only of the core but particularly of the adjacent shell of hydrogen where fusion will then start. Helium will be produced in this shell and rain down on the core, which will continue to contract and heat up. The outer boundary of the hydrogen-fusing shell will migrate outwards. There will have been an increase in energy output, and this will cause the Sun to swell enormously. Its effective temperature T_e will decrease, so that its yellow hue will turn to red. In spite of this decrease in T_e the huge increase in radius R will result in a large increase in luminosity L, in accord with the fundamental equation:

$$L = 4\pi R^2 \sigma T_e^4 \tag{7.1}$$

In this equation $4\pi R^2$ is the surface area of the Sun and σT_e^4 is the power radiated per unit area of the surface, where σ is a fundamental constant called Stefan's constant ($5.67 \times 10^{-8}\ \mathrm{W\,m^{-2}\,K^{-4}}$, see Resources). The Sun will then have become a giant, specifically a *red giant*, a transition from the main sequence that will have taken a few hundred million years.

7.1.2 The giant phase and after

As a red giant, the Sun will initially have a structure like that in Figure 7.2. The tiny core, with a radius about twice that of the Earth, will consist of almost pure helium. It will be surrounded by a thick shell in which hydrogen fusion takes place. But though most of the mass of the Sun will be in the core and shell, most of the volume will be in a bloated envelope of hydrogen and helium that ultimately will reach out to a radius of about 1 AU! The inert helium core will continue to acquire mass from hydrogen fusion, and will therefore continue to contract and rise in temperature.

After a hundred million years or so, the core temperature will reach about 10^8 K and consequently core helium fusion will occur with a sudden onset called a helium

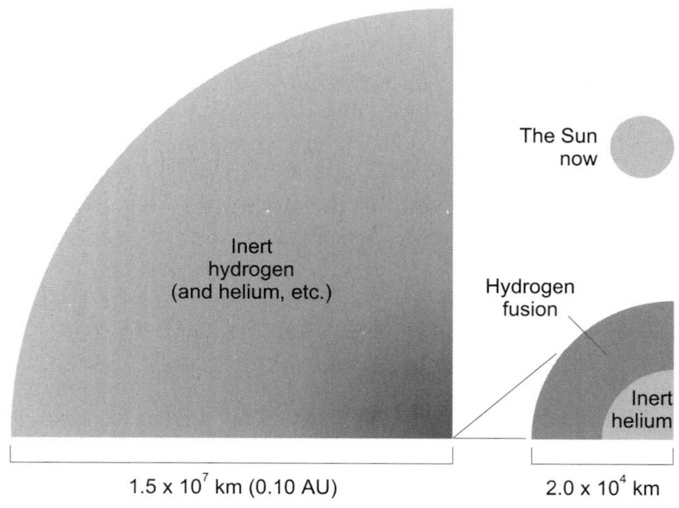

Figure 7.2 The Sun as it will be in its early years as a red giant compared to the Sun as it is now.

core flash. Helium fusion does not occur at the lower temperatures at which hydrogen fusion occurs because each helium nucleus contains two protons, whereas each hydrogen nucleus contains only one. Therefore, the electrostatic repulsion between two helium nuclei is greater, and higher temperatures are then required for the greater collision speeds necessary for the nuclei to approach each other closely enough to undergo fusion. The outcome of helium fusion is the production of carbon and oxygen. First:

$$^{4}\text{He} + {}^{4}\text{He} \rightarrow {}^{8}\text{Be} \tag{7.2}$$

^{8}Be is highly unstable, and on most occasions breaks up into two helium nuclei. But in the very dense helium core of a red giant a third helium nucleus sometimes comes along before this can happen, resulting in:

$$^{8}\text{Be} + {}^{4}\text{He} \rightarrow {}^{12}\text{C} + \gamma \tag{7.3}$$

The carbon nucleus is stable. This is the common isotope of carbon, and it is in giants that most carbon in the Universe has been made. The production of a carbon nucleus from three helium nuclei is called the triple-alpha process, a helium nucleus also being known as an alpha particle. It is an exothermic process, and the heat released establishes a pressure gradient in a giant core that prevents further contraction. Some of the carbon will undergo further fusion:

$$^{12}\text{C} + {}^{4}\text{He} \rightarrow {}^{16}\text{O} + \gamma \tag{7.4}$$

where ^{16}O is the common isotope of oxygen . This reaction is also exothermic, and is the source of most ^{16}O in the Universe.

With the onset of helium fusion, the heat released in the core will cause it to expand. The core will therefore cool slightly, and around it the hydrogen shell will also cool, which will reduce the rate of hydrogen fusion there. The outermost part of the hydrogen envelope will consequently contract, and the radius of the Sun will thus decrease, though it will still be far larger than it was as a main sequence star. The contraction of the outer layers will result in a modest rise in effective temperature – the Sun will take on a yellow hue again – but the contraction is the greater effect and so the luminosity will decrease (Equation 7.1). Since it left the main sequence about 20% of the mass of the Sun will have been lost in an intense solar wind, mainly just before the helium core flash.

Helium fusion in the core of the Sun will last for about 100 Ma. It will end when all the helium there is consumed, at which point the core will consist almost entirely of carbon and oxygen. The core will then contract, and so its temperature will rise yet again. The temperature of the adjacent helium will also rise, and, in a manner analogous to hydrogen, helium fusion will continue in a shell adjacent to the core. As before, the energy output from the shell will cause a huge expansion in the Sun's radius, accompanied by a decrease in temperature – it will again take on a red hue – but overall there will be a large increase in luminosity.

The remaining giant phase evolution will be complex, though short-lived. You will shortly see that the Solar System by now will have become uninhabitable, so the evolutionary details will not concern us. The Sun will end its giant phase by throwing

Figure 7.3 The planetary nebula NGC6369 that has been flung off by a giant star at the end of its life. The stellar remnant (at the centre) is becoming a white dwarf.
NASA and The Hubble Heritage Team (STScI).

off a succession of shells of gas. About 50% of the Sun's present mass will remain, consisting largely of the very hot carbon–oxygen core. Such a remnant will initially have an extremely high effective temperature, but it will quickly cool to about 2×10^4 K. It is then called a white dwarf, and though its temperature will still be high, it will only be about the size of the Earth, so it will have a very low luminosity (Equation 7.1). When it is young, the white dwarf will emit sufficient UV radiation to cause the cast-off shells of gas to fluoresce, and they will be seen as a planetary nebula, such as that in Figure 7.3. (They are nothing to do with planets, but named from the disc-like appearance of some planetary nebulae in low-power telescopes.) There will be no nuclear reactions to generate a supporting pressure gradient. Instead, the white dwarf will be supported by a pressure that arises from a quantum phenomenon that occurs when many electrons share the same space, as they do in the high density of a white dwarf. This is called electron degeneracy pressure (see physics texts in Resources). With no nuclear reactions and no contraction, the white dwarf will cool slowly over billions of years, fading into obscurity.

If any life had somehow survived the earlier phases of solar evolution, it would be unlikely to survive the subsequent frigid conditions under the feeble light of the white dwarf.

7.2 THE EFFECT OF SOLAR EVOLUTION ON SOLAR SYSTEM HABITABILITY

We will now consider what the future holds for the Earth's biosphere and for the habitability of other planets and their satellites as the Sun evolves. But first we consider the fate of the planets themselves.

7.2.1 Planetary orbits

Figure 7.4(a) shows the change in the Sun's radius during and after the main sequence phase – note the scale change on the time axis. The first giant peak is immediately before the onset of helium fusion in the core. The radius starts to grow again after the helium in the core has been consumed, and stays large until the planetary nebula is shed, which is where the radius plunges. Figure 7.4(a) also shows the present semimajor axes of the planets' orbits. If these orbits stayed at their present size then you can see that Mercury and Venus would not survive the first giant peak, and that the case of the Earth is marginal. Fortunately, the Sun does shed mass, and this causes the planetary orbits to expand, as shown. You can see that this gives the Earth a small safety margin. Mars will certainly survive, and so too will the outer planets and their satellites.

A possible consequence of orbital changes is interplanetary approaches close enough for a planet to either be ejected from the Solar System or collide with the Sun. Studies show that this is unlikely, except for Pluto, which could suffer a catastrophe within a few billion years of the Sun becoming a white dwarf. Another possibility is that the Sun will retain too little mass to hold on to a planet. However, studies show that even the white dwarf Sun, at about 50% of its present mass, is likely to retain all the planets that survive consumption by the Sun at the giant phase, and any subsequent loss through interplanetary close encounters.

Survival of a planet as a body does not, of course, mean that it survives as a potential or actual habitat. What will happen to habitability as the Sun evolves?

7.2.2 The habitable zone (HZ)

Recall from Section 4.2 that the distance of the HZ from the Sun depends on the solar luminosity and photospheric temperature. The luminosity has the larger effect. Figure 7.4 shows that the post main sequence events clearly will have a devastating effect on the habitability of the Solar System, but in fact there is trouble a long time before that!

Figure 7.5 shows the HZ in the Solar System during the Sun's main sequence phase for various HZ boundary criteria. The criteria used in Section 4.2 correspond to the intermediate values in Figure 7.5. The semimajor axes of Venus, the Earth,

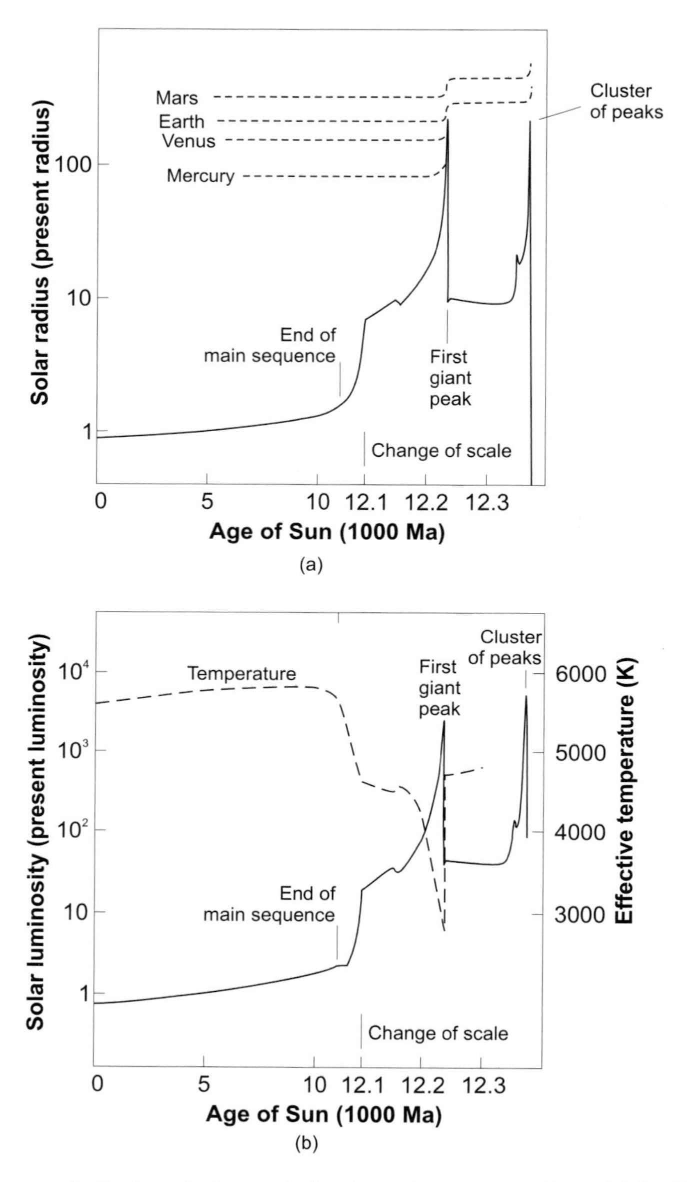

Figure 7.4 Changes in the Sun during and after its main sequence phase. (a) Radius, where the planets orbital semimajor axes are also shown. (b) Luminosity and effective temperature.

and Mars are also shown. It can be seen that, on the most optimistic of these criteria, the Earth will become uninhabitable about 4400 Ma from now. The intermediate criteria give us 2800 Ma. On the most pessimistic criteria, the Earth would be uninhabitable in about 800 Ma. Habitability is brought to an end by an increase in the

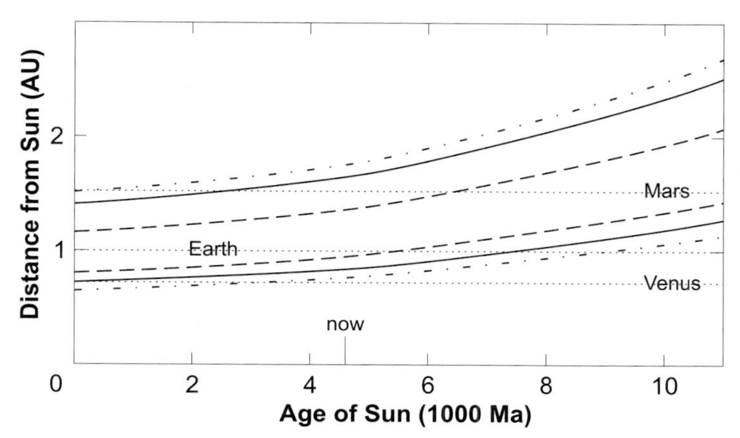

Figure 7.5 The HZ in the Solar System during the Sun's main sequence phase, for three sets of HZ boundary criteria. The intermediate criteria (solid lines) are those adopted in Section 4.2. The dashed and dotted lines are extreme criteria. The semimajor axes of Venus, Earth, and Mars are also shown.

surface temperature that increases the water vapour content of the atmosphere. This enhances the greenhouse effect, which results in further evaporation, and so on, to give a runaway greenhouse effect as outlined in Section 4.2. Even the outer few kilometres of the crust, where life also exists today, would be heated sufficiently to be sterilised. Deeper down the Earth would be no cooler. The fate of the Earth would thus resemble that thought to have been suffered by Venus much earlier in Solar System history. Subsequent increases in solar luminosity, to giant values, will strip the Earth of its atmosphere, leaving it a bare, rocky world, baked during the day, extremely cold at night.

Figure 7.5 shows a future improvement in Mars's prospects throughout most of the Sun's remaining main sequence phase. At present the surface of Mars is uninhabitable, though had it been a more massive planet then it would be habitable (Question 5.2), in accord with its present location within the HZ. In the distant future, the growing luminosity of the Sun should liberate volatiles, notably CO_2 and water, from surface reservoirs, enhancing Mars's greenhouse effect and thus raising surface temperatures considerably. The surface pressure would also rise, making water stable in the liquid phase over much of the surface. If there is subsurface life on Mars today then this could spread to the surface. If Mars is presently barren, then life could emerge. It could then evolve for several billion years, perhaps to the end of the Sun's main sequence lifetime, though the rapid rise in luminosity thereafter would soon extinguish life on Mars as it previously had done on Earth.

Europa's possible habitability does not at present depend on solar radiation but on tidal heating (Section 4.3.3), and as long as the Sun is insufficiently luminous to melt the icy crust this should remain the case. Any aquatic biosphere on Europa should survive into the giant phase, but early in the ascent of the first giant peak the icy crust is likely to melt, and the subsequent loss to space of Europa's water could be rapid, given the low mass of this body – 8% that of Mars. Life could cling on a bit longer beneath what would then be the rocky surface, provided that it was either there already or could colonise the crust rapidly. However, it would be unlikely to survive the first giant peak. Jupiter itself would survive, its atmosphere expanding and contracting in response to the huge changes in solar luminosity.

Titan will become warm enough for liquid water in the giant phase, and this might start the prebiotic processes that would lead to the origin of life, if there were enough time. Unfortunately there will only be a few tens of millions of years before the ascent of the first giant peak will boil the surface, and this is unlikely to be long enough for life to emerge.

The giant phase of the Sun will extinguish any life that still remained in the Solar System. The subsequent ejection of the hot gas that will form the Sun's planetary nebula will only make matters worse, and the rapid drop in luminosity of the remnant Sun as it becomes a feeble white dwarf will freeze the whole system. The only hope would be for life deep in the giant planets' atmospheres, sustained as long as heat flows from their interiors. This is to grasp at a straw.

7.3 SUMMARY

- Supernovae or close encounters with passing stars could sterilise the Solar System, though it is extremely unlikely that such externally caused cataclysms will occur before the evolution of the Sun does the job.
- Though the Sun's radius increases greatly during its giant phase, only Mercury and Venus are likely to be consumed – the Earth will probably escape!
- As the Sun moves through the 11 000 Ma main sequence phase of its lifetime its increasing luminosity will cause a runaway greenhouse effect on the Earth at some time, roughly 1–4 billion years, into the future. Mars will become temporarily habitable as its volatile reservoirs form a dense atmosphere, only to lose its habitability through a runaway greenhouse effect, though perhaps not until the Sun's giant phase, 6000 Ma from now, when the solar luminosity will start to increase enormously.
- The icy shell of Europa is likely to melt as the Sun's luminosity ascends the first giant peak. This low-mass satellite is then likely to lose its water to space, and with it any aquatic biosphere. Any crustal biosphere would perish soon after.
- There is a remote possibility of life flourishing briefly further out in the Solar System, when the Sun's luminosity is in the right range for liquid water to exist there.
- The only possible locations where life could survive through to the Sun's white dwarf stage, some 8000 Ma from now, are in the giant planets' atmospheres, where internal heat sources could sustain sufficiently high temperatures. This is a very remote possibility.

7.4 QUESTIONS

Answers are given at the back of the book.

Question 7.1

Describe two distinct ways in which terrestrial life could sustain itself well beyond when the Earth becomes uninhabitable.

Question 7.2

Estimate by what factor the luminosity of a young white dwarf with an effective temperature of 2×10^4 K is less than the present luminosity of the Sun. Hence describe in qualitative terms where the HZ of a young white dwarf would lie.

8

Potential habitats beyond the Solar System

So far we have restricted ourselves to the Solar System in our search for extra-terrestrial life. In the remaining chapters we look much further afield. We shall focus almost entirely on our Galaxy, and in the main on the nearby regions – the difficulties of finding potential and actual habitats increase the more distant the target because generally the light we receive from them is then more feeble.

We shall continue to concentrate on carbon–liquid water life, and therefore we look to planets and large satellites beyond the Solar System. The kinds of processes that formed the Solar System, described in Section 1.2, are expected to be common, in which case a substantial proportion of stars should have planetary systems. In Chapter 11 you will see observational evidence that this expectation is fulfilled. But which of these many planets might harbour life? This clearly depends on the planet, with *Earth-like planets* being favoured. These are rocky Earth-mass planets endowed with suitable volatiles (an Earth-mass planet having a mass between a few times that of Mars and a few times that of the Earth). It also depends on the planet's location. A protracted period in the HZ is required for *surface* habitability. Surface life is of particular importance in our search for life beyond the Solar System. This is because for the foreseeable future the search will have to be carried out from on or near the Earth, and surface life through its effect on atmosphere and surface, will probably be much more easily detectable from afar than life beneath the surface. Other conditions might also be necessary for life, such as the presence of a giant planet to screen the potential habitat from too many impacts (Section 4.2.2).

But habitability also depends on the type of star the planet orbits, and on where the star lies in the Galaxy, and it is on these factors that we focus in this Chapter starting with the variety of stars.

8.1 THE VARIETY OF STARS

8.1.1 The Hertzsprung-Russell (H-R) diagram

There is a great variety of stars, both by internal constitution and the corresponding external appearance. The latter is demonstrated by the H-R diagram, named after the Danish astronomer Ejnar Hertzsprung and the American astronomer Henry Norris Russell, who independently published the diagram in 1911 and 1913 respectively. It displays the luminosities and effective temperatures of stars (or parameters related to these quantities). Figure 8.1 is a generic H-R diagram showing where the main types of star are concentrated. Note that the luminosity scale is logarithmic and that the unit is the present solar luminosity (3.85×10^{26} W). The effective temperature scale is also logarithmic, and increases from right to left.

Recall from Section 1.1.2 that the effective temperature of the Sun (5780 K) is obtained from the black body spectrum that best fits the solar spectrum, which is approximately of black body shape. The same is true for all other stars. Figure 8.2 shows black body spectra at three temperatures, where the radiance is per unit area of radiator surface. You can see that as temperature increases, as well as a huge increase in the radiated power, the peak wavelength shifts to shorter wavelengths,

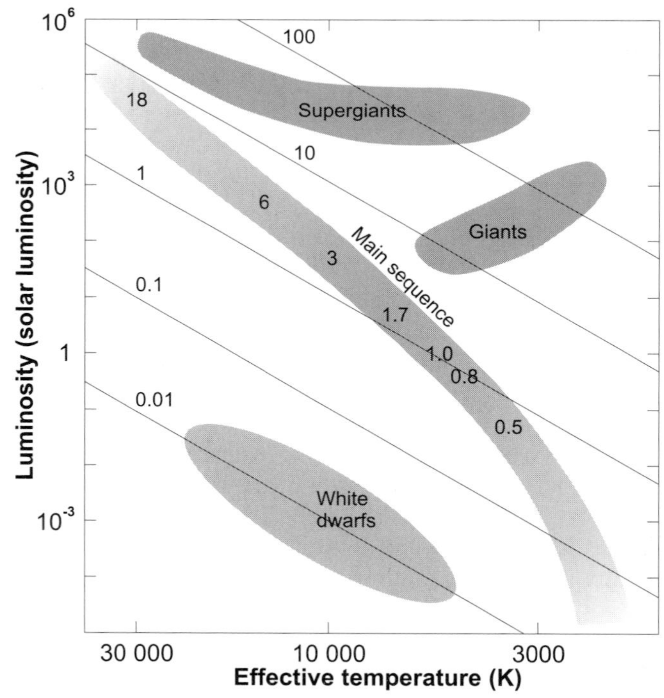

Figure 8.1 The H-R diagram, showing where the main types of star are concentrated. The diagonal lines are constant radii, labelled in solar radii. The values on the main sequence are stellar masses in units of the Sun's mass.

Plates

Plate 1 Comet Hale–Bopp, imaged from Earth on 1 April, 1997. The broad curving, yellowish tail of dust grains is to the right; the straighter, blue plasma tail is to the left.
Akira Fujii.

Plate 2 Mercury, imaged by Mariner 10 on 24 March, 1974, as it approached the planet for the first time. It shows the heavily cratered landscape that dominates the only hemisphere so far imaged.

NASA and US Geological Survey.

Plate 3 Venus, a spacecraft image (visible spectrum), with little structure apparent in the clouds, that are mainly composed of sulphuric acid and which completely obscure the surface. This image was obtained by the Galileo spacecraft in 1990, en route to Jupiter via Venus and the Earth.

NASA.

Plate 4 Meteosat-8 visible wavelength image of the Earth, showing Africa and a cloud-covered Antarctica, acquired from geostationary orbit on 9 May, 2003.

Plate 5 Mars, imaged by the Hubble Space Telescope (HST) on 26 August, 2003, when the planet was 55.7 million kilometres from Earth. The dark, 'shark-fin' shape to the right is Syrtis Major; the horizontal dark lane to the left is Sinus Meridiani. At the bottom of the image, where it is summer, the southern polar cap is shriking. Daytime highs are just above freezing in the Hellas impact basin, the light circular feature below-right of image centre. Many summer dust storms originate in this basin, though it is remarkably clear of dust here.

J Bell (Cornell University), M Wolff (SSI), and NASA/STScI.

Plate 6 Jupiter, a true-colour view composed of four images taken by the Cassini spacecraft on 7 December, 2000. The banded cloud tops are clearly visible, as is the Great Red Spot (lower right) in the southern hemisphere. Jupiter's moon Europa is casting a shadow on the planet (lower left).
NASA/JPL and University of Arizona.

Plate 7 Saturn, two images from the Hubble Space Telescope (HST), captured in 1996 (top) and 2000, which show Saturn's rings opening up from just past nearly edge-on to nearly fully open, as it moves from autumn towards winter in its northern hemisphere. Saturn's equator is tilted to its orbit by about 27 degrees, so as Saturn moves along its orbit, first one hemisphere, then the other is tilted towards the Sun. This cyclical change causes seasons on Saturn.
NASA and the Hubble Heritage Team, STScI/AURA.

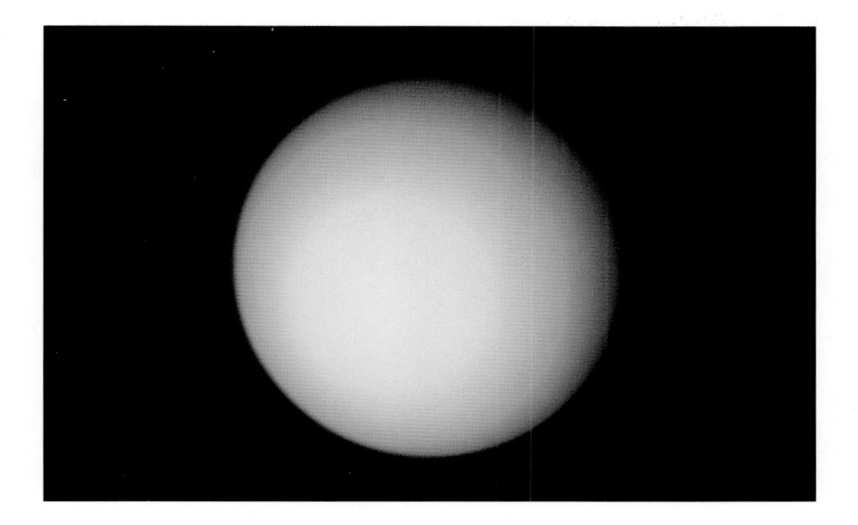

Plate 8 Uranus, imaged by Voyager 2 in January 1986. The planet's pole of rotation is just left of centre. The blue-green colour is caused by methane in the atmosphere. The cloud layers are beyond visibility, deep in the atmospheric haze.
NASA/JPL.

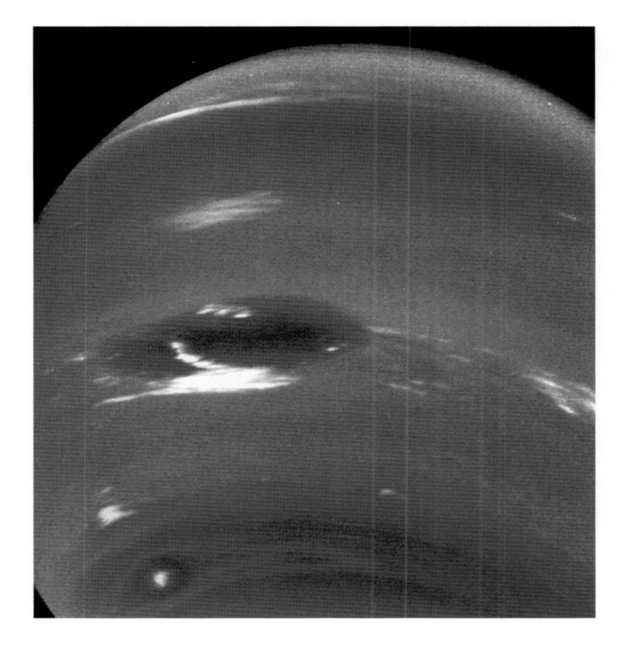

Plate 9 Neptune, imaged by Voyager 2 in August 1989. The North Pole is just out of sight (left of top centre). The atmospheric haze is thinner than on Uranus, so cloud structures can be seen.
NASA/JPL.

Plate 10 Most of a hemisphere of Io, the innermost of the four Galilean satellites of Jupiter. This image was obtained by Voyager 1 in March 1979, and shows an eruption plume from Loki, Io's most powerful volcano, rising above the limb (just above left centre).

NASA and US Geological Survey.

Plate 15 Thin deposits of ground frost in the shadows of boulders at the Utopia Planitia site on Mars, late in the northern martian winter, as imaged by the Viking 2 Lander on 25 September, 1977. The frost layer is extremely thin, only a fraction of a millimetre thick. The frost persisted at temperatures higher than the freezing point of carbon dioxide, and so is probably water ice. The colours have been enhanced.
NASA/JPL.

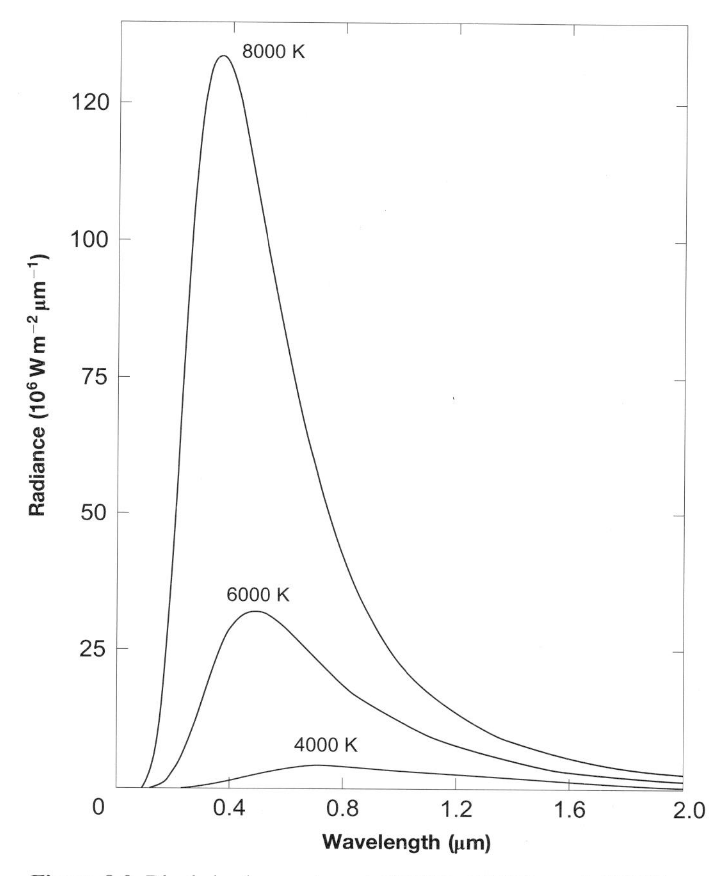

Figure 8.2 Black body spectra at 8000 K, 6000 K, and 4000 K.

from red towards blue. The colour of the black body thus depends on its temperature. The spectra are broad, not concentrated at a specific visual wavelength, and the colours are therefore tints of white rather than pure spectral colours as in a rainbow. The perceived colours are orange-white around 4000 K, yellow-white around 6000 K (including the Sun), and an almost pure white at 8000 K.

Temperatures versus tints are given in Table 8.1. This Table also contains a row of letters, O, B, A, F, G, K, M, called the spectral type of a star. These originate from the use of absorption lines in the star's spectrum to estimate temperature. Each letter corresponds to a range of effective temperatures. The ranges shown in Table 8.1 are for main sequence stars, and are only approximate. Each spectral type is subdivided 0–9, with 0 at the high-temperature end and 9 at the low end. Thus, A9 is slightly hotter than B0 and G4 is slightly hotter than G5. The Sun is a G2 star. For giants and supergiants, though the letter order is the same, many of the corresponding ranges of effective temperature are moderately different. Various mnemonics have been used to recall the letter sequence, such as 'Oh! Be A Fine Guy/Girl, Kiss Me'.

Table 8.1. Effective temperatures (representative values), tints, and spectral types of main sequence stars.

	Effective temperature range (K)						
	>30 000	30 000 to 9800	9800 to 7300	7300 to 5940	5940 to 5150	5150 to 3840	3840 to 2700
Tint of white	blue	blue	white	yellow	yellow	orange	red
Spectral type	O	B	A	F	G	K	M

Luminosity L, effective temperature T_e, and stellar radius R are linked as in Equation (7.1), reproduced here:

$$L = 4\pi R^2 \sigma T_e^4 \tag{8.1}$$

A rearrangement of this equation shows that if L and T_e are known for a star then its radius can be calculated:

$$R = \left[\frac{L}{4\pi\sigma T_e^4}\right]^{1/2} \tag{8.2}$$

Lines of constant radius have been added to Figure 8.1, in terms of the solar radius (6.96×10^5 km).

The mass of a star cannot be simply added to the H-R diagram – a star of approximately fixed mass moves across the diagram as it evolves. The exception is during the main sequence. You can see in Figure 8.1 that main sequence stars occupy a fairly narrow band, making an elongated S-shape running diagonally from low temperature–low luminosity to high temperature–high luminosity. During its main sequence lifetime a star does not move very far across the diagram, and so approximate masses can be attached, and these increase along the main sequence as in Figure 8.1. Measured masses range from about 100 solar masses ($M_\odot$) down to 0.08 $M_\odot$. Lower mass objects are known, but their interior temperatures never become high enough for sustained thermonuclear fusion of hydrogen, so there is no main sequence phase. These low-mass objects are the *brown dwarfs* (Section 1.3.1). Main sequence stars are called dwarfs, because they are smaller than giants, particularly the spectral types cooler than O and B (Figure 8.1). Main sequence stars are denoted by the roman numeral V. Spectral type prefixes this classification, and so, for example, the Sun is a G2V star.

After the main sequence, stars with masses less than about 4 $M_\odot$ evolve to become giants. The star then sheds a planetary nebula and the remnant becomes a white dwarf (Figure 8.1). The particular case of the Sun was described in Section 7.1, but the story applies to other masses too. Stars with masses in the range 4–8 $M_\odot$ also become giants, but when the helium in the core is exhausted, the carbon–oxygen core is too massive to be supported by electron degeneracy pressure – the maximum that can be supported in this way is about 1.4 $M_\odot$. Therefore, there is further core contraction to the point where the temperatures are high enough for carbon to undergo fusion, yielding neon, sodium, and magnesium. However, like the less massive stars, they shed a planetary nebula and the remnant becomes a white dwarf with a sufficiently small mass (less than about $1.4\,M_\odot$) to be stabilised by electron degeneracy pressure. Stars with main sequence masses greater than about $8\,M_\odot$ become supergiants (Figure 8.1). A supergiant starts its life rather like a giant, but it ends its life in a supernova (Chapter 7), leaving as a remnant either a *neutron star* (a 10 km size object with a typical mass about that of the Sun, and made almost entirely of neutrons), or, even more bizarre, a black hole. You will see in Section 8.2 that massive stars are unlikely hosts of inhabitable planetary systems, so we will say no more about their evolution.

Stellar lifetimes, from birth to catastrophe, are very sensitive to stellar mass, but in all cases the great majority of this time is spent on the main sequence. Main sequence lifetimes are the subject of Section 8.2.1.

Evolutionary tracks across the H-R diagram

Figure 8.3 shows the evolutionary tracks across the H-R diagram for stars of various main sequence masses. The initial chemical composition of a star, before thermonuclear fusion has had much effect, has a modest influence on its initial position on the main sequence, on the track, and on the rate that track is traversed. The initial composition is usually given as the proportions by mass of H, He, and all the other chemical elements, namely, the heavy elements. Recall that the heavy element proportion is called the metallicity, and that all the heavy elements are included, not just the metallic ones (Section 1.3.1). For the Sun at birth the proportions were 71%, 27%, and 2%. Other stars could have had several percent more or less He. Metallicities are typically in the range 0.05% to 5%, but can be far less than

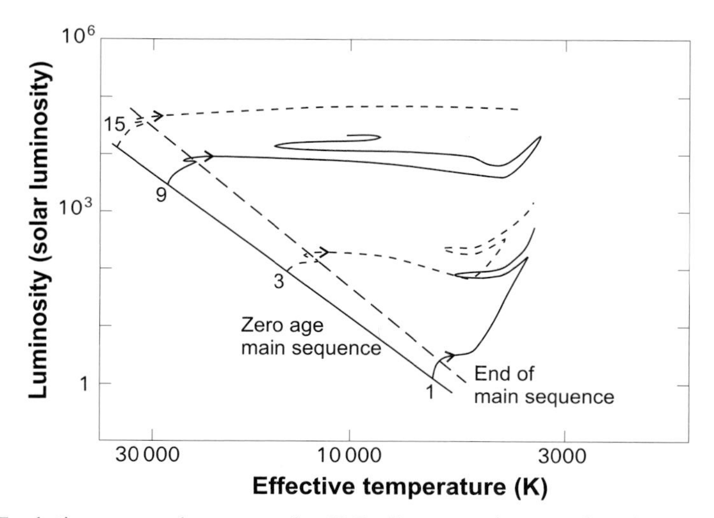

Figure 8.3 Evolutionary tracks across the H-R diagram of stars of various main sequence mass, shown in terms of the Sun's mass. The tracks stop short of the cataclysmic end points.

0.05%. In Figure 8.3 the proportions have been fixed at around the initial solar values, though on the scale of this Figure the tracks for stars with other initial proportions would not be greatly different.

Each track starts at the point where the star commences its main sequence phase – this is called the zero age main sequence or ZAMS point. The main sequence phase ends at the point shown, and you can see that during this phase the stars do not move very far across the H-R diagram. Thereafter the movement is considerable and much more rapid than on the main sequence. The tracks stop short of where giants shed planetary nebulae to leave a white dwarf remnant, and where supergiants become supernovae to leave neutron stars or black holes.

Stars shed little mass during the main sequence phase, but shed considerable proportions thereafter, particularly at the onset of helium fusion in the core and at the planetary nebula/supernova end points. For stars with $1\,M_\odot$ ZAMS mass, the white dwarf is left with about 50% of the mass, but for $7\,M_\odot$ ZAMS mass only about 20% is left. The even more massive stars that end as supernovae can leave an even smaller proportion.

Multiple-star systems

A high proportion of stars are in multiple systems (i.e. in systems where several stars are in orbit around each other). In the solar neighbourhood the proportion is about 70%. In most multiple systems there are just two stars, comprising a binary system. If the pair are closer than a few times the stellar radii, then the stars influence each other significantly, and the evolution just outlined is greatly modified, particularly after the main sequence phase. However, the gravitational disturbance of two stars swinging around each other would destabilise planetary orbits in the HZ, which would lie outside the binary orbit. Only at distances too great for habitability could there be stable orbits. We will therefore not consider further the 40% or so of binaries that have such small separations. If the two (or more) stars are further apart, more than the order of 10 AU, then stable orbits in the HZ of one or both stars are possible. Moreover, such stars do not appreciably influence each other's evolution, so they each follow the evolutionary story outlined above.

8.1.2 Stellar populations

Not all ZAMS masses are equally abundant. The lower the mass the greater the proportion of stars, right down to the $0.08\,M_\odot$ brown dwarf limit. Observational evidence, with some theoretical support, indicates a ZAMS relationship in the Galactic disc of the form:

$$\left.\begin{aligned} \delta n &= k0.035\,M^{-1.3}\,\delta M && \text{when } 0.08 \le M \le 0.05\,M_\odot \\ &= k0.019\,M^{-2.2}\,\delta M && 0.019 < M \le 1.0\,M_\odot \\ &= k0.019\,M^{-2.7}\,\delta M && 1.0 < M \le 100\,M_\odot \end{aligned}\right\} \quad (8.3)$$

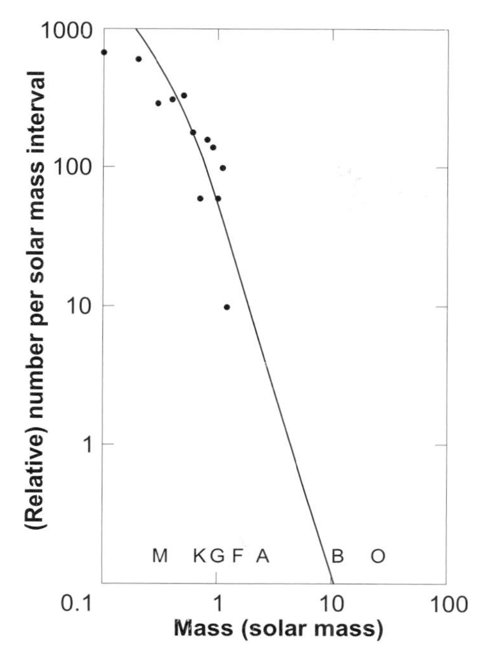

Figure 8.4 The line is the initial mass function of stars in the Galactic disc, with arbitrary vertical scaling. The dots are the numbers per solar-mass interval for nearby stars.

where δn is the number of stars in the small range of mass δM centred on the mass M in units of solar mass $M_{\odot}$. The value of the constant k depends on the volume of space being considered. The relationship between δn and δM is called the initial mass function (IMF). The solid line in Figure 8.4 shows this function with arbitrary vertical scaling. The spectral type is also shown, and with low-mass stars dominating it follows that M dwarfs are the most common, and that O stars are very rare. Figure 8.4 also shows the stars within 30 ly of the Sun. Nearly all of these are main sequence stars so it is appropriate to compare these data with the IMF. You can see that there is fairly good correspondence. The discrepancy at very low masses is probably because some faint stars between 15 and 30 ly have not yet been detected.

8.2 SUITABLE STARS FOR LIFE

In Section 7.2 we described the sterilisation that will befall the Solar System when the Sun becomes a giant. A similar fate awaits any planetary system when its star becomes a giant or a supergiant. Therefore, in considering which stars are suitable hosts for habitable planets, we will concentrate on the main sequence phase.

Table 8.2. Main sequence lifetimes of stars with various ZAMS masses (solar composition).

Stellar mass (ZAMS) ($M_{\odot}$)	25	15	3.0	1.5	1.0	0.75	0.50
Main sequence lifetime (Ma)	3.0	15	500	3000	11 000	15 000	100 000
Spectral type (approximate)	O8	B1	A0	F2	G2	K2	M0

8.2.1 Main sequence lifetime and life detectable from afar

The duration of the main sequence phase is of great importance. It has to be long enough for life to develop a distantly observable effect on the planet's atmosphere or surface. In Chapter 3 it was seen that the Earth, and therefore the Sun, was probably no younger than about 2000 Ma before life had such an effect, notably a significant rise in the oxygen content of the atmosphere (Section 3.2.3). From then, or soon after, this elevated content, and surface effects, could have been detected from afar. We shall take 2000 Ma as the minimum main sequence lifetime necessary for any life to have distantly observable effects, though it is clear that this value is very uncertain.

We therefore exclude stars with main sequence lifetimes less than about 2000 Ma. This does not, of course, mean that the remaining stars will have planets with distantly observable life. Main sequence lifetime is merely the first of many constraints by which we reduce the proportion of stars and planetary systems that might harbour distantly detectable life.

Table 8.2 gives the main sequence lifetimes of stars of various ZAMS masses. The main sequence spectral types (approximate) are also included. Note how rapidly the lifetime decreases as mass increases. The mass of hydrogen is roughly proportional to the stellar mass, so if it were consumed at a rate also proportional to the stellar mass the lifetime would be roughly the same for all stars. Therefore, as stellar mass increases, the rate at which the fuel is consumed must rise rapidly. The reason for this is the increase with mass of the core temperature, coupled to the great sensitivity of the rate of hydrogen fusion to temperature. Table 8.2 shows that, on the 2000 Ma main sequence lifetime constraint, O, B, and A stars do not have sufficiently long lifetimes for any life to be distantly detectable. Fortunately, as you can see from Figure 8.4, this rules out only a small proportion of stars.

A rather larger proportion is ruled out by excluding F, G, K, and M stars that are younger than 2000 Ma. About 10% of these stars are ruled out on this basis.

8.2.2 Metallicity and other considerations

The lifetimes in Table 8.2 are for metallicities around that of the Sun, though the general conclusions apply to a wide range of metallicities. However, we probably have to rule out stars with low metallicity, even if they have sufficiently long main sequence lifetimes. This is because they will have had circumstellar disks with such low abundances of substances like iron, silicates, and water, that sufficiently massive rocky or rocky–icy planets are unlikely to have formed. Metallicities less than about half that of the Sun (2%) might yield rocky planets no more than about 10% the mass of the Earth (i.e. no more than about the mass of Mars). Such planets are likely to lose their atmospheres to space and to the surface within about a billion years, as discussed in Section 4.2, and thus be unlikely to have surface life that can ever be detected from afar.

It is not only necessary to have heavy elements, but *specific* heavy elements. For example, substantial atmospheric losses to the surface arise from a low level of geological activity. To sustain such activity large mass helps, but the interior must also contain long-lived radioisotopes, notably ^{40}K, ^{235}U, ^{238}U, and ^{232}Th, to heat the interior over long periods. Such elements therefore need to be present. Whether this is a significant further constraint is unknown.

Stars with low metallicities had their origin when the Galaxy was young, before about 10 000 Ma ago. At that distant time, the interstellar medium would have had a near-primordial composition, with a low abundance of heavy elements. The stars forming from this medium were comparably depleted. A proportion of G, K, and M dwarfs must be at least as old as 10 000 Ma (Table 8.2) – 20% is one estimate – and this proportion is therefore unlikely to have suitable planets. Subsequently, the short-lived massive stars, in which thermonuclear fusion had increased the proportion of heavy elements, enriched the interstellar medium as they lost mass in their giant, supergiant, and subsequent phases. Younger stars will therefore have been born in clouds enriched in heavy elements, and are thus more likely to have habitable planets, though some of these stars will be too young for life to have emerged in their planetary systems.

Stars not yet ruled out from having planets on which we could detect life from afar are thus the higher metallicity main sequence stars of spectral types F, G, K, and M (i.e. with masses less than about 2 $M_\odot$, and main sequence lifetimes exceeding about 2000 Ma, but older than 2000 Ma). We then have to exclude the proportion of such stars that are in close binaries, for reasons given in Section 8.1.1. Further exclusions will be discussed in Section 8.3.

It has been argued that we should exclude M dwarfs. If we do so, then, these being the most common type (Figure 8.4), the proportion of stars in the Galaxy that are retained falls by about a factor of five. Consider the case for *not* excluding them.

8.2.3 Main sequence M stars (M dwarfs)

Four problems have been raised over the suitability of M dwarfs as hosts for habitable planetary systems: the width of the HZ, synchronous rotation, the small proportion of the star's output at visible wavelengths, and M dwarf variability.

The HZ width

Figure 8.5 shows the HZ of an M0 dwarf with a mass 0.50 $M_\odot$ compared to a G2 dwarf of mass 1.0 $M_\odot$ (for solar metallicity). We have adopted the same two criteria here as we did in Section 4.2 for the Sun, though the precise criteria do not much

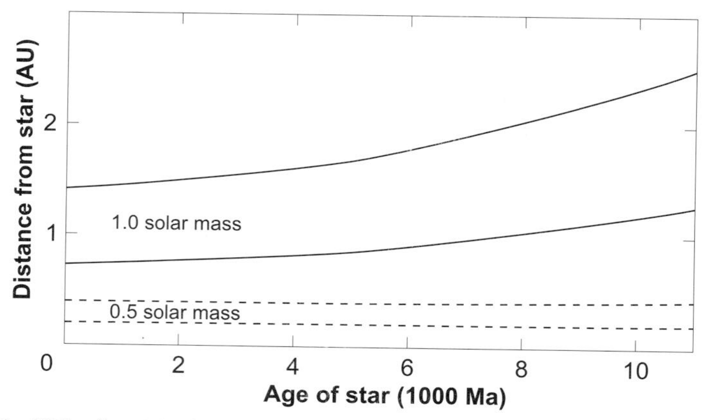

Figure 8.5 The HZs of an M0 dwarf of mass 0.5 $M_\odot$ and a main sequence star of mass 1.0 $M_\odot$ (for solar metallicity).

affect the comparison between the two stars. The zone for the M dwarf is much closer in, because of the star's lower luminosity. It would be even closer if the spectral distribution of its radiation were the same as for the G dwarf, but it is shifted to longer wavelengths, in accord with the lower effective temperatures of M dwarfs (Table 8.1). As a result, a lower proportion of the stellar radiation is scattered back to space, and the planet is warmer than it would otherwise be at a given stellar distance.

As well as being closer in, the HZ of the M dwarf is narrower: at ZAMS it is about 0.2 AU for the 0.50 $M_\odot$ star compared to about 0.7 AU for the 1.0 $M_\odot$ star; at an age of 5000 Ma the values are 0.2 AU and 0.9 AU respectively. This has led some to argue that the probability of finding a planet in the HZ of the M dwarf is reduced. However, it would only be reduced by a small factor, and it would not be reduced at all if the scale of planetary orbit spacing were roughly proportional to the width of the HZ.

Synchronous rotation

Consider a rapidly rotating planet. If it is highly fluid the tidal distortion always lies on the line connecting the planet to the star. Real planets are not highly fluid. Therefore, as the planet rotates with respect to the star the tidal bulges are carried out of alignment, as in Figure 8.6. The gravitational force of the star on bulge A has a component towards the line of alignment, whereas that on bulge B has a component away from it. However, A is closer to the star than B, so the force on A is slightly greater. Therefore, there is a twist, a torque, on the planet tending to bring the bulges into alignment. This torque is opposite in direction to the rotation of the planet. Consequently, the rotation rate slows down (or speeds up if it were initially too slow), and after many orbits the planet is likely to end up with a rotation period equal to its orbital period, and rotating in the same direction. This is synchronous rotation (which you met in Section 4.3.2). In this case it is the result

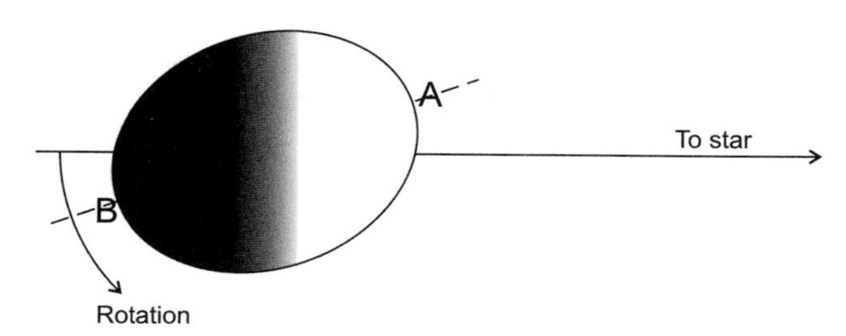

Figure 8.6 The gravitational distortion (the tide) on a planet is carried out of alignment by the rotation of the planet (until synchronous rotation is established).

of what is called tidal locking. The tidal bulge then points towards the star. A consequence of synchronous rotation is that, apart from the slight effect of orbital eccentricity, the planet keeps the same face towards the star. Synchronous rotation is likely for a planet in the HZ of an M dwarf because of the proximity of the planet to the star and the consequent large tidal distortion.

Synchronous rotation has been regarded as inimical to the emergence of life because it has been supposed that the dark side of the planet would act as a cold trap, to the extent that most of any atmosphere would condense there. However, it has been shown that at the sort of radiation levels that occur in the inner HZ of an M dwarf, a 10^4 Pa atmosphere consisting largely of the greenhouse gas CO_2 would prevent atmospheric collapse, and would ensure that liquid water could persist over at least part of the surface, even if, in some cases, only beneath a thin crust of ice on the oceans. The Earth's atmospheric pressure is 10^5 Pa, and though the partial pressure of CO_2 is now only about 35 Pa (Table 2.1), it was surely greater in the past, as discussed in Section 3.2.3. A contributory factor outlined in Section 3.2.3 was the increase in the Sun's luminosity, from 70% of its present value over 4600 Ma. The luminosity increase for M dwarfs in 4600 Ma is far smaller, and so a high level of CO_2 could persist. In any case, water vapour has its own greenhouse effect and would thus reduce the required quantity of CO_2. Synchronous rotation is therefore not necessarily a problem.

Stellar output at visible wavelengths

The third possible problem, the small proportion of the star's output at visible wavelengths, is illustrated in Figure 8.7, where the spectra are shown of an M6 dwarf and the Sun with vertical scaling to match the spectral maxima. The paucity of M dwarf radiation at wavelengths less than about 0.7 μm would deprive Earth organisms of much of the radiation they use for photosynthesis (Section 2.4.2). But even if photosynthesis was absent on the M dwarf planet, life could still be present – on Earth, life can exist independently of photosynthesising organisms (Section 2.5.2). Moreover, some terrestrial bacteria use wavelengths longer than 0.7 μm to perform photosynthesis. Therefore, the weak visible output of M dwarfs need not be a problem.

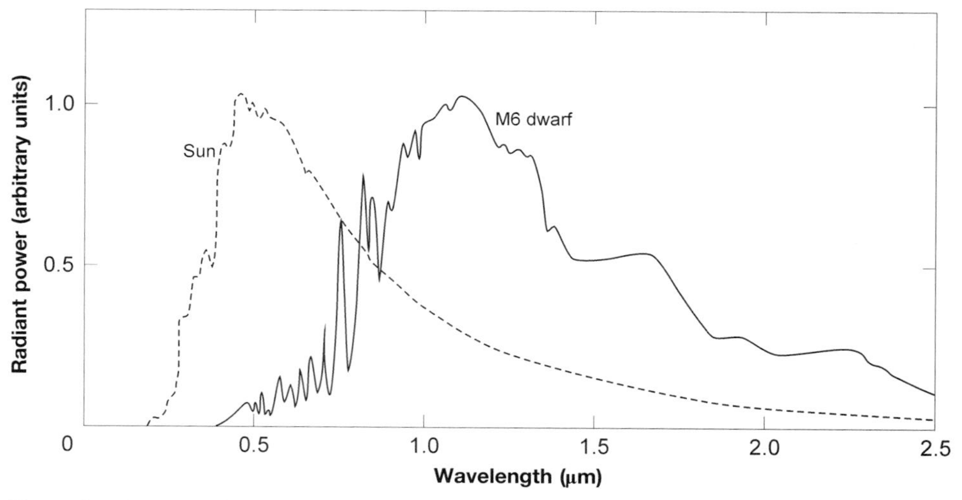

Figure 8.7 Spectra of an M6 dwarf (solar composition) and the Sun with vertical scaling to match the spectral maxima.

M dwarf variability

This final problem also seems non-fatal. All stars flare, including the Sun. A flare is a sharp increase in the emission of electromagnetic radiation and charged particles from a restricted region on the photosphere, often associated with starspots. Flares can last up to a few minutes, though a few tens of seconds is typical, and even the longer flares consist of a peak of short duration flanked by slow rise and fall. During a flare, the increase in the biologically damaging X-ray and UV output can be especially large, though X-rays would not penetrate a planet's atmosphere, and so it is UV that is of particular concern. The UV output can exceed the quiescent UV output of the whole star by up to a factor of 100 or so. Fortunately, the quiescent UV output of an M dwarf is so feeble (Figure 8.7) that even a factor of 100 increase, would, for a planet with an Earth-like atmosphere, only increase the surface flux to a few times the amount the Earth receives from the quiescent Sun.

Though the flare intensity is low, young M dwarfs typically flare far more frequently than the Sun, several times per day in some cases. Fortunately, the frequency declines on a timescale of order 1000 Ma, and so, even if high frequency were a problem, it might only *delay* the emergence of life on a planetary surface. Life in crustal rocks and in the oceans need not be affected.

Another type of variability is in the luminosity of a star due to cool star spots. M stars can have spots proportionately much larger than those on the Sun, so large that they can cause a decrease of a few tens of percent in luminosity for up to several months. However, models of planets with atmospheres show that the temperature drop would not eliminate life, not even at the surface.

Therefore, in the list of stars that are possible hosts of planets on which we could detect life from afar, there seems to be no compelling reason to exclude the abundant M dwarfs.

8.3 THE GALACTIC HZ

Just as a star has a HZ so does the Galaxy. Figure 8.8 is a sketch of the Galaxy showing an edgewise view of the major structural components – the thin and thick discs, the central bulge, and the halo (Section 1.3.2). Note that the thick disc permeates the thin disc, and is recognisable there as a distinct population of stars. The numbers of stars that constitute the thin disc, thick disc, nuclear bulge, and halo are in the approximate ratios 100:20:10:1, and so the thin disc accounts for about three-quarters of the stars in the Galaxy. The Galactic HZ is defined in terms of the likelihood that habitable planets could be present in each structural component.

The metallicity of the medium from which a star and its planetary system formed is of prime importance, as was outlined in Section 8.2.2, with metallicities less than about half that of the Sun perhaps unlikely to yield suitable planets. The thin disc has a long and continuous history of star formation, so that the metallicity of its interstellar medium was raised early in Galactic history and has continued to rise

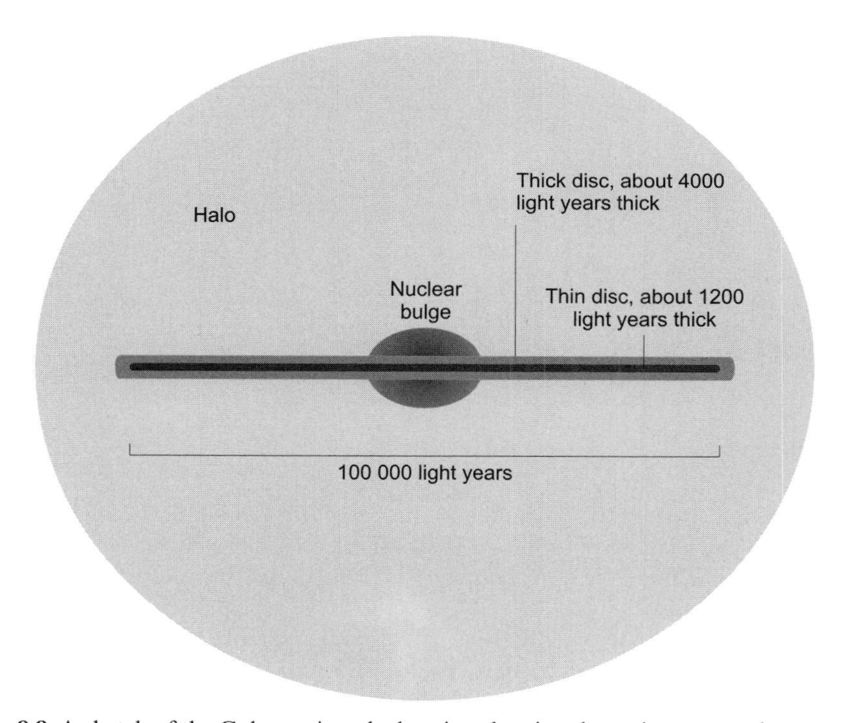

Figure 8.8 A sketch of the Galaxy, viewed edgewise, showing the main structural components. The boundaries are not as sharp-edged as shown here.

since. It is the prime location for stars with habitable planets. In its outer regions it is less enriched, so suitable planets might be scarcer there. The thick disc has a much higher proportion of old, low-metallicity stars, and so habitable planets are probably rarer. The halo is dominated by stars even older than those in the thick disc, and habitable planets are thus likely to be rare. About 1% of the halo stars are in globular clusters (Figure 1.14), found also in the nuclear bulge. In the nuclear bulge, star formation peaked some time ago, but is still continuing. Habitable planets might be common, though the mix of heavy elements differs from that in the thin disc, with uncertain consequences.

In addition to metallicity, there are two other factors that affect planetary habitability – transient radiation events and gravitational disturbance. A proportion of planetary systems will have been sterilised by transient radiation events, such as supernovae, and a further small proportion will have been disrupted by the gravitational disturbance of nearby stars (Chapter 7). Transient radiation events occur throughout the disc, but in the outer disc are well separated and rare. They are more pervasive in the nuclear bulge and the inner disc and probably reduce significantly the habitability there. They must also have reduced the habitability of globular clusters, where massive stars long ago died in supernovae and bathed the cluster with lethal radiation. Gravitational disturbances are also significant in the bulge and in the globular clusters, because they are comparatively densely packed with stars.

Thus, the thin disc is where the greatest proportion of stars are likely to have habitable planets, particularly in an annulus that excludes its outermost and innermost regions. The Sun is in this annulus! With the thin disc accounting for about three-quarters of the stars in the Galaxy we thus have to exclude somewhat more than a quarter. Of the three-quarters, some proportion is unlikely to have planets on which any life is observable from afar, for the various reasons given earlier in this chapter.

As an upper limit, roughly half of the stars in the Galaxy could have planets with life detectable from afar if M dwarfs are included, otherwise 5–10%. It must be emphasised that these are *very* rough figures, and will be revised downwards when further constraints on planetary formation and survival are added in later chapters.

8.4 SUMMARY

- The external appearance of stars, and their evolution, are well shown on the H-R diagram, which displays luminosity versus effective temperature, or parameters related to these quantities, such as spectral type for effective temperature – O, B, A, F, G, K, and M.
- The mass of a star as it embarks on its main sequence phase (a ZAMS star) largely determines its subsequent evolution. Stars up to about $8\,M_\odot$ become giants that then shed planetary nebulae, leaving white dwarfs as the remnant. More massive stars become supergiants, and end their lives in supernovae, leaving a neutron star or a black hole as a remnant.

- The main sequence lifetime decreases rapidly with increasing ZAMS mass, and accounts for the great majority of the time between a star's birth and the planetary nebula/supernova event.
- The abundance of stars decreases from M to O, with M stars by far the most abundant.
- Earth-like planets are the most likely potential habitat for surface life, with a location in the HZ for at least the order of 2000 Ma for surface life to be detectable from afar, through its effect on the planetary surface or atmosphere.
- Stars not yet ruled out from having planets on which we could detect life from afar are the higher metallicity main sequence stars of spectral types F, G, K, and M (i.e. with masses less than about $2\,M_{\odot}$). These have main sequence lifetimes exceeding about 2000 Ma, and need ages exceeding 2000 Ma. From these we exclude close binaries, systems sterilised by transient radiation events, and systems subject to gravitational disturbance. There is no compelling reason to exclude M dwarfs.
- Stars with habitable planets are likely to be concentrated in the thin disc of the Galaxy, particularly away from its outer and inner edges.
- As an upper limit, roughly half of the stars in the Galaxy could have planets with life detectable from afar if M dwarfs are included, otherwise 5–10%. These are *very* rough figures, and will be revised downwards when further constraints on planetary formation and survival are added in later chapters.

8.5 QUESTIONS

Answers are given at the back of the book.

Question 8.1

State, with justification, whether each of the following stars should be ruled out from having planets on which we could detect life from afar (remember that V = main sequence).

(i) An A3V star.
(ii) A binary system in which a solar-mass star is 3 AU from an M dwarf.
(iii) A solar-mass star in a globular cluster.
(iv) A G2V star 1000 Ma old.
(v) An M0V star 5000 Ma old, in the thick disc about half way to its outer edge.

Question 8.2

Some of the stars known to have giant planets have metallicities less than 1%. Explain why this is not at variance with the statement in Section 8.2.2 that such stars are unlikely to have planets with surface life.

9

Searching for planets: direct methods

We now turn to methods of finding planets beyond the Solar System, particularly habitable worlds. In this chapter we shall concentrate on *direct* methods, where we attempt to detect the radiation reflected or emitted by the planet. We could then analyse this radiation to learn about the planet and see whether it is habitable and indeed inhabited. Direct methods are very challenging, more so than *indirect* methods where we infer the presence of a planet from its influence either on the motion of the star it orbits, or on the quantity of radiation we receive from this star or from some background star. We learn less about a planet by indirect methods, but they currently play a very important role in our search, and are described in Chapter 10. In Chapter 11 the results of searches with the various methods are summarised.

9.1 THE CHALLENGE OF DIRECT DETECTION

Consider a planet orbiting a star about 30 ly (about 10 parsecs) away. This is not far in terms of the Galactic diameter of about 100 000 ly (Figure 8.8), but is far enough to be representative of distances in our cosmic neighbourhood, and it has come to be used as a standard distance for comparing detection methods. There are about 500 stars within 30 ly of the Sun, roughly 70% of which are multiple systems, mainly binary.

Viewed from 30 ly, the planets in the Solar System are not particularly faint, and would be well within the grasp of modest Earth-based telescopes. Nevertheless they would at present be undetectable. This is because the Sun is so much brighter than its planets and is so close to them, that its light would drown the far fainter planetary light – it's a bit like trying to detect a candle next to a powerful security light. Figure 9.1 illustrates this problem.

The upper curve in Figure 9.1 shows the solar spectrum, and the lower three curves are the spectra of Jupiter, the Earth, and Uranus. (Note that all of these

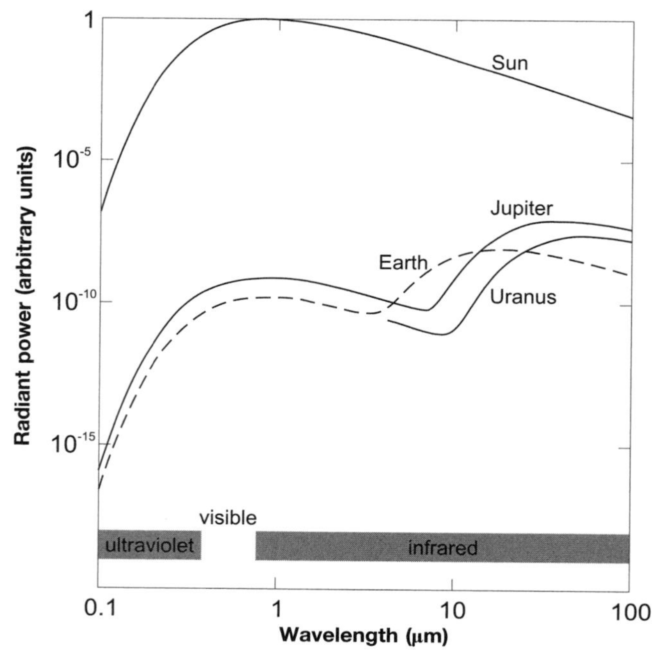

Figure 9.1 The spectrum of the Sun, and, to the same scale, the spectra of Jupiter, Earth, and Uranus (part). All these spectra have been smoothed. Note that the radiant power is per unit frequency interval, and therefore the spectral peaks are at different wavelengths from when the power is per unit wavelength interval.

spectra have been smoothed, so they don't show spectral lines and bands.) The planets shine partly by reflected sunlight and partly through their infrared emission. The reflected light, like its source the Sun, peaks at visible wavelengths. It dominates the Earth's spectrum into the near-infrared (3 μm), and the spectra of Jupiter and Uranus into the mid-infrared (10 μm). At longer wavelengths the infrared emission from the planets takes over, apparent in the broad peaks at long wavelengths in Figure 9.1. You can see that at visible wavelengths the Sun is 10^9–10^{10} times brighter that these three planets, whereas in and beyond the middle infrared the ratio is somewhat smaller, 10^5–10^6. These large contrast ratios would not be a problem if a telescope could produce images sufficiently sharp that the stellar and planetary images were clearly separated. Unfortunately there is a limit to image sharpness, as shown in Figure 9.2.

Figure 9.2(a) shows a reflecting telescope with a circular primary mirror of diameter D forming an image of a distant point object delivering plane wavefronts at the wavelength λ. The image is formed in the image plane across which the position is measured by the angle θ. The image will be something like that in Figure 9.2(b). It consists of a central fuzzy disc plus fuzzy rings (just one ring is shown). The graph beneath the image is a section through the pattern, where the

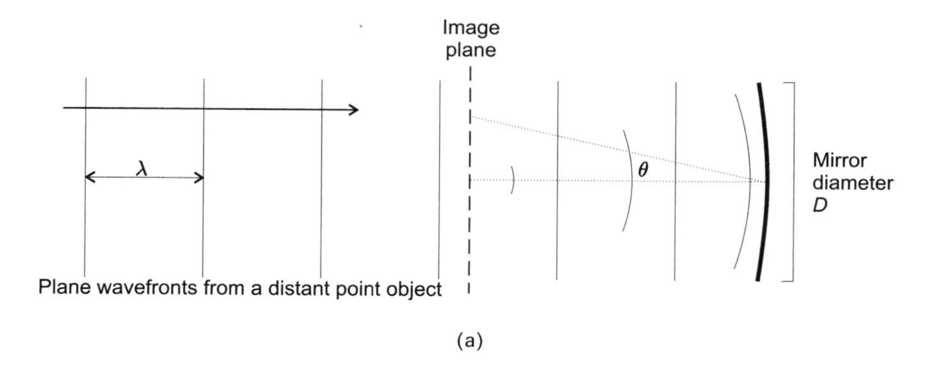

(a)

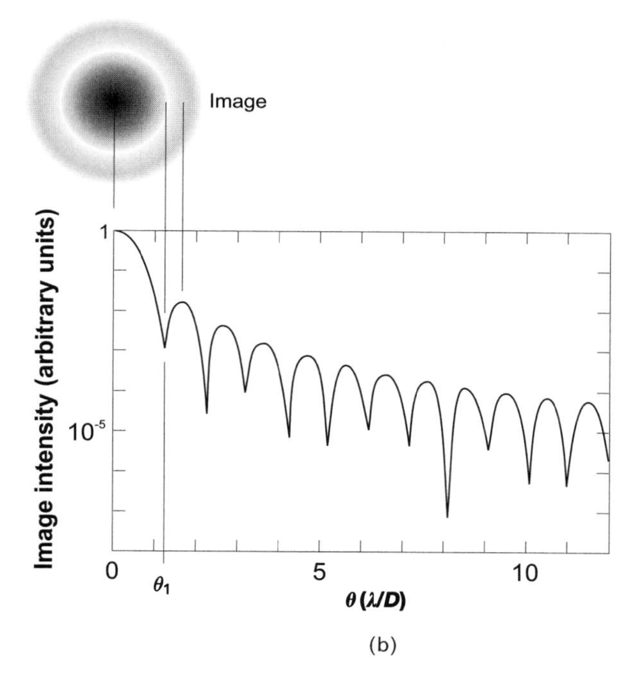

(b)

Figure 9.2 (a) A reflecting telescope with an optically perfect circular mirror of diameter D forming an image of a distant point object. (b) The image of a distant point object produced by the mirror, and the corresponding *psf* of the mirror.

rings now appear as sidelobes. The mirror is optically perfect, and for the time being we neglect any blurring effects of the Earth's atmosphere. You might wonder why the image of a point object is not itself a point at the centre of the image plane (i.e. at $\theta = 0$). This is due to a fundamental optical limit called the *diffraction limit*. It can be thought of as a consequence of the mirror only *sampling* the wavefront from the object – it clearly does not include the light that misses the mirror. The outcome is that the light from a distant point object is spread out to form a pattern. This pattern is called the *point spread function* (*psf*) of the mirror. Note that the *psf* has an angular scale proportional to λ/D, and that the angle θ_1 to the first minimum in Figure 9.2(b) is given (in radians) by:

$$\theta_1 = 1.22\frac{\lambda}{D} \tag{9.1}$$

The inverse, $1/\theta_1$, is a measure of the *resolving power* of the telescope. Consult optical texts in Resources for details.

Here is the difficulty. A distant star is essentially a point object, and thus will give an image like that in Figure 9.2. The image of a planet will be similar, but much lower in brightness and displaced as shown in Figure 9.3, where θ_p is the angular separation in the sky of the planet from the star. The *psf* decreases rather slowly as θ increases, and so the image of the planet is still about 10^4 times fainter than the sidelobes. In this particular example the contrast ratio is for Jupiter and the Sun at visible wavelengths, and the planet is at an angle $\theta_p = 9.0\,\lambda/D$ radians. What kind of telescope would give this outcome? If the Solar System were 30 ly away, then Jupiter and the Sun would be separated by $\theta_p = 0.56\,\text{arcsec} = 2.7 \times 10^{-6}$ radians. Thus, $D = 9.0\lambda/(2.7 \times 10^{-6})$, and at $\lambda = 5.0 \times 10^{-7}\,\text{m}$ (0.5 μm, visible), $D = 1.7\,\text{m}$. Therefore Jupiter would be very hard to detect if viewed from 30 ly with a 1.7 m telescope operating at visible wavelengths.

To improve the situation we could increase D. This narrows the *psf* (Equation 9.1) and thus pulls obscuring starlight away from the planet's image. However, away from the central peak of the *psf* the strength of the sidelobes decreases with increasing angle in proportion to $(\lambda/D)^{1/3}$. Therefore, a large increase in D is required for a significant improvement in performance, leading to very large mirror sizes. At infrared wavelengths the star–planet contrast ratio is less (Figure 9.1), but λ is greater and so the *psf* is more spread out. It would be better if we could reduce the intensity in the stellar image with much less reduction in the planet's image. One way to do this is by coronagraphy.

9.2 CORONAGRAPHY

So far you have met the *psf* of a circular mirror (or lens) that is optically perfect – a diffraction-limited *psf*. The effect of imperfections is to transfer light from the central peak of the *psf* to the sidelobes – *exactly* where we don't want any extra light, because it is among the sidelobes that the image of any planet would lie. Optical perfection means an optically perfect shape to the mirror surface on the large scale and smoothness on a microscopic scale. To achieve smoothness the surface has to be free of dust, scratches, and pits that scatter light, and ideally it also has to be free of microscopic ripples that diffract light. In practice one has to aim at smoothness *considerably* better than a tenth of the wavelength of light at which the image is obtained. The telescope structure should also not scatter light.

Suppose that we have a telescope as near to perfection as is practicable. The next

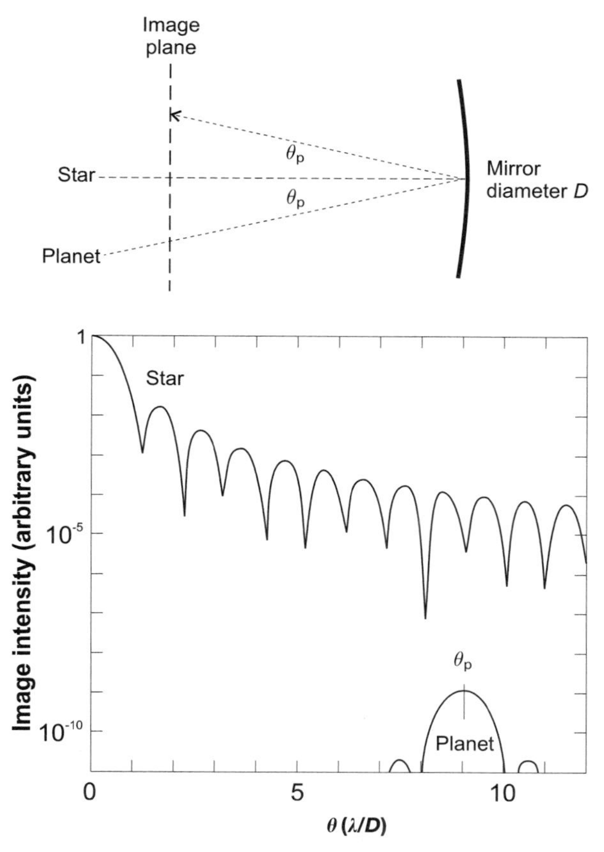

Figure 9.3 Sections through the images of a star and a planet separated by an angle θ_p, produced by an optically perfect circular mirror of diameter D.

step is to reduce the stellar image intensity by a much greater factor than that of the planet. A coronagraph is one device that can do this, but consider first how the advantage of intensity reduction is quantified.

9.2.1 Signal-to-noise ratio (*snr*) in a telescope image

When light interacts with an image detector it does so as a stream of photons (Section 2.4.2). The stream is never steady, so if we accumulate photons from a source for a time Δt, called the *integration time*, we will accumulate different numbers in successive intervals. If the mean number accumulated in Δt is N, then a fundamental result is that the fluctuations in number from one Δt to another are typically $N^{1/2}$ (see statistics texts in Resources).

Suppose that N is the number of photons that has created the star's image in the region of the image plane where the planetary image lies, and that n is the number for the planet's image. It is assumed that the planet's image is separated from the central peak of the star's image so that, in principle, the two objects can be distinguished. $(N+n)$ is accumulated in Δt and the fluctuations are therefore $(N+n)^{1/2}$. It is usual to call n the signal. The fluctuations are called the noise. Whether the signal can be identified is measured by the signal-to-noise ratio, *snr*, defined as:

$$snr = \frac{n}{(N+n)^{1/2}} \tag{9.2}$$

If F is the rate at which stellar photons arrive at the telescope per unit area per unit time (the flux density), then $N = p_N F A \Delta t$ accumulate at the image plane where p_N is the proportion of stellar photons arriving at the aperture that arrive at the position of the planet's image, and A is the area of the telescope aperture $(\pi(D/2)^2)$. Likewise, $n = p_n f A \Delta t$ where f is the flux density from the planet, and p_n is the proportion of photons arriving at the aperture that constitute the planet's image. Equation (9.2) can thus be rewritten as:

$$snr = (A\Delta t)^{1/2} \frac{p_n f}{(p_N F + p_n f)^{1/2}} \propto (A^2 \Delta t)^{1/2} \tag{9.3}$$

The final (approximate) step is not obvious, and comes from the relationship between p_n, p_N, and A when the planet's image is well away from the central peak of the star's image. We will not go into detail, but it is clearly reasonable that as A gets larger p_n/p_N also gets larger. In order to make a reasonable claim that a planet has been discovered after an integration time of Δt, *snr* needs to exceed about three. Therefore, large apertures and long integration times are advantageous.

A further increase in p_n/p_N can be accomplished with coronagraphy.

9.2.2 The effect of one type of coronagraph

A *coronagraph* consists of mirrors (or lenses), and masks (or irises), interposed between the main mirror (or lens) of a telescope and the image plane, in order to reduce the intensity in the image of an object whose image is centred on the image plane. The Lyot coronagraph is one particular type, and is named after the French astronomer Bernard Ferdinand Lyot, who developed it to observe the faint atmosphere of the Sun without the requirement of a solar eclipse to block the bright photosphere. This he achieved in 1931. The Sun's atmosphere is called the corona, hence the name of Lyot's instrument and indeed of all the types of instrument that employ the same general principles. A detailed description of his coronagraph's optical components and their function would take us too far afield (see solar texts in Resources), but its performance is illustrated in Figure 9.4 along with the original *psf* for comparison. You can see that a considerable reduction in the intensity of the stellar image has been achieved right across the image plane, but notably in the sidelobes where a planet might lie. The reduction in the intensity of the planet image would be relatively slight. This is because the planet is off the telescope axis – were the planet on-axis then we would suppress the planet's image instead!

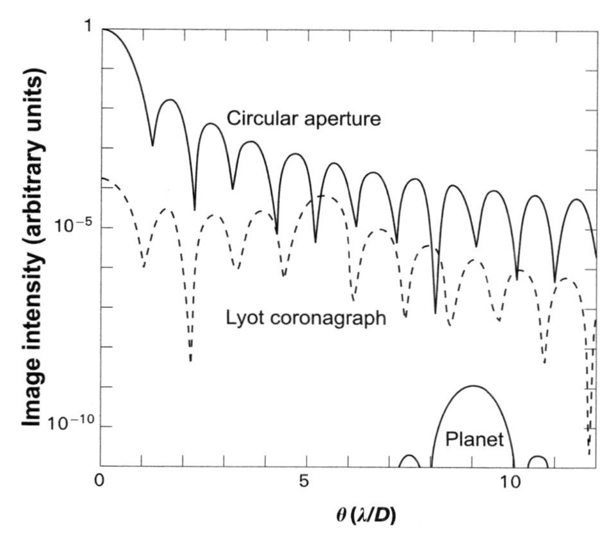

Figure 9.4 A diffraction-limited *psf*, and the *psf* of a typical Lyot coronagraph. θ is the angle from the centre of the image plane, in units of λ/D, where D is the diameter of the mirror and λ is the wavelength.

For the example in Figure 9.4 (where we have shown no effect on the planet), the planet is still fainter than the star, though now by a factor of 10^2–10^3 rather than 10^4–10^5. This does make it feasible to detect Jupiter in the Solar System from 30 ly, provided that we use a large aperture (of order 10 m) and long integration times (a few hours) to collect enough photons, as indicated by Equation (9.3).

For Lyot to image the corona it was also necessary for him to be at a high-altitude observatory, in his case the Pic du Midi observatory in the Pyrenees. This was to reduce the effect of light scattered by the atmosphere. The only way to eliminate atmospheric effects completely is to go into space, where coronagraphs are particularly effective. On the ground we have to do our best to reduce atmospheric effects, for coronagraphy and otherwise, and this reduction is the subject of the next section.

9.3 ATMOSPHERIC EFFECTS AND THEIR REDUCTION

The Earth's atmosphere has four deleterious effects on telescope images that prevent us from obtaining images of exoplanets. First, it *absorbs* radiation from space, reducing the quantity reaching the ground and thus reducing the signal. Figure 9.5 shows the proportion absorbed down to sea level at various wavelengths in cloud-free conditions for a source well above the horizon. Most of the absorption is due to atmospheric gases. You can see that the atmosphere is fairly transparent at visible wavelengths, but less so in the infrared where much of the absorption is due to water

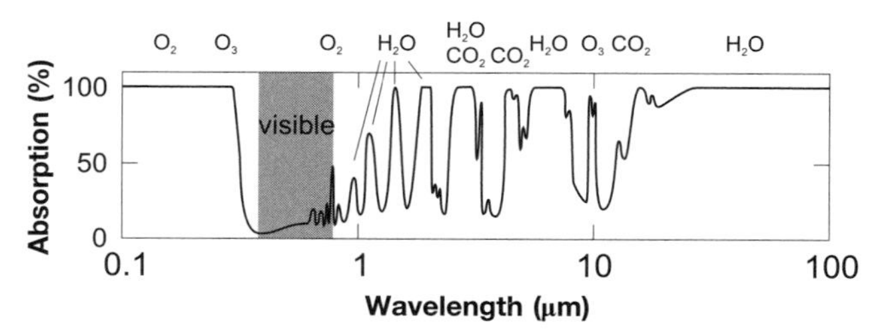

Figure 9.5 The absorption spectrum of the Earth's atmosphere at sea-level in cloud-free conditions for a source well above the horizon. The H_2O features vary with humidity.

vapour in the lower atmosphere. It is completely opaque in much of the UV, due largely to O_2 and O_3. It becomes transparent again at radio wavelengths, particularly in the approximate wavelength range 10 mm to 10 m. (Beyond 10 m the ionosphere is highly reflecting.)

Second, the atmosphere *emits* radiation. Most of the emission arises from the thermal motions of the atmospheric gases. The wavelength range of this thermal emission depends on the temperature of the atmosphere, and for the Earth it is in the mid-infrared, 5–100 μm. At these wavelengths the atmosphere seems to glow, night and day. This emission adds to the noise (Equation 9.2) and thus makes it more difficult to discern faint celestial objects. By placing infrared telescopes at high altitudes the problems of emission and absorption in the mid-infrared can be reduced, but only by going into space can they be completely overcome.

The third deleterious effect is the *scattering* of radiation. This is where radiation is redirected rather than absorbed. The molecules in the atmosphere are responsible for some of the scattering, the most obvious example being the blue sky, which is scattered solar radiation. It is blue because the intensity scattered by air molecules increases as wavelength decreases, in fact as $1/\lambda^4$, in what is called Rayleigh scattering. There is also scattering from aerosols. An aerosol is a suspension of tiny liquid or solid particles, and though aerosols supplied in cans are familiar to you, most aerosols in the atmosphere are of natural origin, and include water droplets, ice crystals, dust, and organic particles. An increasingly troublesome artificial aerosol is contrails (ice crystals) from high-flying aircraft. The Sun is such a bright source that scattering of sunlight from air and aerosols obscures the stars beyond. At night the sources of atmospheric illumination include ground-level lighting, causing light pollution. Another source of light pollution, in the present context, is the Moon. Scattering degrades images in two ways. First, it introduces light that adds to the noise. Second, some of the radiation from a celestial object is scattered back to space, thus reducing the signal. By siting a telescope away from artificial lighting and at high altitude these problems can be reduced, but again can only be eliminated by going into space.

The fourth deleterious effect of the Earth's atmosphere is caused by wavefront distortion. This arises from point-to-point variations in refractive index and their changes with time. This effect is measured by what is called the atmospheric 'seeing'. This is discussed in the next section, followed by ways in which it can be combated.

9.3.1 Atmospheric 'seeing' and its effects

Recall that the *psf* of a mirror is the light distribution across its image of a point object at a large distance. The assumption is made that there is nothing between the mirror and the object, in which case the radiation wavefronts are planes at the mirror as shown in Figure 9.2(a). If the mirror is optically perfect then the *psf* has the diffraction-limited form in Figure 9.2(b). Unfortunately for a star there *is* something between object and mirror – the interstellar medium and the Earth's atmosphere. The interstellar medium is extremely thin and can be discounted for present purposes. The Earth's atmosphere is far denser, particularly at lower altitudes, and it is in the lower 10 km or so, roughly corresponding to the troposphere, that the trouble arises.

If the Earth's atmosphere had a refractive index n that varied only with altitude then it would have a negligible effect on the image of a star. Unfortunately, n varies from point to point – turbulence in the troposphere is associated with point-to-point density variations, which cause the variations in n. The wave speed is proportional to $1/n$ and so a plane wave after traversing the atmosphere will be distorted, some parts having travelled slower than other parts. It will no longer be plane at the mirror. It is as if the *psf* of the mirror has been degraded, though it is more appropriate to regard the new *psf* as a composite of the mirror and the Earth's atmosphere – the mirror has not changed. Good seeing corresponds to low turbulence.

At sea level, in good seeing, the wavefronts at a wavelength of 0.5 μm remain fairly flat on a scale of about 0.1 m. Therefore, mirrors up to about 0.1 m diameter display their diffraction-limited *psf*s. Temporal changes in the turbulence, and winds, cause the flat areas to arrive in varying orientations, known as tip and tilt, so the *psf* moves randomly over the image plane. However, this can be compensated (see below). Wavefront distortion is less at longer wavelengths, and it can be shown that the scale over which flatness is preserved is roughly proportional to $\lambda^{1.2}$. Thus, at a wavelength of 2 μm the wavefronts are fairly flat on a scale of about 0.5 m, but this is still a modest aperture. The position improves somewhat at high-altitude observatories, but it remains the case that to obtain images of exoplanets we must use telescopes so large that their *psf*s are compromised by atmospheric turbulence. Fortunately, adaptive optics helps us to combat the effect of turbulence.

9.3.2 Adaptive optics

Figure 9.6(a) shows the essential features of a typical *adaptive optics* system. The wavefront sensor evaluates the distortion of the wavefront from a reference source by comparing the actual wavefront received from it with the wavefront that would have been received in perfect seeing. It then feeds a signal to the tip–tilt mirror to remove the average tilt of the wavefront across the whole area of the mirror. It feeds another signal to the deformable mirror that corrects the smaller scale distortions of the wavefront. The overall effect is that the wavefronts from the reference source are tilted and flattened so that it is as if there is no atmosphere present. Figure 9.6(b) shows an example of an outcome.

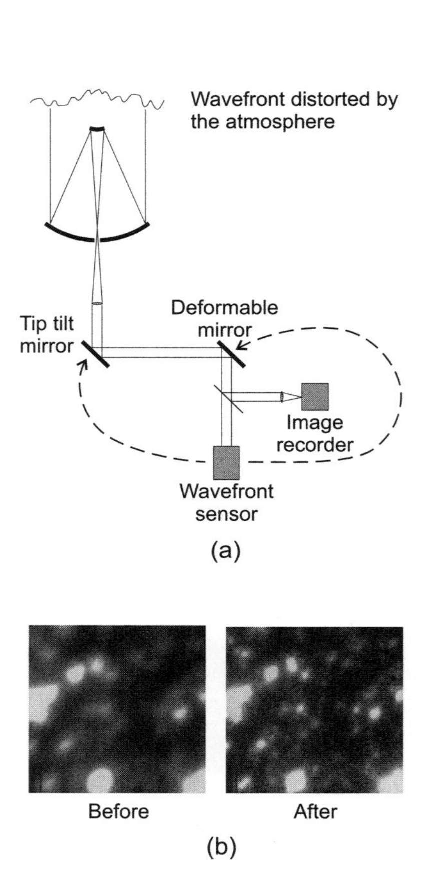

Figure 9.6 (a) A typical adaptive optics system. (b) The outcome of adaptive optics applied to a particular star field.

(b) Isaac Newton Group of Telescopes/Astronomy Technology Centre, Durham University.

The reference source can be a bright single star, though the chance of a sufficiently bright star lying close enough to the object of interest is small. Therefore artificial sources produced high in the Earth's atmosphere are also used. At altitudes 85–100 km there are sodium atoms that can be made to fluoresce at 0.589 μm by a laser system that operates at this wavelength. These artificial stars however suffer from the drawback that they are unable to provide the tip–tilt correction. This is because any wavefront tilt impressed on the laser beam when it is travelling upwards is removed when it is travelling downwards, so we learn nothing about the atmospheric effect. Thus, to make the tip–tilt correction a real star has to be used, and this can be the object of interest. This object cannot, however, be used to correct the smaller scale wavefront distortions – this would require us to know its detailed structure, and this is what we are trying to find out!

The correction so far described is for the atmosphere at one particular instant. But the atmosphere is not frozen in place. The point-to-point density variations are carried along by winds. They also evolve, so that a parcel of low density air can become high density, and vice versa. The former effect is the more important in determining the timescale in adaptive optics, requiring that the deformable mirror is reconfigured several hundred times per second at visible wavelengths. In the infrared the scale of the wavefront distortion is larger (Section 9.3.1) and so the adjustments can be made less often.

The adaptive optics correction is never perfect, even for the reference source. This is because it takes time to integrate enough light from the reference to enable the wavefront distortion to be evaluated. During the integration time the distortion changes, even during the short times that are possible with a bright reference source. Also, the deformable mirror is made up of a finite number of elements (perhaps a few tens), and so does not correct the wavefront on a scale smaller than the elements. Moreover, the correction gets worse with increasing angular separation between sources and reference. This is because the light from the two sources has taken increasingly different paths through the atmosphere to reach the mirror, and thus arrives at the mirror increasingly differently distorted.

Overall, adaptive optics corrections at visible wavelengths enable diffraction-limited *psf*s to be approached over fields of view a few arcseconds across centred on the reference image, and rather larger fields in the infrared. To do better than this we have to get above the Earth's atmosphere, and use telescopes in space. What can be achieved at ground level when adaptive optics is used on the largest optical telescopes, and what can be achieved with optical telescopes in space?

9.4 LARGE OPTICAL TELESCOPES

9.4.1 Ground-based telescopes

Optical telescopes operate at visible and/or infrared wavelengths. Currently, the largest ground-based optical telescopes have main mirrors 8–10 m diameter. Prominent among these are the four 8.2 m telescopes that constitute the European Southern Observatory's VLT group at an altitude of 2640 m on Cerro Paranal in Chile. 'VLT' stands for 'Very Large Telescope'! Another example is the two 9.8 m telescopes that constitute the Keck group at an altitude of 4150 m on Mauna Kea in Hawaii. Consider their potential for obtaining images of exoplanets.

At the VLT, in a few years, it should be possible to image exoplanets in the visible and in what is called the near-infrared, which extends from the long-wavelength visible-light limit at 0.78 μm to about 5 μm. To achieve this imaging with one of the VLT telescopes, adaptive optics and coronagraphy will have to be used. But these alone are not enough. In addition, differential techniques will have to be employed. In such techniques one image is subtracted from another, and the images are chosen so that this suppresses the residual light from the star that would otherwise obscure the planet. The details can be found at the VLT website (see Resources). In this way, a Jupiter-twin could be imaged at a range of 15 ly in one night of integration. A Jupiter-twin is a planet like Jupiter, 5 AU from a solar-type star that is also about the age of the Sun. Age is important because giant planets are very hot when they are young, and therefore very much brighter and easier to image. The giant planet around Epsilon Eridani should be within range at 10.5 ly (Chapter 11).

On one of the Keck telescopes (Keck 2) there is already a suitable camera that operates in the near-infrared (NIRC2). This camera, and its predecessor (KCAM) have the capability to image young giant planets, less than about 100 Ma old. Such young giants would be very luminous in the near-infrared. For example, a young Jupiter-mass planet 19 AU from its star could be seen within about 200 ly. To achieve this, the Keck adaptive optics system would have to be used, plus a differential technique called unsharp masking that enhances point-like objects (like planets) at the expense of diffuse scattered light. A survey has been carried out for 'young Jupiters', but none has yet been detected.

We would of course image fainter planets, or planets closer to their stars, if the mirror were considerably larger – recall that, as Equation (9.3) shows, with other things being equal, the *snr* ratio is larger the larger the aperture.

Whereas US proposals for extremely large telescopes (ELTs) centre on 20–30 m diameters, Europe is thinking bigger, at its biggest in a proposal centred on the European Southern Observatory for a 100 m telescope called OWL (OverWhelmingly Large Telescope!). Figure 9.7 is an artist's impression of this monster. A funding proposal might be made in 2004 for a 10-year project costing €900 million at 2002 prices. This is only about six times more than the cost of an 8 m telescope, yet the mirror area is nearly 160 times greater, and is in fact greater than the combined area of all professional optical telescopes that were in use in 2003. The relatively low cost stems from the use of mass production. The mirror would not be a single piece of glass but would consist of 2000 2 m hexagonal segments. This approach has been used on the two Keck telescopes on Mauna Kea. Adaptive optics would be essential, but with it OWL would be capable of imaging a Jupiter-twin up to hundreds of light years away, and might even be capable of imaging an Earth-twin out to several tens of light years (Earth mass, 1 AU from a solar-type star). In a few cases it might allow us to investigate any such planets spectroscopically, to see if they were capable of supporting life, and indeed to see whether biospheres were present, as will be discussed in Chapter 12.

9.4.2 Telescopes in space

Space has the advantage of avoiding all the problems posed by the Earth's atmosphere, which can only be partially compensated at ground level. The best known space telescope is probably NASA's Hubble Space Telescope (HST), from which many fine images have been obtained, including some of those in this book. But with a 2.4 m diameter mirror its exoplanet imaging capabilities are severely limited. In 2010 its successor is scheduled to be launched, the James Head Space Telescope

Figure 9.7 An artist's impression of the European Southern Observatory proposal for an OWL.

European Southern Observatory.

(formerly the Next Generation Space Telescope). With a 6 m mirror, and a coronagraph, it will be able to image planets in the infrared. For example, it would be able to see a Jupiter-twin within about 25 ly with three hours integration, and more luminous (younger and/or larger) planets further out. Spectroscopic studies of some planets might be possible.

Earth-twins will be beyond the capabilities of the James Head Space Telescope. They would not be beyond one possible design of NASA's Terrestrial Planet Finder (TPF). In this design, TPF would consist of a single telescope with a mirror about 8 m diameter, and fitted with a coronagraph and other devices to reduce diffracted light. The technical challenges are huge.

There is another design under consideration for TPF, with equal potential for planet-imaging, and NASA will decide between the two in 2006, for launch in the middle of the next decade. In this design TPF would be a giant interferometer.

9.5 INTERFEROMETERS

9.5.1 The basic principle of interferometry

To understand how an interferometer works it is first necessary to recall the wave nature of electromagnetic radiation. Electromagnetic radiation travels as a wave,

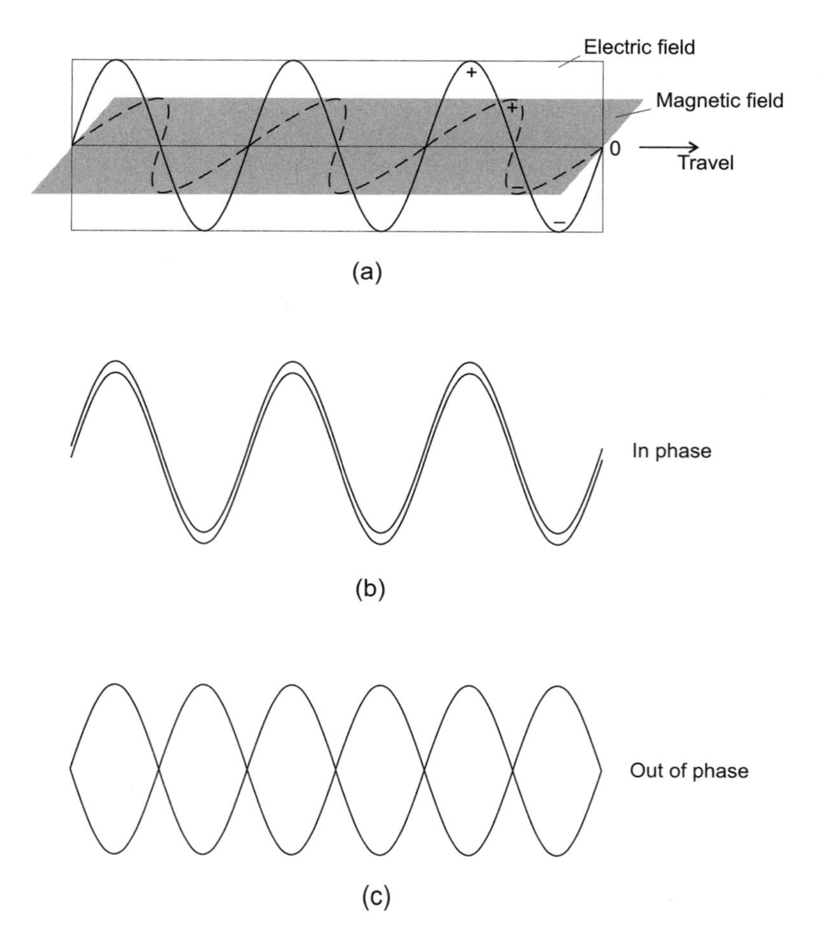

Figure 9.8 (a) An electromagnetic wave. (b) Waves reinforcing each other. (c) Waves of equal amplitude cancelling each other.

and if there is just a single wavelength present then the waveform is sinusoidal, as in Figure 9.8(a). The sinusoids display the strength of the electric and magnetic fields in the wave at each point in space at a particular instant in time. The electric and magnetic fields are at right angles to each other and to the direction of wave travel. Negative values are where the field points in the opposite direction to positive values. When we detect a wave we normally detect the intensity, which is proportional to the square of the fields. Thus, intensity is always positive. Figure 9.8(b) shows two waves travelling in phase, that is peak coincides with peak, trough with trough. Only the electric field is shown, but as the magnetic peaks and troughs in a single wave coincide with the electric peaks and troughs (as in Figure 9.8(a)) there is no loss of generality. In phase, the fields reinforce and the intensity is consequently enhanced. Figure 9.8(c) shows the out-of-phase case (i.e. peak coinciding with

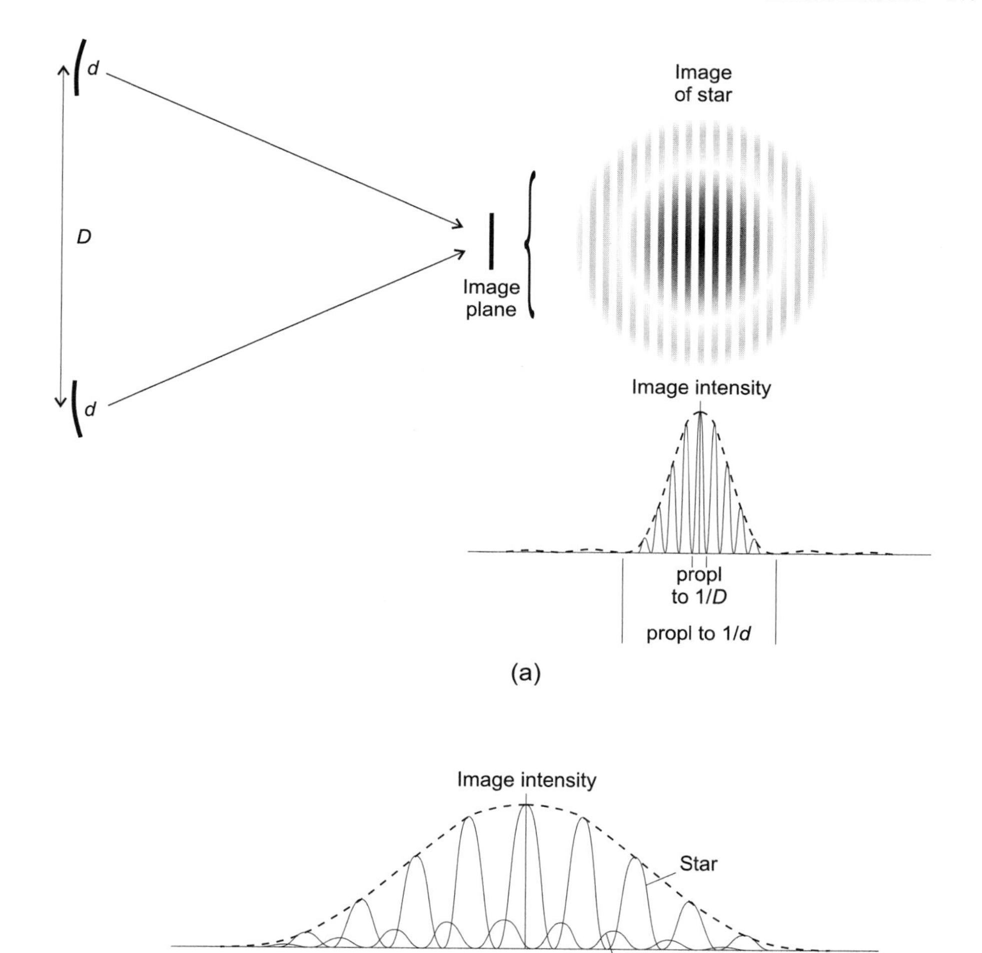

Figure 9.9 (a) A simple two-mirror interferometer imaging a star. (b) The same interferometer imaging a planet and the star (horizontal scale expanded with respect to (a)).

trough). In this case, for two equal amplitude waves, the fields cancel and the net intensity is zero.

Figure 9.9(a) shows a simple interferometer consisting of two identical perfect mirrors that each produce an image of a distant object on the image plane. If the object is a star that can be regarded as a point, then (if there is no atmosphere or if adaptive optics is fully effective) each mirror will produce an image which has the form of the diffraction-limited *psf*. This is the envelope shown lower right in Figure 9.9(a), and its width is proportional to $1/d$ where d is the diameter of the mirror. With both mirrors feeding light from the star into the image plane the outcome is modified. At the centre of the image plane, if the star is on-axis, light from the mirrors arrives in phase and so the waves reinforce each other. Away from the centre the light has travelled further from one mirror than from the other. If this path difference is half a wavelength ($\lambda/2$), then the waves arrive out of phase, so the waves cancel, and there is a dark strip on the screen. This is also the case at $3\lambda/2$, $5\lambda/2$, and so on. Across the screen there is thus a set of *interference fringes*, as shown under the envelope. The path difference is proportional to $1/D$ and hence the fringe spacing is also proportional to $1/D$, where D is the separation of the two telescopes.

The advantage of interferometry becomes apparent when the source is other than a point. Suppose that it is a star with a planet. The outcome is shown in Figure 9.9(b) (on a horizontally-expanded scale). There are now two sets of fringes, one from the star, the other from the planet. Compared to the star, the planetary fringes are lower in intensity (much lower than shown) and are slightly displaced in the image plane. Suppose that each telescope working alone is too small to detect this planet – its *psf* is too broad. The interference fringes from the two telescopes working together might nevertheless be discernable. In Figure 9.9(b) the planetary fringes are shown with maxima near to where the star's fringes are minima. The overall effect would be a lowering in the contrast of the fringes. By varying D and observing how the contrast changes, we could in principle tell that there was a planet present and how far it is from its star. We would then have gained information about the source that depends on D rather than on the much smaller mirror size d. The resolving power of the interferometer is thus much greater than that of its individual telescopes. It is nevertheless important to maximise d in order to increase the *snr*.

The information content from a two-telescope interferometer would be rather small. We can do much better with an array of mirrors, and of course were we to entirely fill a circle of diameter D with mirrors we would be back to an ELT. Well short of this we could in principle use several much smaller mirrors to obtain the information that could be obtained with an ELT. The penalty is the time taken to build the image. We would have to vary the relative positions of the mirrors so that the $\pi(D/2)^2$ aperture were eventually filled, and use integration times such that we collected the same number of photons that we would have collected with an ELT. Observations would have to be completed in a time smaller than that in which the planet changed appreciably. In practise a limited number of relative positions are used. Thus, an image constructed from this information would be 'fringy' and resemble, somewhat vaguely, a picture of the star plus planet. Nevertheless, the existence of a planet could be established. We could also get its (projected) distance from its star, and follow its orbital motion. The light from the planet could be analysed to determine atmospheric temperatures and composition (Chapter 12).

9.5.2 Imaging interferometers

In the previous section you read that the alternative design for TPF is an interferometer. This would be a multiaperture interferometer capable of yielding

Figure 9.10 An artist's impression of a possible design of ESA's proposed interferometer, Darwin, in a version using six mirrors, each about 1.5 m diameter.
Alcatel/ESA.

(fringy) images of exoplanets, including Earth-twins. ESA is also considering a space-based interferometer with a similar capability. This is called Darwin, and Figure 9.10 shows one possible design. This particular version consists of six telescopes each with a mirror a few metres in diameter. They lie on the circumference of a circle up to about 100 m in diameter, and move around this circumference and in and out to vary the relative positions. The light is fed to a central hub where an image can be reconstructed. Darwin would operate in the mid-infrared (around 10 μm). This is preferred over the visible because the contrast between planet and star is better (Section 9.1), and it is an informative spectral region about life on an exoplanet (Chapter 12). Also, the atmosphere precludes ground-based infrared interferometry – the Earth's atmosphere emits strongly in the mid-infrared, and the infrared absorption bands block much of our infrared view of space (Section 9.3). It is thus extremely advantageous to place an infrared interferometer in space. Space has the further advantage, perhaps surprisingly, that it is easier to maintain the interferometer configuration than on the ground, which vibrates. A configuration

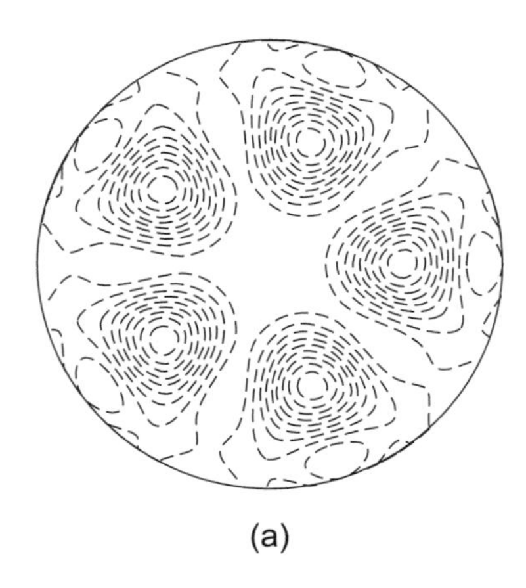

(a)

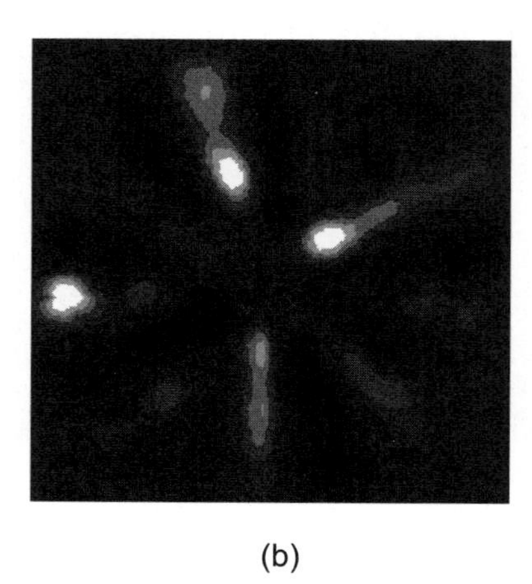

(b)

Figure 9.11 (a) Attenuation contours in the image plane when nulling interferometry is used with a five-telescope interferometer. The five 'hills' are places of minimum attenuation. (b) A possible image from Darwin of the Solar System, showing Venus, Earth, and Mars.
(b) Mennesson and Mariotti.

either needs to be maintained to a small fraction of the wavelength of the radiation used (i.e. a small fraction of 10 μm), or we need to measure where each telescope is to this accuracy, and use optical delay lines to correct for positioning errors. Darwin would achieve this through laser monitoring of the telescope positions and a set of rockets on each telescope.

Darwin would make use of *nulling interferometry*. By adjusting the path differences from each telescope to the hub it is possible for the waves from an object on the axis of the mirror array to cancel or be nulled, as illustrated in Figure 9.11(a), for a five-telescope array. The contours build five 'hills' where the attenuation is least for a particular configuration. A star on-axis would be greatly suppressed (not entirely so, because of the finite size of the star). A 2-D image is built by repositioning the telescopes radially and around a circle a sufficient number of times to obtain the required detail. An image might look something like that in Figure 9.11(b). This is what the Solar System would look like – the three bright objects are Venus, Earth, and Mars. The other structures are spurious, arising ultimately from the incomplete filling of the aperture achieved by the telescope separations.

One or both of Darwin and TPF (which might operate at visible wavelengths) could be in space by the middle of the next decade, and either of them would be able to discover Earth-twins out to several tens of light years with a few hours of integration time, and with several days or weeks of integration time to investigate them for life (see Chapter 12). There are also nulling interferometry plans for ground-based telescopes, notably at the VLT and Keck. These will pave the way for Darwin and TPF respectively, but should have some limited planet-detecting capability. However, at present, the direct detection of exoplanets is unlikely in space or from the ground. Therefore, we have to rely on indirect methods to discover and investigate exoplanets. These methods are described in the next chapter.

9.6 SUMMARY

- The direct detection of exoplanets is challenging because the planet is close to a far brighter object – its star.
- Coronagraphy makes it easier to see planets by reducing the intensity of the point spread function (*psf*) of the star far more than that of the planet.
- The Earth's atmosphere absorbs certain wavelengths, particularly at infrared and UV wavelengths. It also emits, mainly in the mid-infrared, and scatters at all wavelengths. These processes all reduce the signal-to-noise ratio in images and thus make it harder to detect planets.
- The atmosphere also degrades images through atmospheric turbulence – it is as if the telescope *psf* is broadened. It can be partially compensated by adaptive optics for small fields of view centred on a reference 'star'.
- The largest ground-based optical telescopes (e.g., VLT and Keck), have very limited capabilities for the direct detection of exoplanets now, but by 2010 the VLT should be able to image a Jupiter-twin out to about 15 ly. In about 2010 the James Head Space Telescope will have a somewhat better capability. ELTs (of order 100 m diameter mirrors), equipped with adaptive optics, would be capable of imaging Jupiter-twins out to distances of hundreds of light years, and might be able to image Earth-twins within a few tens of light years. These are planned for about 2014.
- Interferometers in space, such as Darwin and one design for the TPF, would be able to image Earth-twins out to several tens of light years, and investigate them for life. Realistic launch dates are around 2015.
- Though there is an increasing capability, at present the *direct* detection of exoplanets is unlikely in space or from the ground.

9.7 QUESTIONS

Answers are given at the back of the book.

Question 9.1

(a) Describe, qualitatively, how the contrast ratios of the Earth (Figure 9.1) might change if the Earth were moved closer to the Sun.
(b) State the factors that would make direct detection of the Earth (i) easier and (ii) harder if it were moved closer to the Sun.

Question 9.2

Explain why a coronagraph is useful to a space telescope while adaptive optics is not.

Question 9.3

Explain why size matters:

(i) for a single telescope; and
(ii) for the spacing of telescopes in an interferometric array.

10

Searching for planets: indirect methods

Though direct methods can tell us more about an exoplanet than indirect methods, it will probably be some years before we will be able to detect planets orbiting other stars in this way, and so at present we have to rely on the comparatively less challenging indirect methods. In these methods we infer the presence of a planet from its influence either on the motion of the star it orbits, or on the quantity of radiation we receive from this star or from some background star.

10.1 DETECTING A PLANET THROUGH THE MOTION OF ITS STAR

10.1.1 The effect of a planet on its star's motion

Figure 10.1(a) is the usual way of showing the orbit of a planet around a star, in which the orbit is shown with respect to the star. An alternative point of reference is the centre of mass of the system, around which the star and the planet orbit, as shown in Figure 10.1(b). The centre of mass is located at a point such that:

$$M \times r_{\mathrm{s}} = m \times r_{\mathrm{p}} \tag{10.1}$$

where M and m are the masses of star and planet, and r_{s} and r_{p} are their instantaneous distances from the centre of mass. This equation applies at all points in the orbit, and so it also applies when r_{s} and r_{p} equal the semimajor axes of the orbits (a_{s} and a_{p}) with respect to the centre of mass. Thus:

$$M \times a_{\mathrm{s}} = m \times a_{\mathrm{p}} \tag{10.2}$$

The eccentricity e and periods P of the three orbits in Figure 10.1 are the same. Also, just as the planet moves at a non-uniform speed around its orbit in Figure 10.1(a), with maximum speed at periastron (nearest the star), and minimum speed at apastron (furthest from the star), in accord with Kepler's second law (Section 1.1.1),

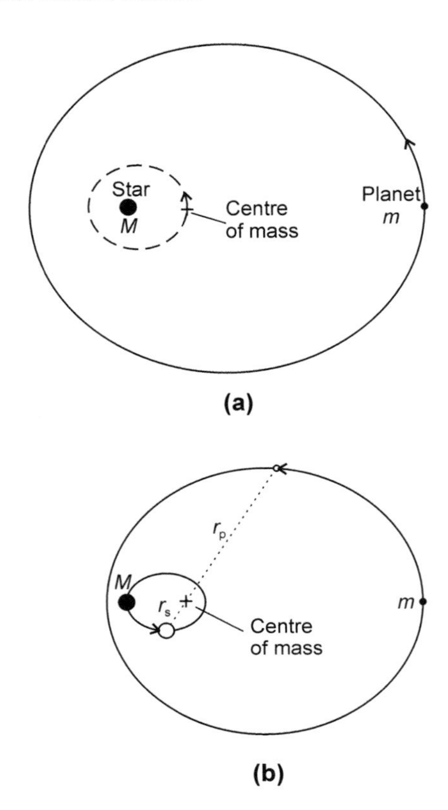

Figure 10.1 (a) The orbit of a planet with respect to its star. (b) The orbit of a star and planet with respect to their centre of mass.

so too do planet and star in their centre of mass orbits in Figure 10.1(b), with maximum speeds at the pericentres (nearest the centre of mass) and minimum at the apocentres. Note that $a = a_s + a_p$. For stars and planets $M \gg m$, so $a_p \gg a_s$, and therefore $a \approx a_p$.

Thus any star with a planetary companion will be moving around a small orbit. This is superimposed on the star's steady drift across the sky, called its proper motion. If we can detect the orbital component of the star's overall motion then we can infer the presence of the planet and learn something about its mass and its orbit. Astrometry and Doppler spectroscopy are two ways in which the orbital motion can be measured.

10.1.2 Astrometry: principles

In *astrometry* the position of the star is measured with great precision at a series of times. Its overall motion is then analysed to see if there is any orbital component. Whether this component can be detected depends on the angular size β of the star's orbit. To obtain β we rearrange Equation (10.2) as $a_s = (m/M)\, a_p$. If the distance from which we observe the star is d, then:

$$\beta = \frac{a_s}{d} = \frac{m}{M}\frac{a_p}{d} \approx \frac{m}{M}\frac{a}{d} \tag{10.3}$$

where β is in radians and where we have used the approximation $a \approx a_p$. This equation assumes that the system happens to be presented to us so that the semimajor axis of the star's orbit lies in the plane of the sky, otherwise the excursion of the star across the sky would be less than a_s.

We can express β in terms of the orbital period P of the system, by expressing a in terms of P. From Newton's laws of motion and law of gravity it can be shown that:

$$a = \left(\frac{GM}{4\pi^2}\right)^{1/3} P^{2/3} \tag{10.4}$$

From Equations (10.3) and (10.4), it follows that:

$$\beta \approx \left(\frac{G}{4\pi^2}\right)^{1/3} \left(\frac{P}{M}\right)^{2/3} \frac{m}{d} \tag{10.5}$$

(Note that Equation (10.4), when applied to the Solar System, gives Kepler's third law, Section 1.1.1.)

Equation (10.3) shows that the astrometric signal β is greater:

- the larger the ratio m/M;
- the larger the value of a (and so the larger the value of P); and
- the smaller the value of d.

Thus, massive planets in large orbits around low-mass stars, observed from small distances, are the easiest to detect, just as one would expect. However, large orbits have large orbital periods, and we need to accumulate data for at least a large fraction of an orbit. This might be many years.

Figure 10.2 quantifies $\beta \times d$ in a few specific cases. To get a feel for the data in Figure 10.2, consider the Solar System, assuming that the only planet is Jupiter. In this case the Sun would go around an orbit with a semimajor axis $a_s = 7.4 \times 10^8$ m, just a bit greater than the Sun's radius of 6.96×10^8 m. Jupiter has an orbital period of 4.33×10^3 days, so, reading from the appropriate line in Figure 10.2, $\beta \times d = 1.6 \times 10^{-2}$ arcsec ly. From a distance of 30 ly this is $\beta = 5.3 \times 10^{-4}$ arcsec, or 530 microarcsec (μas). This is a very small angle, equivalent to viewing the width of a human finger from a distance of about 7000 km! Clearly it is not easy to find planets with astrometry.

This example assumed that the semimajor axis a_s of the star's orbit is in the plane of the sky. It gets worse if this is not so, because then we see only some fraction of a_s. Consider the cases in Figure 10.3. If the orbit is presented to us face-on (Figure 10.3(a)), or in any other orientation with a_s in the plane of the sky (Figure 10.3(b) and (c)), we see the full $2a_s$ excursion of the star. But if it is presented from any other

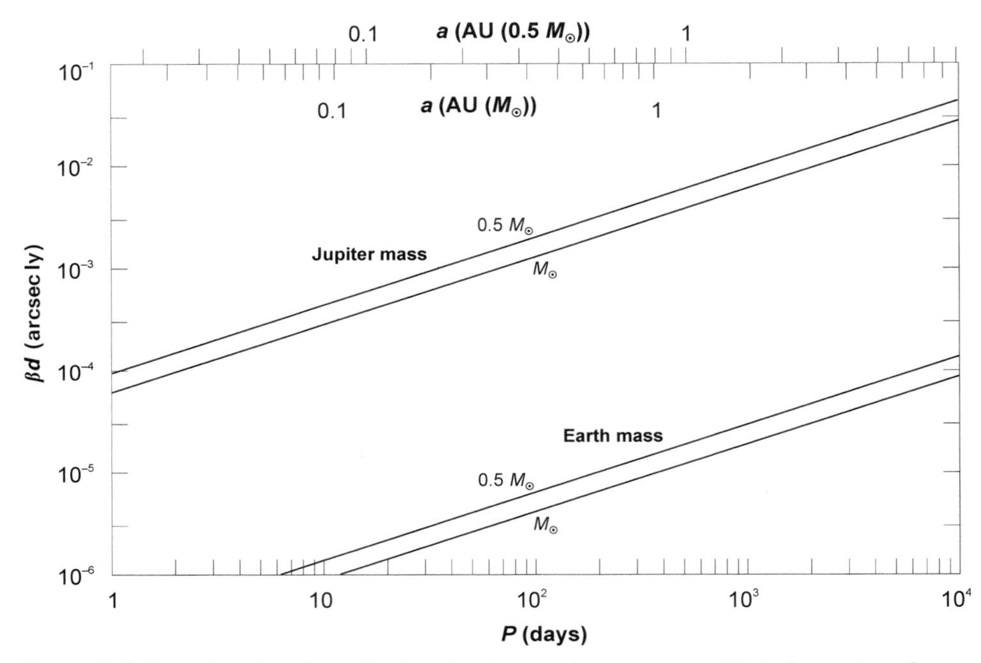

Figure 10.2 How $\beta \times d$ varies with P and a for a solar-mass star ($M_\odot$), for a star of mass $0.5\,M_\odot$, and for Jupiter-mass and Earth-mass planets. Note that a is the semimajor axis of the *planet's* orbit.

angle we see less than the full excursion. The worst case is shown in Figure 10.3(d). Here the semimajor axis is at 90° to the plane of the sky, so we see only the narrowest dimension of the elliptical orbit (twice the semiminor axis).

The question arises of whether the true orbit can be deduced when it is not presented face-on to us. The answer is that it can. Consider the case in Figure 10.3(b). If this were the true orbit then it would have a much higher eccentricity than the actual true orbit in Figure 10.3(a), and the centre of mass would be further to the left. In this case the star would move around its orbit in a different way. For example, the difference in orbital speed between pericentre and apocentre would be much higher. By measuring the time intervals between various points in the star's observed orbit the true orbit can be deduced. This is true for all orientations. We can thus infer the value of β in Equations (10.3) and (10.5), which corresponds to the full motion a_s.

Planetary masses and orbits

One way of obtaining the mass of the planet is from a rearrangement of Equation (10.5):

$$m \approx \beta d \left(\frac{M}{P}\right)^{2/3} \left(\frac{4\pi^2}{G}\right)^{1/3} \tag{10.6}$$

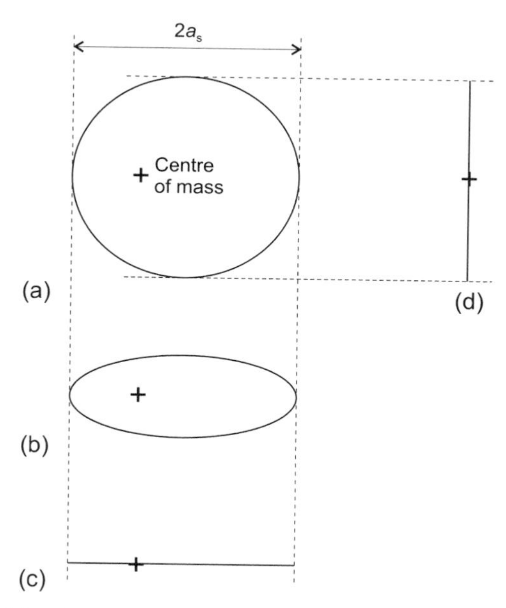

Figure 10.3 (a) The orbit of a star seen face-on. (b) The same orbit seen between face-on and edgewise. (c) The same orbit seen edgewise, with the semimajor axis in the plane of the sky. (d) The same orbit seen edgewise with the narrow dimension (semiminor axis) in the plane of the sky.

We obtain β and P from observations of the star's motion, and its distance d from other measurements on the star, such as its trigonometric parallax (Section 1.3.1). The star's mass M can be estimated as follows. From its spectrum we can identify it as a main sequence star, which we are concerned with. We can then use its measured luminosity with the relationship between mass and luminosity illustrated in Figure 8.1 to obtain its mass. (This mass–luminosity relationship has been established by measuring the masses of the two stars in a binary system from their motions around each other – see the astronomy texts in Resources.) From P and M we can get the semimajor axis a of the planet's orbit from Equation (10.4). The eccentricity of the planet's orbit is the same as that of the star's orbit (Figure 10.1), and as we can deduce the true orbit of the star this eccentricity is known. In practice, various orbits are tried, and varied until the best fit to the data is obtained. There is always some uncertainty in the measurements, and this leads to corresponding uncertainty in planetary masses and orbits.

So far we have considered just one planet in orbit around a star. When there is more than one, the star's motion is more complicated. This is illustrated in Figure 10.4, where the actual motion of the Sun is shown, as would be seen face-on to the Earth's orbit from a distance of 30 ly. Though Jupiter has the greatest effect, the other planets, particularly Saturn, cause large departures from a simple elliptical motion of the Sun. It would be possible, given enough measurements of sufficient accuracy, to disentangle the effects of the different planets, and determine the masses

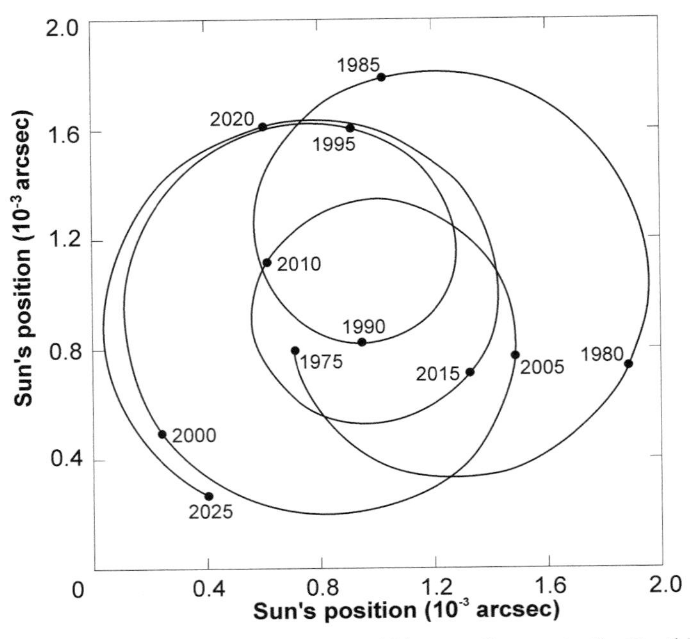

Figure 10.4 The actual motion of the Sun, as would be seen face-on to the Earth's orbit from a distance of 30 ly.

and orbits of Jupiter and Saturn, and, with greater difficulty because of their smaller effects, the masses and orbits of the other planets too.

10.1.3 Astrometry: practice

Central to the practice of astrometry is the use of a reference frame against which the motion of the target star can be measured. The reference frame is provided by other stars, and so the position of the target star with respect to each of a set of reference stars has to be measured many times. Relative positional changes will arise from the different proper motions of the stars, from any orbital motion of a reference star, and from the different parallax shifts when observations are made with the Earth at different times in the year. The data have to be processed to see if there is a residual orbital motion of the target star.

The first major attempt to detect planets through astrometry was by the Dutch astronomer Peter van de Kamp. He began his search when he joined the Sproule Observatory in Pennsylvania in 1937. He used a refracting telescope (i.e. one using lenses rather than mirrors), with a main (objective) lens diameter D of 0.61 m. At a wavelength λ of 0.5 μm the angular radius of the central disc in the diffraction-limited *psf* is then 0.2 arcsec ($1.22\lambda/D$ radians, Equation (9.1)). The ultimate positional resolution will be some fraction of this. He recorded positions on photographic plates, for several target stars over several decades. He became convinced that he had detected planets around a few stars, including Barnards' Star, which, at a distance of 5.94 ly is the closest to the Sun after the triple system comprising Proxima Centauri and Alpha Centauri A and B. A major overhaul of the telescope in 1949 caused him to exclude from analysis the plates he had obtained before that date, but further telescope adjustments in 1957 caused more uncertainties. His positive results were challenged by others, partly on the basis of the 1957 adjustments, and partly because of problems caused by his underexposure of photographic plates and his use of rather few reference stars. Nevertheless, he went to his grave in 1995 believing that Barnard's Star has at least one planet. This claim has not been confirmed by others, though the residual data do show some indication of oscillatory motion.

Others have attempted astrometry from the ground using single aperture techniques like van de Kamp, more recently replacing the photographic plate with electronic devices. No definite discoveries have been made.

Another approach is to use an interferometer (Section 9.5) to obtain greater positional resolution as illustrated in Figure 10.5. Suppose that there is a single star in the field of view. The light from the two apertures comes together in the image plane with various phase differences to create a pattern of light and dark fringes. Where the waves come together with the same optical path length in each arm the waves are in phase and the pattern is at its brightest. If there is a second star in a different direction its pattern will be brightest at a different place in the image plane, but can be brought to the same location as the reference star by adjusting one of the optical paths in the interferometer. The amount of adjustment gives the angular separation of the two stars. In general the baseline between the two telescopes will not be parallel to the stars' separation, so either the baseline has to be rotated, or there needs to be further telescopes with baselines in different directions. By repeated measurements the motion of an orbiting star with respect to relatively fixed background stars can be observed.

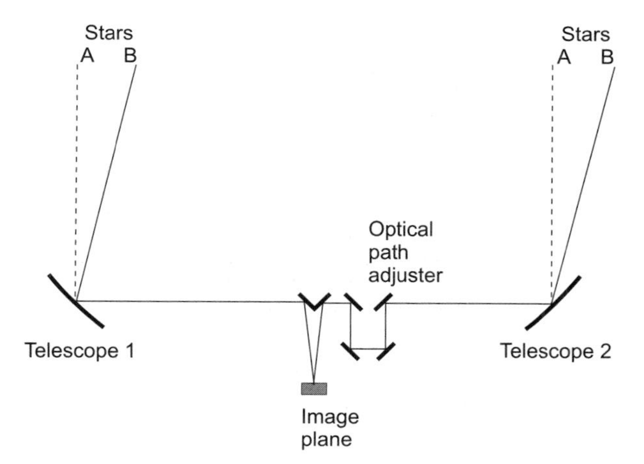

Figure 10.5 An interferometer being used to measure the angular separation between two stars (schematic).

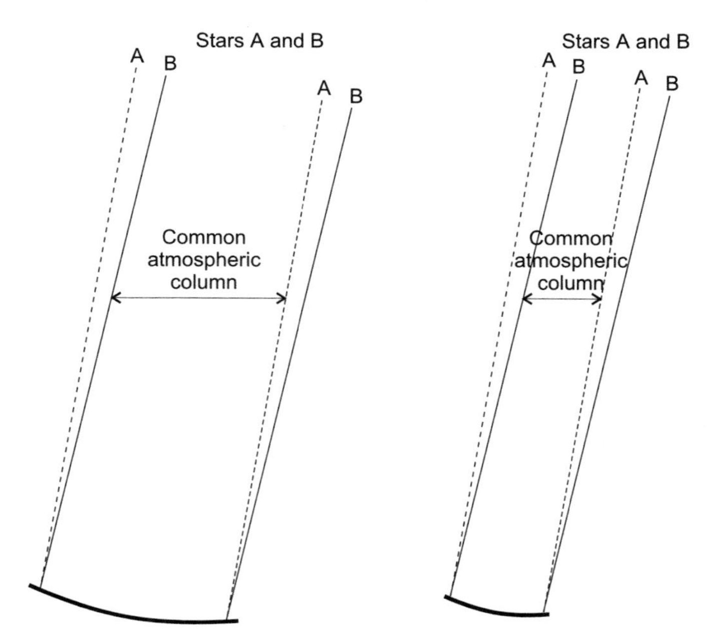

Figure 10.6 The advantage of a large aperture in combating the effect of the atmosphere.

The Earth's atmosphere poses two big problems for ground-based astrometry, which so far has not been able to detect the tiny stellar orbits that would be produced by planets. First, atmospheric refraction depends on the altitude of an object above the horizon, and therefore introduces distortion across an image that varies with the altitude at image centre. Second, atmospheric turbulence (Section 9.3.1) causes a jitter in the positions of star images that limits the precision of measurement of star separations to about 0.1 arcsec even at high-altitude observatories where the effect of the atmosphere is reduced.

There is a reduction in jitter if large apertures are used. Consider two stars with a small angular separation, as shown in Figure 10.6. The large aperture collects starlight that has traversed much the same column of atmosphere for each star. Therefore the jitter affects the two images in much the same way. With a small aperture the two columns are more distinct. They are also more distinct if the angular separation of the two stars is larger. The area in the image plane where the jitter effects are much the same is called the isoplanatic patch. For a 10 m telescope on a high-altitude site the patch will be 50–100 arcsec across. Within this patch angular precisions of 100 μas or better can be obtained for the separation between two stars (unless they are very faint). This assumes integration times of the order of an hour to accumulate sufficient photons to locate the *psf* centres with sufficient precision. Jitter is further reduced if adaptive optics is used. Note from Section 10.1.2 that 100 μas is good enough to detect the effect of a Jupiter-twin at a distance of 30 ly (i.e. a planet like Jupiter about 5 AU from a solar-type star, more specifically in the context of astrometry a Jupiter *mass* planet about 5 AU from such a star).

Forthcoming astrometry on the ground and in space

The two Keck telescopes on Mauna Kea and the four VLT telescopes in Chile are 8–10 m telescopes at high altitudes, and are capable of operating interferometrically. In principle they could detect planets astrometrically. Further developments are needed at the Keck telescopes before substantial surveys for planets can be made. At the VLT, it is hoped that by 2005 astrometry will be performed with the four telescopes, each using adaptive optics, in a system called Phase Referenced Imaging and Microarcsecond Astrometry (PRIMA). PRIMA would have an astrometric precision of 10 μas. The semimajor axis of the star's orbit would have to be about twice this for the orbit to be determined with useful accuracy. PRIMA would then have the capability to detect the stellar motion caused by a Jupiter-twin out to about 800 ly (Equation (10.3)). Earth-twins will not be detectable, but Earth-mass planets several AU from nearby low-mass M dwarfs would be detectable, though several AU is well beyond the HZ (Figure 8.5). Further in the future, around 2010, an array of radiotelescopes operating at millimetre wavelengths will have the interferometric capability to discover planets astrometrically. This array, the Atacama Large Millimetre Array (ALMA), will consist of 64 antennas and be located in Chile. It will have an astrometric precision of about 100 μas, which will give it the capability of detecting the effect of a Jupiter-twin out to many tens of light years.

Though there as yet have been no confirmed ground-based detections of exoplanets, by interferometry or otherwise, there has been one success in space. In 2002 the HST detected the astrometric motion of the star Gliese 876 due to the giant planets orbiting it. This was a single aperture rather than interferometric observation, and was facilitated by the low mass of Gliese 876, which is an M dwarf about one-third of the mass of the Sun, and by the knowledge from Doppler spectroscopy (Section 10.1.4) that giant planets were present.

Space has the huge advantage for astrometry that the deleterious effects of the Earth's atmosphere are absent. The potential of space-based astrometry has been demonstrated by ESA's Hipparcos spacecraft, which was launched in 1989 on a 4-year mission. Its best precision was about 500 μas, on the threshold of being able to detect the effect of Jupiter at a distance of 30 ly.

In the near future several space missions should be capable of astrometric detection of planets. These include the NASA mission Space Interferometry Mission (SIM) and the ESA mission Gaia, each under study for launch around 2010, both with two telescopes. SIM will be a targeted mission, and will be able to achieve precisions of a few microarcseconds even for faint stars, corresponding to the Sun at a few thousand light years. These faint stars would require an integration time of several hours per single observation, and a few tens of observations per star to get the orbit. Jupiter-twins could be detected out to a few thousand light years,

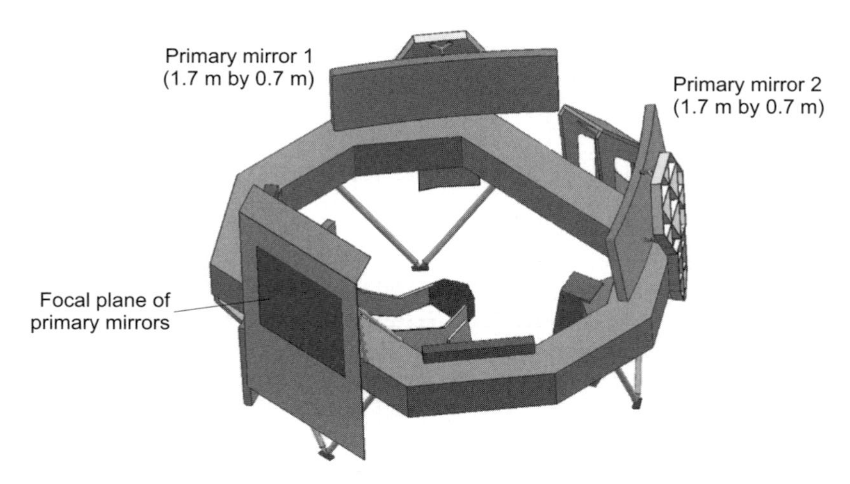

Figure 10.7 The optical heart of Gaia, showing the two primary mirrors 1.7 m by 0.7 m used for astrometry.
ESA.

and Earth-twins to a few tens of light years, but relatively few stars could be targeted per year.

Gaia is a sky survey mission, rather than a targeted mission. The optical heart of Gaia is shown in Figure 10.7. There are two astrometric telescopes, each with primary mirrors 1.7 m by 0.7 m. These will each image small areas of the sky 106° apart. As the spacecraft rotates, very accurate separations between stars will be obtained in the direction of rotation. To fix accurately the separation in another direction, and hence get the overall separation, the star field will be recorded again, with the rotating spacecraft pointing in a different direction. Over the nominal 5-year mission the same field will be recorded about 80 times. For stars brighter than visual magnitude 10 (a solar-type star at a distance of 350 ly), the semimajor axis of the star's orbit (as projected on the sky) will need to be about 40 μas for the orbit to be measurable with reasonable accuracy. This is sufficient for a Jupiter-twin to be detected out to about 350 ly, and to detect planets several times the mass of the Earth at 1 AU from M dwarfs out to a few tens of light years. Earth-mass planets will be beyond detection.

A third telescope will be used to measure stellar velocities along the line of sight. Though Gaia will not attain sufficient precision to detect planets this way, this is the basis of the Doppler spectroscopy technique, to which we now turn.

10.1.4 Doppler spectroscopy: principles

Whereas in astrometry the orbital motion of a star is revealed by accurate measurements of its position at a series of times, in *Doppler spectroscopy* the orbital motion is revealed by repeatedly measuring the radial velocity of the star (i.e. that component

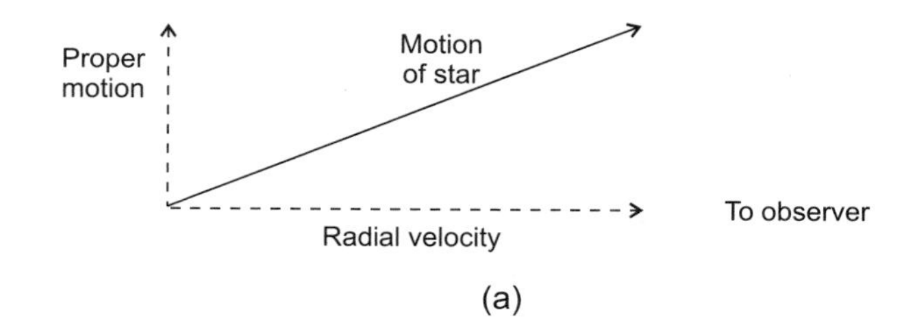

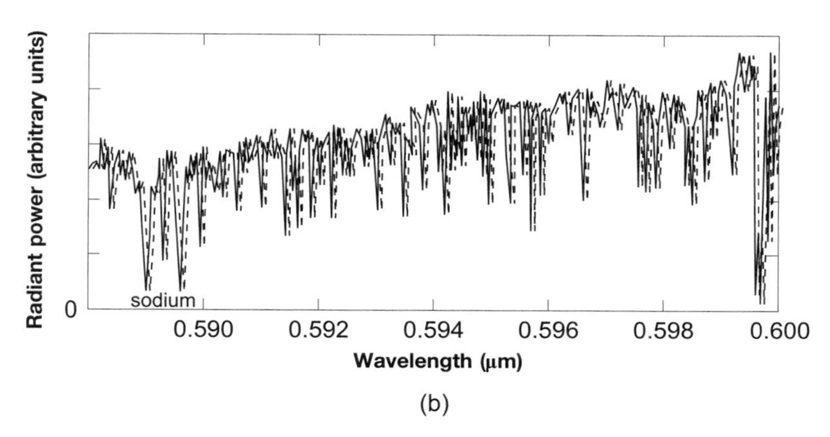

Figure 10.8 (a) The radial velocity. (b) The absorption spectrum of a star, without a Doppler shift, and (dashed line) with a Doppler shift (some of the low amplitude fine structure is noise).

of its overall velocity with respect to us that lies along our line of sight, as shown in Figure 10.8(a)). Consequently, the method is sometimes called the radial velocity method. If the radial velocity exhibits cyclic changes then we can infer orbital motion of the star, and hence that it has a companion.

The radial velocity is measured using the spectral lines of the star. Figure 10.8(b) is an example of a stellar spectrum, displaying many absorption lines, narrow ranges of wavelength where there are deficits of radiation. These lines are formed in the stellar atmosphere, which consists mainly of the elements hydrogen and helium plus small but significant quantities of the other elements. Most of the elements are present as single atoms, a proportion of which will be ionised, depending on the element and the temperature. In cool atmospheres there will also be simple molecules. Radiation wells up from the interior of the star, and passes through the atmosphere. As it does so, the atoms of each element absorb radiation at a set of well-defined wavelengths, characteristic of the element. Much of the absorbed energy is re-emitted at the same wavelengths, but in all directions. Some is emitted at longer wavelengths and the rest is extracted from the atoms when they collide with each other. This collisional de-excitation helps sustain the atmospheric temperatures. Collisions also *excite* atoms, causing them to emit at their characteristic wavelengths. The net effect for an element can either be absorption lines, as shown, or emission lines where there is an excess of radiation over narrow wavelength ranges, in both cases at the characteristic wavelengths. In Doppler spectroscopy it is absorption lines that are commonly utilised, because they are abundant.

We can obtain the radial velocity from the spectral lines through the Doppler effect. For any source of radiation the wavelengths measured by an observer will depend on the source's radial velocity. Thus, the wavelengths of the stellar absorption lines as measured by us are changed, as illustrated by the displaced spectrum in Figure 10.8(b) (dashed line). The change is only in the wavelengths we *measure*, not in the values at the source, which are unchanged by its motion. Note that the proper motion (the transverse velocity component) has no effect. The radial velocity v_r is related to the change in observed wavelength by:

$$v_r = c\frac{(\lambda_{observed} - \lambda_{source})}{\lambda_{source}} \tag{10.7}$$

where c is the speed of the waves (the speed of light in this case), λ_{source} is the wavelength at the source and $\lambda_{observed}$ is the observed wavelength. If $\lambda_{observed} > \lambda_{source}$ then the source is moving away from the observer, and Equation (10.7) shows that in this case v_r is regarded as positive. We shall not derive Equation (10.7) – this can be found in standard works on optics (see Resources) – we will instead put it to work. We shall ignore the average radial motion of the star, and focus on the orbital variation.

Circular orbits

First, consider the case of the star in a circular orbit of radius a_s presented edgewise to us, as in Figure 10.9(a). When the star is at A its radial velocity *away* from us is a maximum, and so $(\lambda_{observed} - \lambda_{source})$ is a maximum. At B its radial velocity *towards* us is a maximum, and so $(\lambda_{observed} - \lambda_{source})$ has its greatest negative value. As the star goes around its orbit we would observe $(\lambda_{observed} - \lambda_{source})$ going through the sinusoidal changes shown in Figure 10.9(b). The existence of an orbiting companion is thus indicated.

Figure 10.9(b) also shows the sinusoidal changes in radial velocity. These have an amplitude denoted by v_{rA}. When the orbit is presented edgewise, v_{rA} is also the star's orbital speed v. For a star with a mass M to be moving at a speed v around a circular orbit with a radius a_s, there must be a force directed towards the centre of the circle given by Mv^2/a_s (physics texts in Resources). In this case the force is provided by the gravitational attraction of the planet on the star, which is GMm/a^2, where m is the mass of the planet and a is the distance between the planet and the star. Therefore:

$$\frac{Mv^2}{a_s} = \frac{GMm}{a^2} \tag{10.8}$$

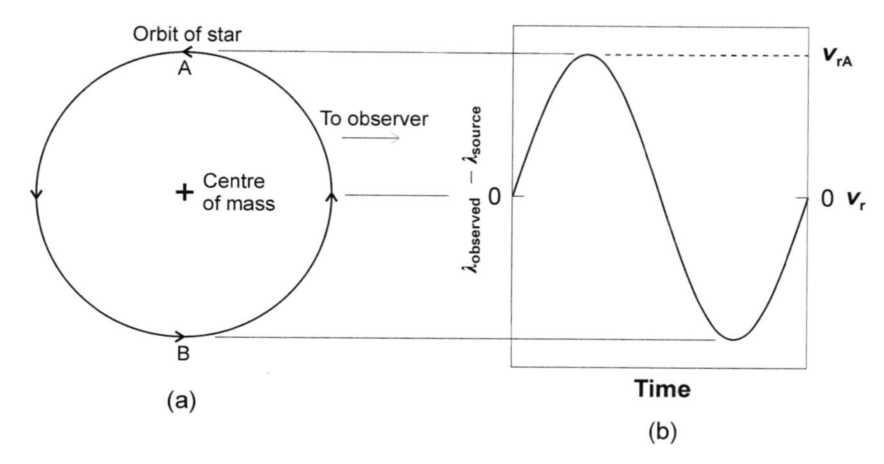

Figure 10.9 (a) A star in a circular orbit presented edgewise to an observer. (b) The observed sinusoidal changes in $(\lambda_{\text{observed}} - \lambda_{\text{source}})$, and the corresponding cyclic changes in v_r.

From Equations (10.8), (10.2) and (10.4), plus the approximation $a \approx a_p$, we get:

$$v \approx m\left(\frac{G}{aM}\right)^{1/2} = \frac{m(2\pi G)^{1/3}}{P^{1/3}M^{2/3}} \tag{10.9}$$

where $v = v_{rA}$. Equation (10.9) shows, as might be expected, that the Doppler signal v_{rA} is greater:

- the larger the value of m and the smaller the value of M; and
- the smaller the value of P (and hence the smaller the value of a).

That the signal is greater the smaller the value of P means that data in favourable cases can be acquired quickly. This is opposite to the astrometric case, where the signal is greater the larger the value of P. Also, the Doppler signal is independent of the distance of the star, though nearby stars have the great advantage of providing greater flux at our telescopes, and therefore it is easier to measure the Doppler shifts in their spectral lines.

Figure 10.10 quantifies v_{rA} in a few specific cases. It can be seen that a Jupiter-twin ($P = 430$ days) will produce $v_{rA} = 12.5\,\text{m s}^{-1}$. From Equation (10.7), with $c = 3.0 \times 10^8\,\text{m s}^{-1}$, the maximum $(\lambda_{\text{observed}} - \lambda_{\text{source}})/\lambda_{\text{source}} = 4.2 \times 10^{-8}$. This shows that the spectroscopic precision required for Doppler spectroscopy is demanding. For an Earth-twin the values are $0.09\,\text{m s}^{-1}$ and 3.0×10^{-10}.

The sin(i_0) *problem in measuring masses of planets*

Consider the case of a star in a circular orbit presented *face-on* to us. The radial component of the star's orbital velocity is now zero, and so v_r is constant. The Doppler shifts in the spectral lines are also constant, and so we have no sign of an

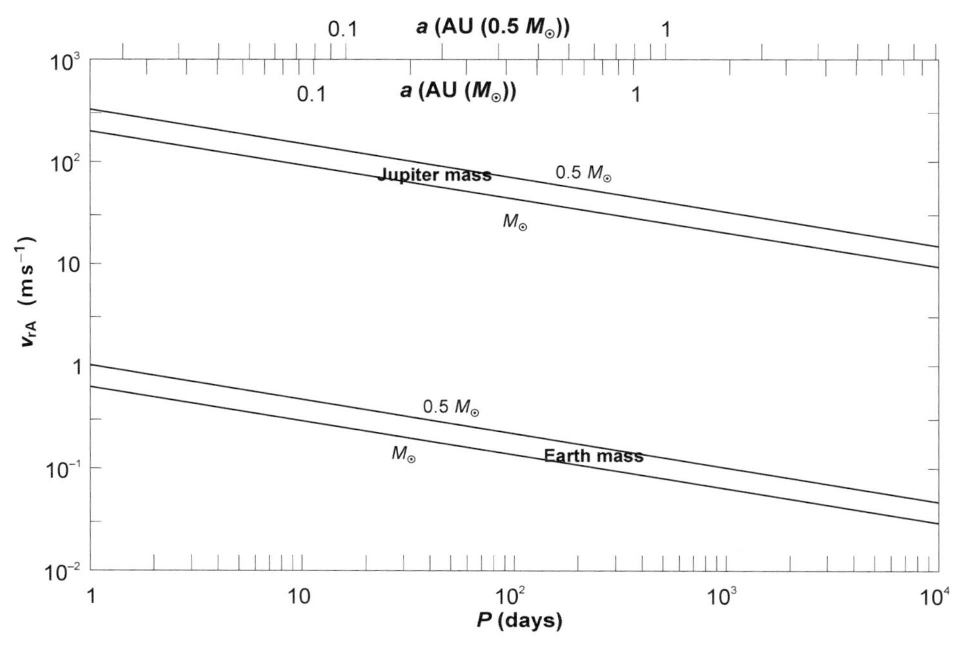

Figure 10.10 How v_{rA} varies with P and a for a solar-mass star ($M_\odot$), for a star of mass $0.5\,M_\odot$, and for Jupiter-mass and Earth-mass planets. Note that a is the semimajor axis of the *planet*'s orbit.

orbiting companion. The general case of the circular orbit is shown in Figure 10.11, where the orbit is inclined at an angle of i_0 with respect to the plane of the sky of the observer. If the speed of the star in its orbit is v, then the amplitude of the radial component is given by:

$$v_{rA} = v \sin(i_0) \tag{10.10}$$

The limiting cases already discussed correspond to $i_0 = 90^\circ$ (edgewise), and $i_0 = 0^\circ$ (face-on). From observations of the sinusoidal changes $(\lambda_{\text{observed}} - \lambda_{\text{source}})$ it is not possible to obtain the value of i_0. This means that we can only obtain the *lower limit* to the mass of a planet. This can be seen if we rearrange Equation (10.9) as:

$$m = v \frac{(P^{1/3} M^{2/3})}{(2\pi G)^{1/3}} \tag{10.11}$$

and use Equations (10.10) and (10.11) to obtain:

$$m \sin(i_0) = v_{rA} \frac{(P^{1/3} M^{2/3})}{(2\pi G)^{1/3}} \tag{10.12}$$

Even if all the quantities on the right are known, the astronomer can only calculate m $\sin(i_0)$, not m. There is a $\sin(i_0)$ ambiguity – we don't know if v_{rA} as measured by us is due to a low-mass planet in an orbit viewed nearly edgewise (i_0 large), or a much

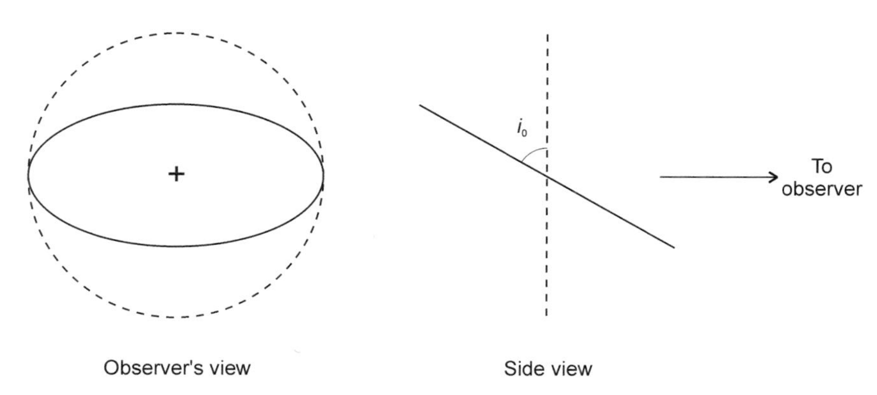

Figure 10.11 A star in a circular orbit inclined at an angle of i_0 with respect to the plane of the sky of the observer.

more massive planet in an orbit viewed nearly face-on (i_0 small). Equation (10.12) gives only a lower limit to m.

Values for i_0 can be estimated in several ways. First, if the planet is seen to transit its star then we know that i_0 is very close to 90° (see Section 10.2). Second, if the star has a circumstellar disc of dust or gas then it is likely that the planet's orbit is near or in the disc plane (this follows from the way that planetary systems are thought to form, as outlined in Section 1.2). Consequently, if the disc has been imaged, then i_0 can be obtained on the assumption that, if it appears elliptical to us, it is actually circular. Third, observations of a star can provide an estimate of the inclination of its rotation axis with respect to our line of sight. Assuming that the planet's orbit is in the equatorial plane of the star, we then get an estimate of i_0. Finally, astrometric measurements can provide an upper limit on m, and hence on i_0, because if i_0 were larger, then m, in order to produce the observed Doppler signal, would also be large enough to cause an observable astrometric signal.

Planetary masses and orbits in the general case

In the case of the circular orbit considered so far, the observer measures v_{rA} and the orbital period P, and can obtain the stellar mass M from the mass-luminosity relationship. A value of $m\sin(i_0)$ can then be obtained from Equation (10.12). From P and M we can obtain the planet's semimajor axis a from Equation (10.4).

Figure 10.12 shows an *elliptical* orbit presented edgewise. The associated cyclic changes in v_r are now non-sinusoidal. Comparison with the edgewise circular case in Figure 10.9 shows the effect of the non-uniform speed at which the star moves around its orbit, with the maximum speed at pericentre, and the minimum at apocentre. The eccentricity could be deduced, though the details will not concern us. We could also obtain $m\sin(i_0)$ (in this edgewise case equal to m) and a as before. Note that, for a given value of a, v_{rA} is greater the higher the eccentricity. This is because of the larger speeds at the pericentre. This somewhat favours the detection of high-eccentricity orbits of given a.

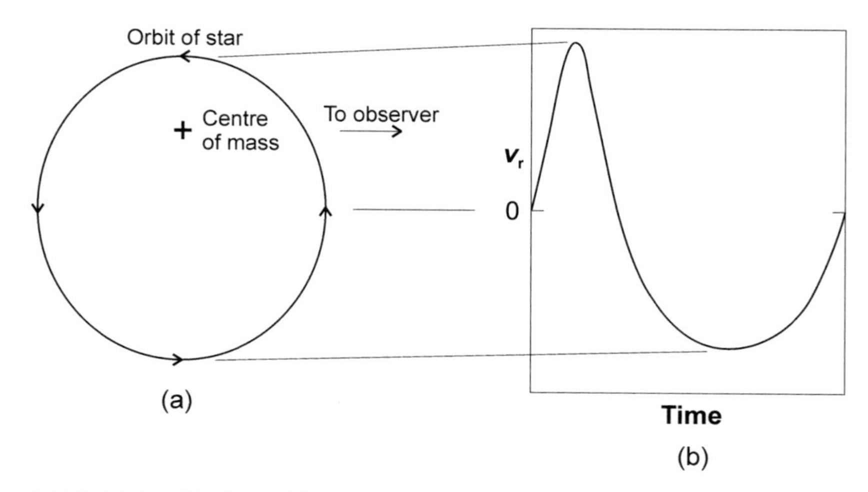

Figure 10.12 (a) An elliptical orbit presented edgewise to an observer. (b) The associated non-sinusoidal cyclic changes in v_r (schematic).

In the actual case of an unknown arbitrary orientation of a possibly elliptical orbit, it is usual practice to invent an orbit and vary it until there is the best fit to the v_r graph. In this way, as well $m\sin(i_0)$, we can get the eccentricity and semimajor axis of the planet's orbit (and other orbital parameters too). The $\sin(i_0)$ ambiguity can be removed in some cases in the ways discussed above. The best fit to the v_r graph can be subtracted to leave a set of residuals. These can be further analysed to see if there is any evidence of a second planet. If there is then we return to the complete data and simultaneously fit orbits for both planets to find the best fits. If the best fit for both orbits is subtracted from the graph this leaves a new set of residuals, and if these are significantly smaller than those when just the first orbit is subtracted, then the existence of the second planet is confirmed. This procedure can be extended to more than two planets. Individual measurements of v_r are subject to uncertainty, so there is some scatter in the v_r graph, as in Figure 10.13. This leads to uncertainty in the planetary masses and orbits.

10.1.5 Doppler spectroscopy: practice

In order to obtain a star's spectrum, the light collected from the star by a telescope has to be passed through a device that disperses different wavelengths into different directions so that they fall in different positions on a detector, as illustrated in Figure 10.14(a) in the simple case when a source has just three well-separated absorption lines. A familiar disperser is a glass prism, though it does not spread apart wavelengths widely enough to be useful. A better device is called a diffraction grating, which consists of many parallel rulings with spacings the order of the wavelengths of interest (see optics texts in Resources). The optical layout is usually such that the spectrum is imaged as a strip, as shown in Figure 10.14(b), crossed by spectral lines.

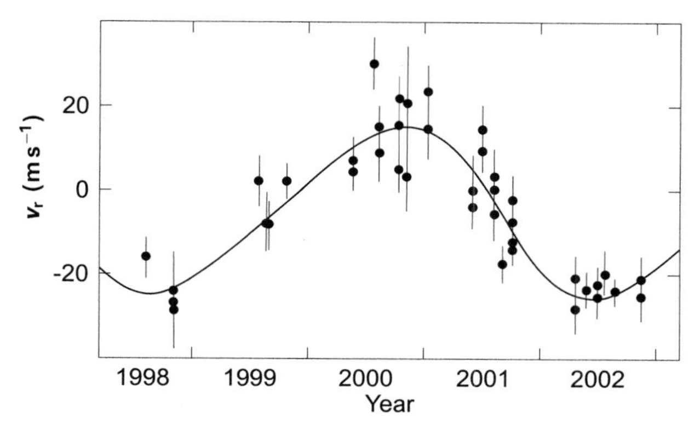

Figure 10.13 A graph of actual measurements of v_r, showing scatter. (This is for the star Tau[1] Gruis.)

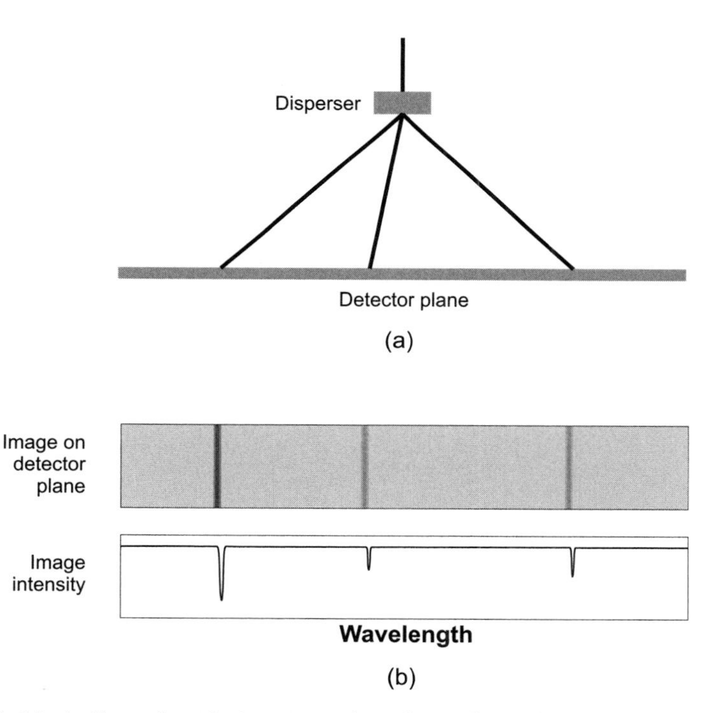

Figure 10.14 (a) A dispersing device separating three absorption lines. (b) The spectrum displayed on the detector plane and the image intensity obtained by scanning across these lines.

A scan across the lines gives the graph of image intensity versus wavelength also shown in Figure 10.14(b).

Once the stellar spectrum is produced we have to repeatedly measure the wavelength of at least one spectral line to detect any variations in radial velocity. In practise very many lines are measured to increase accuracy. The stellar wavelengths are measured by comparing their positions on the detector with those from a reference source in the observatory that has numerous narrow, accurately known, and stable spectral lines over a wavelength range where the star also has many lines. A commonly used source is a gas of molecular iodine (I_2). This has many extremely narrow spectral lines over the range 0.5–0.6 μm that are very stable in wavelength.

From the varying differences in wavelength between the iodine lines and those of the star, the variations in radial velocity are derived. But there is one further step – we have to subtract the Earth's varying radial velocity with respect to the star, arising from the Earth's orbital motion. When this is done we have the radial motion of the star with respect to the centre of mass of the Solar System, which moves through space at a sufficiently constant speed to obviate the need for further adjustments.

There are other approaches. A recently developed technique is externally dispersive interferometry. In the briefest of terms, a set of narrow interference fringes is locally produced at the detector plane over a wide range of wavelengths, as in Figure 10.15. The stellar absorption lines traverse this plane making a small angle with respect to the fringes, which creates a Moiré pattern. Any slight shift in the wavelength of the lines due to a radial velocity change causes a slight sideways shift in their position but a much larger shift in the position of the Moiré pattern. This allows very small radial velocity changes to be measured.

Radial velocities are currently being measured with a precision approaching $2\,\mathrm{m\,s^{-1}}$, corresponding to shifts much smaller than the line widths in Figure 10.8(b). This is certainly enough to detect a Jupiter-twin (Section 10.1.4), provided that the observations are made for at least the order of the 11.9 year orbital period. It is unlikely that this can be pushed to the $0.1\,\mathrm{m\,s^{-1}}$ needed to detect Earth-twins, though an Earth-mass planet in the HZ of a low-mass M dwarf could produce a variation of several $\mathrm{m\,s^{-1}}$, and so would be detectable. This is because of the low mass of the star and the proximity of the HZ, thus reducing M and P in Equation (10.9). The ultimate limit might be set by stellar activity. For example, the convective cells in a star's outer regions cause Doppler shifts, and the convective activity will

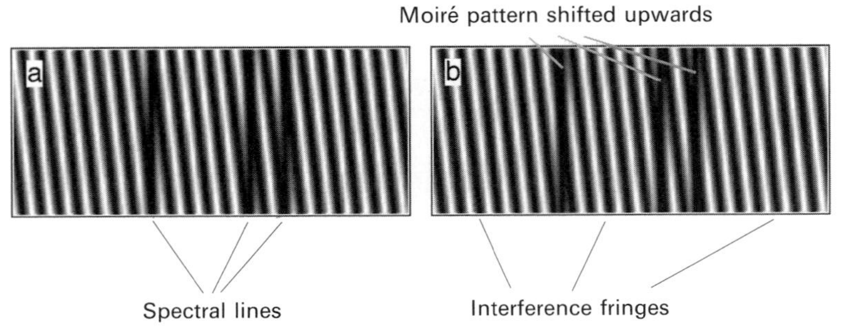

Figure 10.15 The output from an externally dispersive interferometer, showing the shift in the Moiré pattern when there is a slight change in the radial velocity of a star from (a) to (b). Dave Erskine.

vary with the stellar cycle, thus giving the appearance of a varying radial velocity. For a solar-type star the variation will be of the order of $1\,\mathrm{m\,s^{-1}}$, and this might stand in the way of unambiguous interpretation. For many M dwarfs, notably when they are young, the variation is probably higher, which would hamper the detection of Earth-mass planets around such stars.

For planets down to about one-tenth of the mass of Jupiter, Doppler spectroscopy has been very fruitful, as you will see in Chapter 11. Solar-type stars are bright enough for this technique to be used out to thousands of light years.

10.2 TRANSIT PHOTOMETRY

If the orbit of a planet is presented to us sufficiently close to edgewise, then once per orbit, for a small fraction of its orbital period, the planet will pass between us and the star – the planet will be observed in *transit*. The planet emits negligible radiation at visible wavelengths, so the visible radiation we receive from the star is reduced. Even at infrared wavelengths, where the planet emits radiation, the brightness of the planet is far less than the area of the star it obscures, and so the transit is likewise detectable. If the surface of the star has uniform brightness, and if, as in Figure 10.16, the whole of the planet passes across the stellar disc, then the maximum fractional reduction f in apparent brightness of the star is the ratio of the cross-sectional areas of the planet and star i.e.:

$$f = \frac{\pi R_p^2}{\pi R_s^2} \tag{10.13}$$

where R_p is the radius of the planet and R_s is the radius of the star. The radius of the

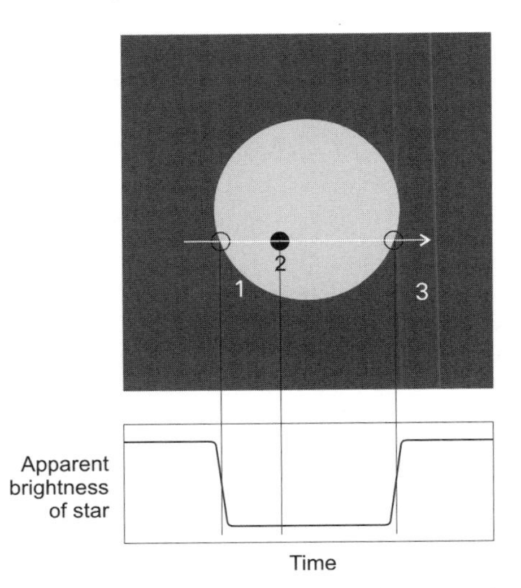

Figure 10.16 A planetary transit, and the associated light curve (neglecting limb-darkening of the star).

star can be estimated, for example, from its luminosity and effective temperature (Equation 8.2), and so we obtain R_p, which we cannot derive from astrometry or Doppler spectroscopy. These other techniques do however provide the mass m of the planet, even in the case of Doppler spectroscopy because we know that $i_0 \approx 90^\circ$ for a transit to occur, so we get m not $m \sin(i_0)$. Thus, if a planet is observed in transit and also by one of these other two methods then we get m and R_p. Hence the planet's mean density is known, which places a constraint on its composition.

In fact, stars dim slightly towards the limb (edge of disc). This is limb-darkening, and if the transit is near the centre of the stellar disc it makes f slightly greater than $(\pi R_p^2)/(\pi R_s^2)$. If it is near the limb, but the planet still fully overlaps the star's disc, f will be slightly smaller than $(\pi R_p^2)/(\pi R_s^2)$. However, the shape of the light curve is different in each case. Therefore, if we could obtain sufficiently precise curves we could estimate the limb-darkening and thus obtain an improved estimate of R_p/R_s.

Another possibility is the grazing transit of a dim companion star where the companion appears to nip the edge of the star's limb. This could be confused with a non-grazing planetary transit, though the light curves again differ, and so we could again distinguish these cases in principle. Another problem arises if the star undergoing a transit (by a planet or a star) has its light contaminated by a foreground star near the line of sight or by a close companion star. Care has to be taken to detect such contamination and eliminate its effects.

If Jupiter were to be observed to transit the Sun then f would be about 0.01, and so to obtain a useful measurement the radiation from the Sun would have to be measured with a precision of a few parts in 10^3. Such photometric precision is readily achievable with ground-based telescopes. If the Earth were to be observed to transit the Sun then f would be about 8×10^{-5}. In this case the photometric precision would need to be a few parts in 10^5. Unfortunately, the photometric disturbance of the Earth's atmosphere limits precision to about one part in 10^4, and so the detection of Earth-size planets around solar-type stars will have to be attempted in space. Only for M dwarfs less than order of one-tenth the radius of the Sun, would the detection of Earth-size planets from the ground be feasible.

The ultimate limit is imposed by stellar variability that mimics the effect of a transit. There is the possibility of a dip in the star's luminosity for the few hours that are typical of a transit. If only a single dimming is observed then for solar-type stars this limits transit photometry to planets with radii greater than about $R_E/2$ in a 1 AU orbit, where R_E is the Earth's radius. In a smaller orbit, smaller planets could be detected because solar variability on the shorter timescale of such transits is less. For M dwarfs, planets down to about $R_E/3$ (Mercury-size) could be detected out to a few tenths of an AU from the star. If several dimmings are observed then this helps distinguish a transit from stellar variability, and this can push the size limits further downwards. Another approach is to detect colour variations. The limb is darker because it appears cooler, and therefore it is also redder. In a transit the starlight will therefore be slightly less red when the planet obscures the limb, and slightly more red when it is more central. Any colour variations due to stellar variability would be different. This approach would however require an order of magnitude improvement in photometric precision.

Table 10.1. Comparison of indirect detection methods.

	Astrometry	Doppler spectroscopy	Transit photometry	Gravitational lensing
Mass of planet	yes	$m\sin(i_0)$	no	yes
Radius of planet	no	no	yes	no
Semimajor axis of orbit	yes	from period	from period	projected value
Period of orbit	yes	yes	yes	no
Eccentricity of orbit	yes	yes	no	no

Note: for each technique, to get the mass of the planet, the mass of the star needs to be known.

The interval between dips gives the orbital period P of the planet, and from an estimate of the star's mass the orbital semimajor axis a can then be calculated. Table 10.1 compares the properties of a planet and of its orbit that we can obtain from transit photometry with what we can obtain from the other detection methods described in this chapter.

If, as is surely the case, the orbits of exoplanets have random inclinations i_0 with respect to the plane of the sky, then we can estimate the proportion of exoplanetary systems that should present transits. For the centre of the planet to lie within the diameter of the star the proportion is approximately R/a, where R is the stellar radius and a is the semimajor axis of the planet's orbit. Clearly, the closer the planet is to the star, and the larger the star, the greater the proportion. For planets about 0.05 AU from a solar-radius star the proportion is about 10%. At about 5 AU (Jupiter distance) from a solar-radius star the proportion has fallen to about 0.1%. Moreover, at 0.05 AU the transits are spaced by a few days, whereas at 5 AU they are spaced by about ten years. However, solar-type stars are bright enough for transits to be detected out to many thousands of light years.

Most search programmes for planetary transits have been launched only recently, numbering a few tens. Although only a few discoveries have so far been made (2003), many more are expected soon. Two space missions are likely to discover many planets through transit photometry. One of these, COROT, is a French mission due to launch in 2004. It will be able to detect planets down to a few Earth-radii. Another, a NASA Discovery mission called Kepler, will launch in 2006. A third, Eddington, is a possible later ESA mission. Kepler will push transit photometry towards its limits, and in surveying hundreds of thousands of stars could discover tens of thousands of planets, including Earth-twins and Mars-size planets closer in.

10.3 GRAVITATIONAL MICROLENSING

This method relies on the effect of an exoplanetary system on the light we receive from a more distant star, called the source star, or background star. Figure 10.17(a) shows (side-view) an exact alignment between a source, an interposed star that might

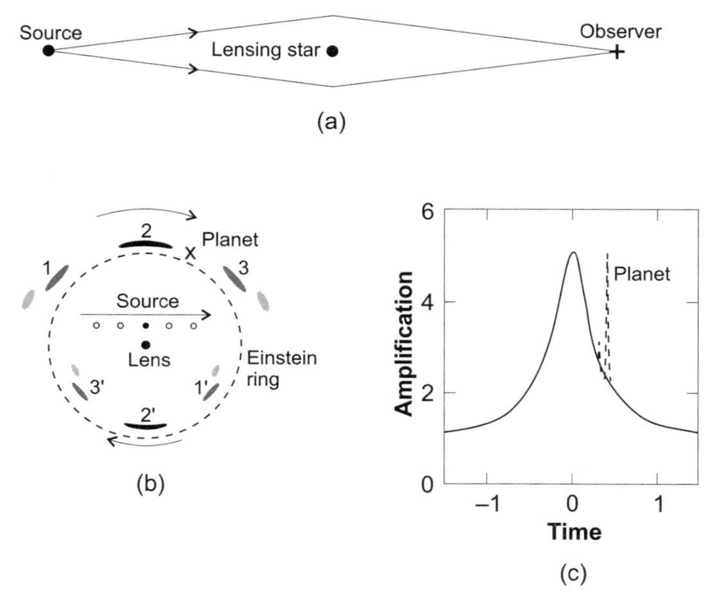

Figure 10.17 (a) Gravitational lensing with exact alignment (side-view). (b) Gravitational lensing with inexact alignment (face-on), where corresponding pairs of source star images are shown at (1, 1′), (2, 2′), and (3, 3′). (c) The light curve from the inexact alignment where the dashed line is the effect of a planet at '×' in (b).

have a planet, and us as observers. You might think that the source would be hidden from view. Instead, the gravitational field of the interposed star bends the light that passes near it. Some of the bent light reaches us in the manner shown, and we see the source as a ring of light around the interposed star. The interposed star has thus acted as a gravitational lens, and is called the lensing star. The ring is called the Einstein ring, after Albert Einstein, whose theory of general relativity (1916) explains the bending in detail. The ring shown face-on in Figure 10.17(b) has an angular radius:

$$\theta_E = \left[\frac{4GM}{c^2}\frac{(d_s - d_l)d_l}{d_s}\right]^{1/2} \tag{10.14}$$

where M is the mass of the lens, d_l is its distance from us, and d_s is the distance to the source star. For typical values of the parameters, θ_E is of the order of a few hundred microarcseconds. This is to be compared with the angular radius of the central disc in the diffraction-limited *psf*, given by Equation (9.1) as $1.22\,\lambda/D$ radians. At a wavelength λ of 0.5 μm and a telescope aperture D even as large as 10 m, this is 1.3×10^4 μas, too large for the telescope to image the ring. The lensing can however be detected because we receive more light from the lensed star than when it is unlensed. Thus, as the stars' proper motions produce the alignment we would observe an apparent brightening of the lensed star. Detection of gravitational lensing via apparent brightening is called *gravitational microlensing*.

Alignments are never exact. Figure 10.17(b) shows a sequence where the two stars' angular separation on the plane of the sky is a fraction x of θ_E – call this fraction x. The lensing produces a sequence of pairs of distorted images (e.g. at (1, 1′), (2, 2′), (3, 3′)). Note that these are not far from the Einstein ring. The amplification of the light received from the lensed star, the light curve, is shown by the solid line in Figure 10.17(c). The amplification A is given by:

$$A = \frac{x^2 + 2}{x(x^2 + 4)^{1/2}} \tag{10.15}$$

It thus depends only on x, and will peak at closest alignment. Alignment well beyond θ_E, where $x \gg 1$, produces negligible amplification (i.e., $A \approx 1$). The encounter lasts for a duration of order:

$$t_E = \frac{\theta_E}{\mu} = \frac{\left[\dfrac{4GM}{c^2}\dfrac{(d_s - d_1)d_l}{d_s}\right]^{1/2}}{\mu} \tag{10.16}$$

where μ is the proper motion of the lens with respect to the source (in angular measure per unit time). For typical parameters t_E is a few days.

A planet, indeed any mass, can also produce lensing. Moreover, the peak amplification depends only on x (Equation 10.15), and so a planet could produce as large an amplification as a star if x were small enough. Unfortunately, θ_E is proportional to the square root of the mass of the lens (Equation 10.14), and therefore the target area is small for a planet, and the likelihood of a sufficiently good alignment is correspondingly small. The duration is also dependent on the square root of the mass (Equation 10.16), and so even if there is excellent alignment the lensing is over in a matter of hours or minutes, and is therefore easily missed.

The practical approach is therefore to detect the beginning of lensing by a star, and use this as an alert to look for modification produced by a planet. This will only be appreciable if the planet is located near the path of one of the source images (e.g., at '×' in Figure 10.17(b)). In this case the planet will deflect the light from the image and modify the light curve as shown by the dashed lines in Figure 10.17(c) that form narrow peaks. The height of the peaks is greater the closer the planet is to the image trajectory, and is independent of the mass of the planet (just as the light amplification in Equation (10.15) is independent of the mass of the lensing star). A peak could thus be tall. Its duration t_{Ep} is of order θ_{Ep}/μ where θ_{Ep} is the angular radius of the planet's Einstein ring. It thus follows that:

$$\frac{t_{Ep}}{t_E} = \left(\frac{m}{M}\right)^{1/2} \tag{10.17}$$

where m is the mass of the planet. The duration is thus greater the larger the mass of the planet.

Given that the planet has to be located near the path of one of the source images, and that these are near the Einstein ring, it is clear that the angular separation of a detectable planet from its star has to be such that, when projected on the plane of the sky, it is roughly θ_E from the star, and in the right direction. Therefore, only a small proportion of exoplanets will be discovered by this method.

Nevertheless, microlensing is an important additional technique, for several reasons. First, it can find planets even further away than Doppler spectroscopy and transit photometry, and around intrinsically less luminous stars, to tens of thousands of light years. Second, we can get the mass of the planet, even small masses – Equation (10.17) shows that if we can estimate M we can get m. Moreover, because the duration is proportional to $m^{1/2}$, the peak width for an Earth-mass planet is only $318^{1/2} = 18$ times less than that of a Jupiter-mass planet. Gravitational lensing could thus detect Earth-mass planets. Third, we do not have to make observations for times of the order of a period of a planet's orbit – a planet in a 5 AU Jupiter orbit could be detected in days rather than years. Finally, satellites of a planet could be detected via subsidiary peaks.

There are, however, limitations. First, we learn little about the planet's orbit. We know that the angular separation on the sky of the instantaneous star–planet distance is roughly θ_E, which is a distance of $R_E = d_1\theta_E$, where R_E is the radius of the Einstein ring. Typically R_E is a few AU, so all we learn is that the planet, at some instant, was at a projected distance of a few AU from its star. Second, lensing is one-off – follow-up lensing events are extremely unlikely, though if the system is near enough it could be explored by some of the other methods described earlier.

A number of microlensing surveys has taken place, in directions in space in which there are many background stars to act as sources and many closer stars to act as lenses, such as towards the nuclear bulge of the Galaxy (Section 1.3.2). Very many lensing events have been recorded, for example, in the Optical Gravitational Lensing Experiment (OGLE) survey that for some years has been using the 1.3 m Warsaw telescope at Las Campanas Observatory in Chile. Early alerts are particularly important so that many observations of any planetary event can be made. The systems for implementing such alerts are now being put in place. In the earlier absence of such systems we have only tentative, unconfirmable evidence for a few planets.

10.4 OBSERVATIONS OF CIRCUMSTELLAR DISCS AND RINGS

Planets are believed to form from a circumstellar disc of gas and dust (Section 1.2). That such discs are found to be common around young stars supports this view. Very young discs are dominated by gas, and structures in the gas indicate the presence of planets, perhaps in the process of forming. All but the very youngest discs are dominated by dust, the gas having been dissipated. Some of this dust will be primordial, but the existence of old dust discs indicates the continuing generation of dust by collisions between cometary and asteroidal bodies. Telltale signs that planets are present in some dust discs include a central hole in the disc a few tens of AU across, and warping of the disc due to the gravitational forces of giant planets. A few cases will be outlined in Section 11.4.3.

10.5 SUMMARY

- Indirect methods of detecting planets rely on the effect of a planet either on the motion of the star it orbits, or on the quantity of light we receive from this star or from a background star.
- Methods that rely on the star's motion include astrometry and Doppler spectroscopy. In astrometry a star's orbital motion is sought by repeatedly measuring its position. Interferometry at the VLT telescopes should, by about 2005, be able to detect Jupiter-twins out to about 800 ly, and Earth-mass planets several AU from nearby low-mass M dwarfs. The Keck telescopes might acquire a similar capability. The space telescope Gaia (about 2010) will be able to detect Jupiter-twins out to about 350 ly, and planets of a few Earth masses 1 AU from M dwarfs out to a few tens of light years. The space telescope SIM (about 2010) will have a better capability through being a targeting mission rather than a sky survey.
- In Doppler spectroscopy we attempt to detect the orbital motion of the star from cyclic changes in the Doppler shift of the spectral absorption lines of the star. Precisions approaching $2\,\mathrm{m\,s^{-1}}$ are currently being achieved, and are sufficient for Jupiter-twins out to thousands of light years, and for Earth-mass planets in the HZs of nearer M dwarfs.
- In transit photometry we seek reductions in light from the star when a planet makes a transit across it. Photometric precision of a few parts in 10^3 is required for Jupiter-size planets around solar-size stars, and a few parts in 10^5 for Earth-size planets. The former precisions are readily obtainable from the ground, but the latter only from space, where COROT (2004), Kepler (2006), and perhaps Eddington are expected to make many discoveries, the latter two down to Earth-twins even at many thousands of light years.
- Gravitational microlensing occurs when the relative proper motions bring a background star to a direction that is within a few hundred microarcseconds of an exoplanetary system. The light from the background star is amplified, and a planet can, if the alignment is right, cause an easily observable distortion in the stellar light curve.
- See Table 10.1 for a comparison of what we learn about an exoplanetary system from the various methods. Additionally, gravitational microlensing is the only method that for now and in the near future could detect Earth-mass planets even in large orbits around stars that are tens of thousands of light years away.
- Doppler spectroscopy has so far been the most productive technique in discovering exoplanets (see Chapter 11).
- The structure of circumstellar discs of gas (young stars) and the existence and structure of discs of dust (older stars) can indicate the presence of planets.

10.6 QUESTIONS

Answers are given at the back of the book.

Question 10.1

The astrometric system PRIMA intended for the VLT would have a precision of 10 μas.

(a) Show that this is sufficient to determine the orbit of a Jupiter-twin out to about 800 ly.
(b) Calculate the maximum range at which PRIMA could determine the orbit of an Earth-mass planet 1 AU from a very low-mass M dwarf one-tenth the mass of the Sun.

Question 10.2

The measured wavelengths of the spectral absorption lines of a solar-mass star are going through sinusoidal changes that cover a range of 2.5×10^{-7} of the wavelength at the source. The cycle has a period of 20 days. Deduce as much as you can about the mass (in Earth masses) and orbit of the companion to this star. Could there be a second companion?

Question 10.3

Suppose that a gravitational microlensing event is observed that lasts for 60 hours, and that a tall spike appears in the light curve for about 9 minutes. The lensing star appears extremely faint but is believed to be an M dwarf with a mass about half that of the Sun.

(a) Estimate the mass of the object that produced the spike (in Earth masses).
(b) Outline the difficulties in obtaining further data from astrometry, Doppler spectroscopy, and transit photometry.

11

Exoplanetary systems

The previous two chapters reviewed the methods being employed in the search for exoplanets. This chapter summarises the results of the searches so far, and considers what sort of exoplanetary systems might await discovery. The next chapter concentrates on how to find life on exoplanets.

11.1 THE DISCOVERY OF EXOPLANETARY SYSTEMS

In 1992, after decades of disappointment and false hope, the first exoplanets were discovered. The US astronomers Alex Wolszczan and Dale Frail announced that they had detected two planets in orbit around a rare type of star called a pulsar. Each planet had a mass just a few times that of the Earth. The claim that a pulsar had planets was greeted with considerable surprise by other astronomers, but it has withstood further investigation. The surprise stems from the way that a pulsar is formed, as the remnant of a massive star after it explodes in a supernova at the end of its life. The remnant can either be a black hole or a neutron star (Section 8.1.1). A *pulsar* is a neutron star that we observe by the beacon of electromagnetic radiation that it sweeps across us as it rotates, giving us a series of regular pulses. The planets were detected through the apparent modulation of pulse spacing resulting from the Doppler effect as the pulsar orbited the centre of mass of the system. It had not been anticipated that planets could survive a supernova explosion, and perhaps any pre-existing planets didn't. Instead, the planets could have formed from debris left by the explosion, or maybe they were captured from a companion star as the pulsar travelled near it. More importantly for us, life could not have survived the explosion, and even if the planets formed afterwards, they would still be uninhabitable so close to the pulsar, with its deadly radiation. We will therefore discount the handful of pulsar planets, and confine our attention to planets around stars more like the Sun.

It was in October 1995 that two Swiss astronomers, Michel Mayor and Didier Queloz of the Geneva Observatory, announced the discovery of the first non-pulsar planet. It is in orbit around the solar-type star 51 Pegasi (star number 51 in the constellation of Pegasus). The result was soon confirmed by others, and by early 1996 astronomers knew that the long drought in exoplanetary discoveries had ended. A steady trickle of discoveries has followed, and continues today. By September 2003, 117 non-pulsar planets were listed in one catalogue, in 102 planetary systems, 13 of which are multiple-planet systems (see Schneider website in Resources), though not all in the list are beyond reasonable doubt. In addition there are a hundred or so candidates for planetary status, as yet unconfirmed, mainly from transit and gravitational microlensing surveys. Exoplanets are named after their stars, by using the letter 'b' to denote the first planet in the system to be discovered, 'c' the next, and so on. Thus the planet of 51 Pegasi is named 51 Pegasi b. If several in a system are discovered at the same time, then the lettering starts with the innermost and works outwards, as in Upsilon Andromedae b, c, and d.

All but one of the 117 non-pulsar exoplanets have been discovered using Doppler spectroscopy. Figure 11.1 shows data for three stars, along with best-fit smooth curves. From the sinusoidal fit for 51 Pegasi in Figure 11.1(a) it can be deduced that the orbit has a low eccentricity (Section 10.1.4). The different-shaped curves in the other two cases indicate more eccentric orbits, and the eccentricity can be deduced, though the details will not concern us. The only confirmed discovery by another method is of a planet orbiting the star OGLE-TR-56. This star was unnamed before the discovery, and its name carries that of the survey – OGLE (Section 10.3). In spite of the survey name, the discovery was made by transit photometry. This is because a survey looking for the light amplification in gravitational microlensing can also detect changes in apparent stellar brightness due to a transit. That OGLE-TR-56b has planetary mass has been confirmed by Doppler spectroscopy.

OGLE-TR-56 is about 5000 ly away. All the other stars with planets are much closer, the furthest, HD47536b, being at 401 ly, which is only about 0.5% of the diameter of the disc of the Galaxy in which we live, and so all these other exoplanets are in our cosmic backyard (HD denotes a particular star catalogue). This is because at close range we receive more stellar radiation. The spectral lines are then clearer, and the tiny Doppler shifts caused by planetary-mass bodies are consequently easier to measure. (Remember that the Doppler shifts themselves are independent of the distance to the star – Section 10.1.4.) The detection of OGLE-TR-56b by Doppler spectroscopy was aided by the very short period of the planet, only 1.2 days, and the consequent large radial velocity amplitude.

One system discovered by Doppler spectroscopy, that of the star Gliese 876 (Gliese is another star catalogue), has since been detected astrometrically. This detection was aided by the low mass of Gliese 876, about one-third that of the Sun, and its proximity, 15.4 ly, leading to a large astrometric signal.

Another exoplanet discovered by Doppler spectroscopy has also been detected by another technique. This is HD209458b, which has subsequently been observed in transit. The transit was observed with a small telescope of just 0.1 m aperture, the

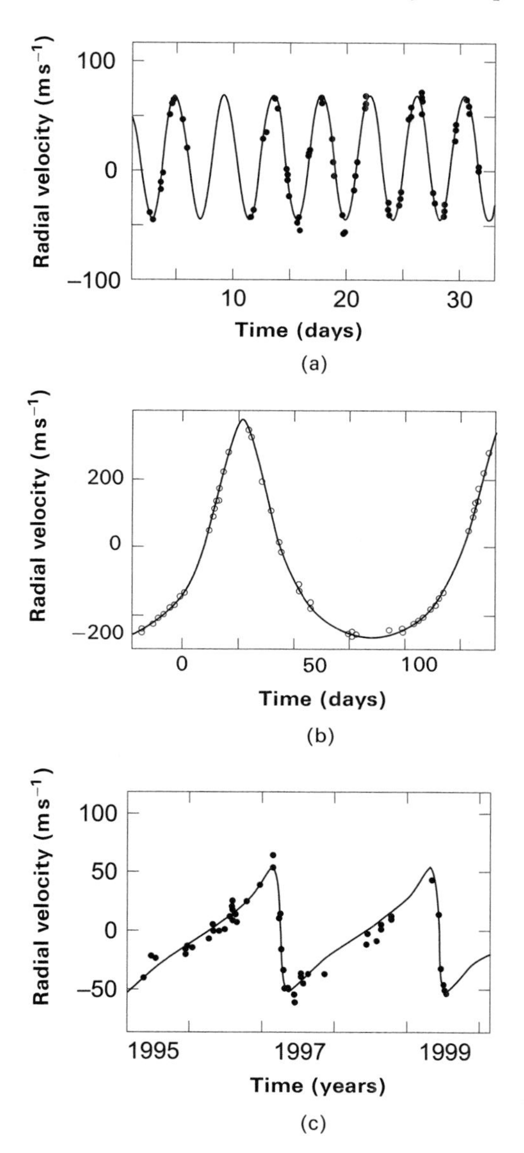

Figure 11.1 The variation of the radial velocity of some stars with planets, as obtained from the Doppler shifts in the spectral lines. (a) 51 Pegasi (planet in a low-eccentricity orbit – data from late 1995). (b) 70 Virginis (planet in an eccentric orbit). (c) 16 Cygni B (planet in an eccentric orbit).

instrument of the Stellar Astrophysics and Research on Exoplanets (STARE) project in Boulder, Colorado. In August and September, 1999, HD209458 was observed on 10 nights, and on two of these (9 and 16 September), evidence of a transit was obtained, as shown by the light curve in Figure 11.2, where the dots are the

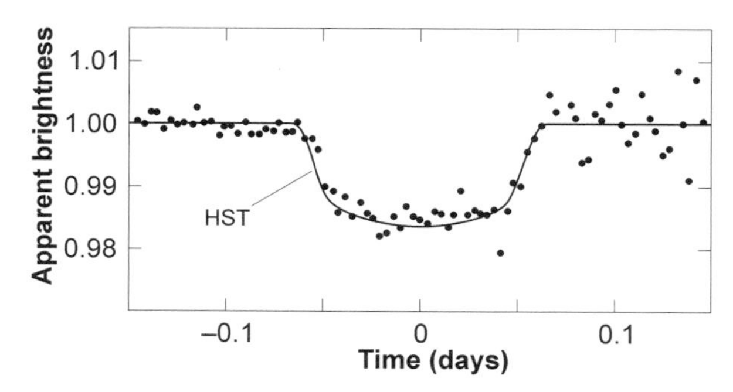

Figure 11.2 The dots show the first averaged light curve for the HD209458 system, from the 0.1 m STARE telescope. The line shows the later light curve from the HST. The scatter of the HST measurements is about the thickness of the line.

STARE data. These dates are separated by the known orbital period of the planet. Since then the transit has been observed many times, and the line in Figure 11.2 shows the copious and far less noisy photometric data from the HST, due to its greater aperture and location in space, free of background light from the atmosphere.

To observe a transit the planet's orbit has to be presented to us very nearly edge-on. Exoplanet orbits are oriented randomly on the sky. Calculations show that in this case, on average, out of the 100 or so confirmed exoplanets, we can only expect to observe transits for one or two, so the rate of success is about right.

We will now look at the confirmed exoplanetary systems in more detail, including the stars, the exoplanets as bodies, their orbits, and the ways in which the various systems might have formed. Remember that we are excluding pulsar planets.

11.2 THE KNOWN (NON-PULSAR) EXOPLANETARY SYSTEMS

11.2.1 The stars that host the known exoplanetary systems

The stars that host the planets are predominantly main sequence stars of spectral type F, G, or K (Section 8.1.1) (i.e., they are predominantly solar-type stars). This is because such stars have attracted most of the search effort by Doppler spectroscopy. They suit this technique because they have plenty of sharp spectral lines, good surface stability, they are fairly luminous, and they are not rare, so there are plenty of bright examples. In relation to the search for life it is fortunate that they also have sufficiently long main sequence lifetimes for any life to have effects we might be able to observe, as discussed in Section 8.2.1. M dwarfs have even longer lifetimes, and they are also abundant, more so than all F, G, and K stars put together (Section 8.1.2).

So, why have M dwarfs not been favoured in the exoplanet searches? The main reason is their low luminosities. There has also been a belief that any planets might not be habitable. This was discussed in Section 8.2.3, where it was concluded that this belief is not very secure. Therefore, it might be the case that M dwarfs have been unduly neglected. More of them are now being scrutinised.

Among the 102 stars with planets, seven of the stars are members of binary systems. The planet orbits one of the stars. For example, in the binary star Gamma Cephei, a giant planet is in a 2.15 AU orbit around a 1.6 solar mass star. The second star, 0.4 solar masses, is at 21.4 AU. It might have been the case that the second star would have prevented planetary formation or ruled out stable orbits, but this is not so. Recall that a high proportion of stars are in multiple systems, mainly binary systems – about 70% of stars in the solar neighbourhood are in such systems. That such systems can have planets increases considerably the potential number of stars with planets.

Roughly 10% of the nearby solar-type stars have giant planets within 4 AU of the star. Nearly all of these stars have metallicities exceeding 0.5%, some exceeding the Sun's 2%. Such metallicities are comparatively high. This might indicate that planetary formation is favoured by increasing the proportion of condensable materials in the stellar nebula (Section 1.2), though comparisons with stars apparently without planets is very incomplete. The fraction of nearby solar-type stars known to have planets can only grow as the more difficult discoveries are made, such as planets with the longer periods characteristic of larger orbits. To these must be added planets that will be discovered around other types of star, particularly the abundant M dwarfs. Whether the majority of nearby stars have planets is unknown, but astronomers are hopeful that the proportion is substantial.

Gravitational microlensing surveys reach further out, and favour the nuclear bulge of the Galaxy (Section 10.3), where the number of stars makes microlensing more frequent than locally. Though there are some unconfirmed detections, the surveys indicate that fewer than about one-third of the stars towards the bulge have Jupiter-mass planets at projected distances in the range 1.5–4 AU from the star, and fewer than about one-half have rather more massive planets in the range 1–7 AU. Note that the typical star observed is an M dwarf.

Transit surveys also reach further out, and again there are unconfirmed detections. However, a survey of the open cluster NGC6819, which consists of high-metallicity stars, failed to reveal any transits of Jupiter-size planets, in spite of the expectation of a few based on statistics from the solar neighbourhood. This might indicate that the solar neighbourhood is particularly well endowed, or that the Doppler spectroscopy method, which has yielded the local discoveries, favours the sort of stars that have planets. A survey of the globular cluster 47 Tucanae for planetary transits also found none, though these ancient stars have low metallicities, and the comparatively high spatial density of stars in such clusters might prevent planets forming, or eject them. However, a planet of a few Jupiter masses discovered in a binary system consisting of a pulsar and a white dwarf in the globular cluster M4 might indicate otherwise. The subject of exoplanets is very young, and there are many uncertainties.

The number of confirmed exoplanets is likely to increase greatly in the next few years. Quite what sort await discovery is the subject of Section 11.4. First, let's examine the ones we *have* discovered.

11.2.2 Exoplanet masses

Figure 11.3 shows the number of measured values of $m\sin(i_0)$ in the mass intervals 0–0.49 m_J, 0.50–0.99 m_J, and so on, where m is the actual mass of the planet and i_0 is the inclination of its orbit with respect to the plane of the sky – remember that Doppler spectroscopy gives $m\sin(i_0)$ not m (Section 10.1.4). The units of $m\sin(i_0)$ are m_J the mass of Jupiter (318 times the mass of the Earth, m_E). The least massive (non-pulsar) planet with a confirmed discovery at the time of writing (September 2003) is 0.12 m_J or 38 m_E, a bit less than half the mass of Saturn and a bit more than twice the mass of Neptune. At the other extreme, we are not concerned with bodies of mass greater than 13 m_J, because they are better classified as stars than as planets. Such massive bodies will be of roughly stellar composition, made largely of hydrogen and helium. For such a composition 13 m_J is the approximate threshold above which the deep interior becomes hot enough for thermonuclear fusion to occur, involving the rare isotope of hydrogen, deuterium (^{2}H). This fusion process can last the order of 1000 Ma, because even though ^{2}H is a scarce isotope it is consumed at a low rate

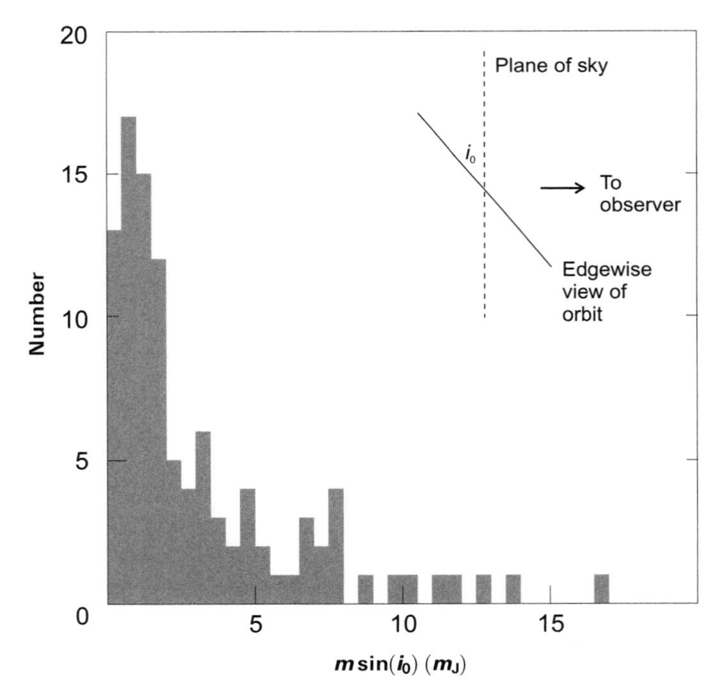

Figure 11.3 The distribution of $m\sin(i_0)$ for the known exoplanets. Inset is an illustration of $\sin(i_0)$.

at the modest temperatures in these comparatively low-mass objects. Recall that only above about 80 m_J is there sustained fusion of ^{1}H, with the star then on its protracted main sequence phase. Hydrogen-rich objects in the approximate mass range 13–80 m_J are the brown dwarfs (Section 8.1.1). A large number have been discovered, particularly towards the upper end of this mass range.

Thermonuclear fusion is only one of several criteria that can be used to distinguish between a planet and a star, and although there is no consensus on which criterion is best, this one is useful because there seems to be a paucity of bodies with masses around 13 m_J (Figure 11.3).

The question arises of how many of the planets in Figure 11.3 are in fact brown dwarfs masquerading as planets, as they would if they have such low values of i_0 that the true masses are above 13 m_J. For example, if $m\sin(i_0) = 4.5\,m_J$ and $i_0 = 5.7°$, the actual mass $m = 4.5\,m_J/\sin(5.7°) = 4.5\,m_J/0.099 = 45\,m_J$. HD209458b is certainly no brown dwarf. This is because it has been observed to transit its star, and so we must be seeing the orbit nearly edge-on, corresponding to $i_0 = 86.1°$ and thus to $\sin(i_0) \approx 0.998$. With $m\sin(i_0) = 0.69\,m_J$, its actual mass is also 0.69 m_J. We can also be fairly certain about Epsilon Eridani. This star probably has a planet in which case it is the one of the very few exoplanetary systems so far observed to have a dust ring. The dust ring is seen to be elliptical, and so, presuming it to be actually circular, and that the planet orbits in the ring plane, a value of i_0 of about 46° has been obtained. The true mass would then be $1/\sin(46°)$ times the 0.86 m_J value of $m\sin(i_0)$, that is 1.2 m_J. Gliese 876b, from astrometry, has $i_0 \approx 37°$, and so $m = 3.3\,m_J$. From the transit of OGLE-TR-56b $i_0 \approx 86.2°$, and subsequent Doppler spectroscopy then gives $m = 0.9\,m_J$.

Useful estimates of i_0 can be obtained in some further cases from observations of the star that yield the inclination of its rotation axis with respect to our line of sight. As it is likely that the planets orbit close to the equatorial plane of the star, we then get an estimate of i_0. In only a few of these cases is i_0 so small that the true mass exceeds 13 m_J. Overall, only a very small proportion of the bodies in Figure 11.3 could be brown dwarfs. This conclusion is supported on statistical grounds. If the orientations of the orbits of exoplanets are random, then it can be shown that the vast majority will have actual masses within a factor of two of their $m\sin(i_0)$ values.

The known exoplanets thus have masses of the order of the mass of Jupiter. But do they also resemble Jupiter in composition?

11.2.3 Exoplanet composition

Astronomers know from detailed observations and modelling that Jupiter has a composition not hugely different from the Sun when it was young, before nuclear fusion altered its constitution. By proportion of mass, the young Sun was 73% hydrogen, 25% helium, and 2% all the other elements (the heavy elements). Jupiter might have a composition fairly close to this, though it is likely to be enriched to the level of 5–10% of its mass by heavy elements, depending on its mode of formation (Section 1.2.1). You have seen that the (non-pulsar) exoplanets all have (minimum) masses of the order of that of Jupiter, and thus considerably greater than the mass of the Earth. But could they nevertheless have Earth-like compositions, dominated by silicates and iron?

An important indicator of composition is the mean density ρ of the planet (i.e. its mass m divided by its volume V, which is $4\pi R^3/3$ for a spherical body with a radius R). Thus,

$$\rho = \frac{m}{4\pi R^3/3} \tag{11.1}$$

Suppose that an exoplanet has the same mass as Jupiter ($318\,m_E$). Assuming that its mean density is the same as that of the Earth, then, from Equation (11.1):

$$\frac{318\,m_E}{4\pi R^3/3} = \frac{m_E}{4\pi R_E^3/3}$$

where R_E is the Earth's radius. Therefore:

$$R = (318)^{1/3}\,R_E = 6.8\,R_E$$

In fact, at the greater mass the pressures inside the super-Earth would be much greater than in the Earth, and so the interior would be compressed to a greater density. The actual radius for a Jupiter-mass planet with an Earth-like composition would be more like 5–6 R_E. In comparison, Jupiter's radius $R_J = 11\,R_E$. The difference between 5–6 R_E and 11 R_E is because (at comparable pressures) silicates and iron are much denser than hydrogen and helium, so a planet of silicates and iron will have a smaller size that a hydrogen–helium planet of the same mass. Therefore, one way to infer composition is to measure the diameter of an exoplanet of known mass.

So far, this has only been possible for HD209458b and OGLE-TR-56b. From Doppler spectroscopy and its transits, the true mass of HD209458b is known to be $0.69\,m_J$. The amount of apparent dimming of the star during the transits enables the radius of the planet to be fixed at $1.42\,R_J$. This huge size but modest mass certainly rules out a silicate–iron composition, but raises the question, why is it 1.42 times the radius of Jupiter but only 0.69 times Jupiter's mass, and thus only 24% of Jupiter's mean density? The main reason is its proximity to its star, only 0.045 AU. You might think that thermal expansion of its atmosphere is responsible, but this is a minor effect. More important is the reduction in the rate of cooling of the planet due to the heat of its star. If the planet has been close to its star from soon after its formation then we have a plausible explanation of its bloated size today. From its size we can thus tell not only that it is a hydrogen–helium-rich body, but also that it has been close-in from soon after it formed. We shall return to this latter point in Section 11.3.

From analogous data for OGLE-TR-56b we have a radius $(1.3 \pm 0.3)\,R_J$ and a mass $(0.9 \pm 0.3)\,m_J$. These values give a mean density about 40% that of Jupiter. This planet is even closer to its star than HD209458b, and presumably it is distended for much the same reasons.

Another way to obtain the composition, at least of atmospheres, is to examine the imprint the atmosphere makes on the light of its star. In this way sodium vapour has been detected in the atmosphere of HG209458b. As yet, the more abundant species are at the limit of detection capabilities.

For the other exoplanets we have only indirect evidence for their composition. However, if they are silicate–iron bodies then the circumstellar discs from which they formed must have been highly enriched in heavy elements to make planets amounting to hundreds of Earth masses. The degree of enrichment required is unrealistically large given what we know about the composition of the sort of interstellar clouds from which stars and their planetary systems have been born. In any case, such enrichment would be seen as very high metallicity in the parent stars. The stellar spectra do indeed indicate some enrichment, but nowhere near large enough to support the hypothesis that the giant exoplanets are silicate–iron.

There is thus reasonable evidence that the Jupiter-mass exoplanets so far discovered have compositions that resemble Jupiter in that they are dominated by hydrogen and helium. At the lower end of the exoplanet-mass range, the masses are more like that of Saturn, about one-third the mass of Jupiter, but Saturn too is largely hydrogen and helium, so the conclusion applies also to smaller giants. Only at the current extreme of $0.12\,m_J$, about twice the mass of Neptune, might hydrogen and helium not dominate, just as they do not in Uranus and Neptune (Section 1.1.3).

11.2.4 Exoplanet orbits

Figure 11.4 shows $m\sin(i_0)$ versus the orbital semimajor axis a for each exoplanet. The dashed line is of constant radial-velocity amplitude for fixed stellar mass, and therefore provides an approximate boundary between the exoplanet discoveries that were relatively hard to make (below the line) from those that were easier to make (above the line). If the orbit has an eccentricity e greater than 0.1 then an elliptical symbol has been used, otherwise a filled circle. The choice of 0.1 as the discriminating value is because it is just larger than the eccentricity of Mars, which has a fairly large eccentricity by Solar System standards, exceeded only by Mercury and Pluto among the planets. The inset to Figure 11.4 shows an orbit with an eccentricity of 0.1. It is not dramatically non-circular, but the offset of the star from the centre of the orbit is considerable. An orbit with an eccentricity of 0.67 is also shown. This is for the planet of the star 16 Cygni B, and is one of the largest exoplanet eccentricities known. The orbit has a clearly non-circular shape.

A striking feature of Figure 11.4 is how small many of the semimajor axes are – note that the scale is logarithmic. Jupiter and Saturn are shown for comparison – they are much further out. The planet closest to its star is OGLE-TR-56b, at 0.0225 AU, which is only about 6% of the semimajor axis of Mercury, also shown in Figure 11.4. This proximity would be much less remarkable if the exoplanets were low-mass iron–silicate bodies, as in the Solar System, but they are hydrogen–helium giants. To see why proximity is then remarkable you need to recall from Section 1.2 two current models for the formation of giant planets in the Solar System. In either the two-stage or one-stage process they form around

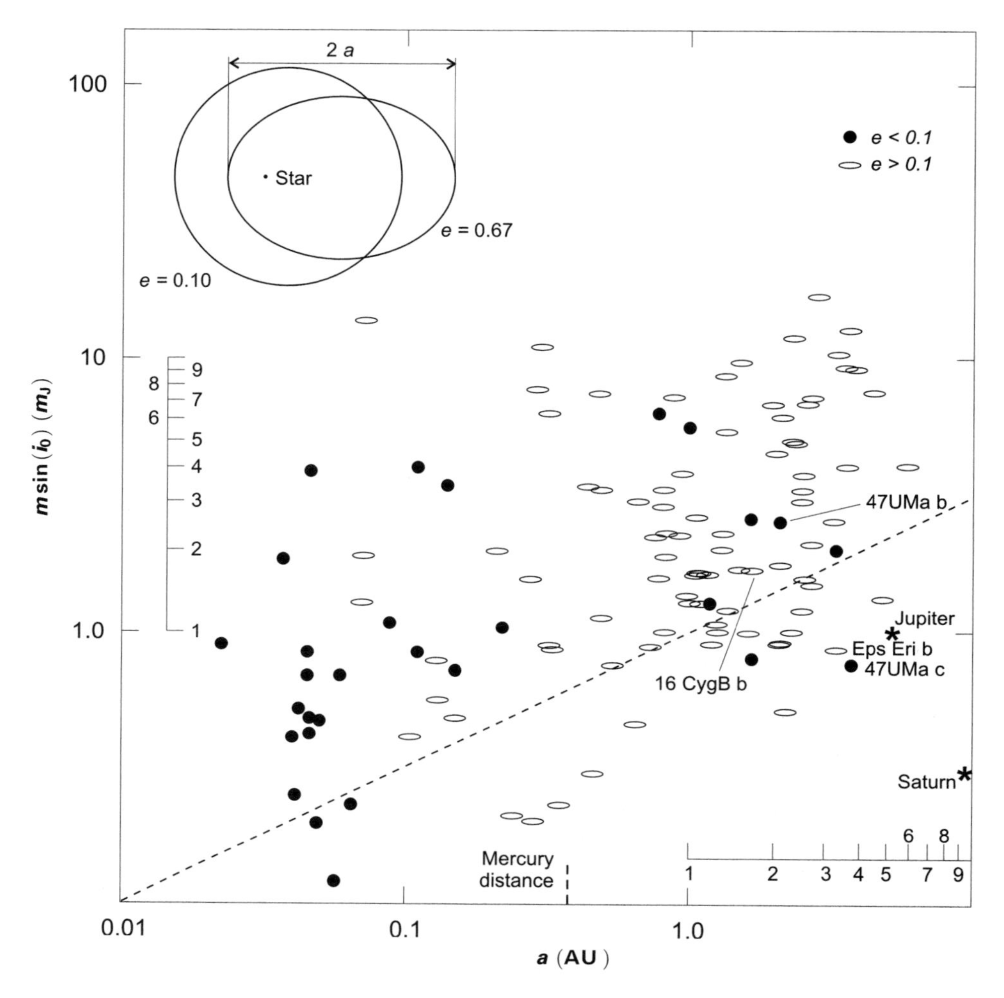

Figure 11.4 $m \sin(i_0)$ versus orbital semimajor axis a for the known exoplanets. Inset (top left) shows orbits with eccentricities of 0.1 and 0.67 (16 Cygni Bb) with the same semimajor axis a.

or beyond the ice line, about 4 AU in the case of solar-type stars, and though it could be that hydrogen–helium giant exoplanets formed close to their stars, this is not yet a well established possibility. Therefore, if the giant exoplanets formed some way out, then those that are now well within the ice line of their star must have subsequently moved inwards. These *hot Jupiters* as they are called, thus indicate planetary migration. That theoreticians have found reasonably convincing migration mechanisms that can produce hot Jupiters, adds weight to the view that the hot Jupiters, like all Jupiter-mass exoplanets, are predominantly hydrogen–helium in composition.

11.3 MIGRATION OF GIANT EXOPLANETS AND ITS CONSEQUENCES

The discovery in the mid-1990s of hot Jupiters, led within months to plausible mechanisms by which planets could migrate inwards. Theoreticians were not being entirely 'wise after the event' – migration had been predicted over a decade earlier, in research that was largely overlooked. Migration mechanisms are outlined in Section 11.3.1. Then, in Section 11.3.2, we examine the implications of giant planet migration for the formation and survival of Earth-mass planets.

11.3.1 Migration mechanisms and consequences for giants

The key to migration is the gravitational effect of the giant planet on the circumstellar disc of gas and dust in which it is embedded and from which it has formed. The details are complex so only a qualitative outline is given here. At first the disc is symmetrical about an axis perpendicular to it and running through the protostar at its centre. But as the mass of the embryonic giant grows, its gravitational field produces spiral structures in the disc that destroy this symmetry, as in Figure 11.5 (note that the disc is not modelled close to the star, hence the 'black hole', which is an

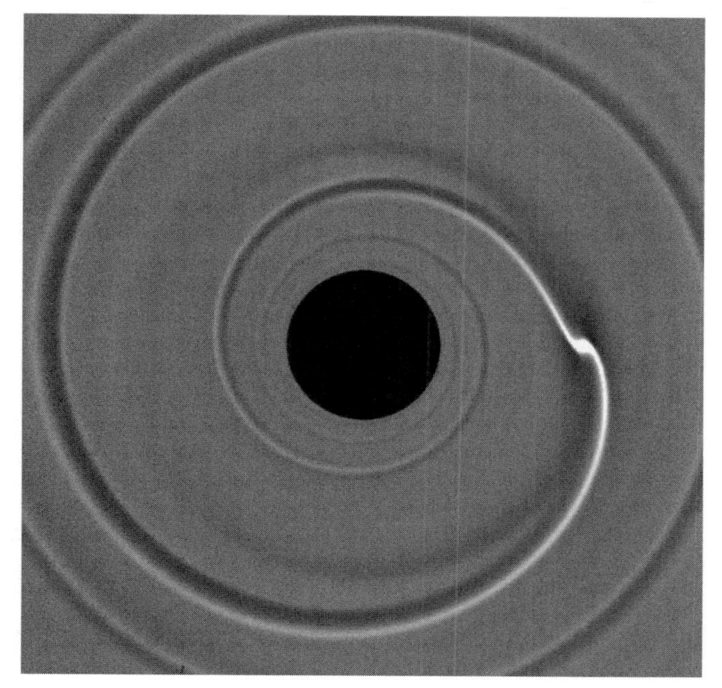

Figure 11.5 A circumstellar disc with a spiral structure created by a planetary kernel or embryo. The disc is not modelled close to the star, hence the 'black hole', which is an artefact. Frederic Masset.

artefact). Consider the two-stage formation of a giant at the point where its icy–rocky kernel has a mass less than m_E. The spiral structure in the disc interior to the kernel has a gravitational influence on the embryo's orbit tending to push the kernel outwards, whereas the spiral structure exterior to the disc exerts a gravitational influence that tends to push it inwards. For any plausible disc model the inwards push is the greater and so the net effect is inward migration. The rate of migration is proportional to the mass of the disc and also to the mass of the kernel, so as it grows it migrates inwards ever more rapidly. This is *Type I migration*, which is comparatively rapid. Note that the disc itself is also migrating inwards, but always more slowly than the kernel.

Type I migration continues until the kernel has grown to sufficient mass to open up a gap in the disc, as illustrated in Figure 11.6. The gap causes a major change in the migration. It now slows dramatically, by 10–100 times, until the kernel and the disc are migrating inwards at the same rate. This is *Type II migration*. The kernel mass at which the transition takes place depends on various properties of the disc (its density, thickness, viscosity, temperature, and so on), and on the distance of the kernel from the star. An approximate range is 10–100 m_E, and so it is likely that there is a fully-fledged giant kernel plus some captured gas at this transition. Gap formation reduces the rate at which the kernel acquires mass from the disc, but does not halt it, and so planets up to several m_J can form. If giant planets form in

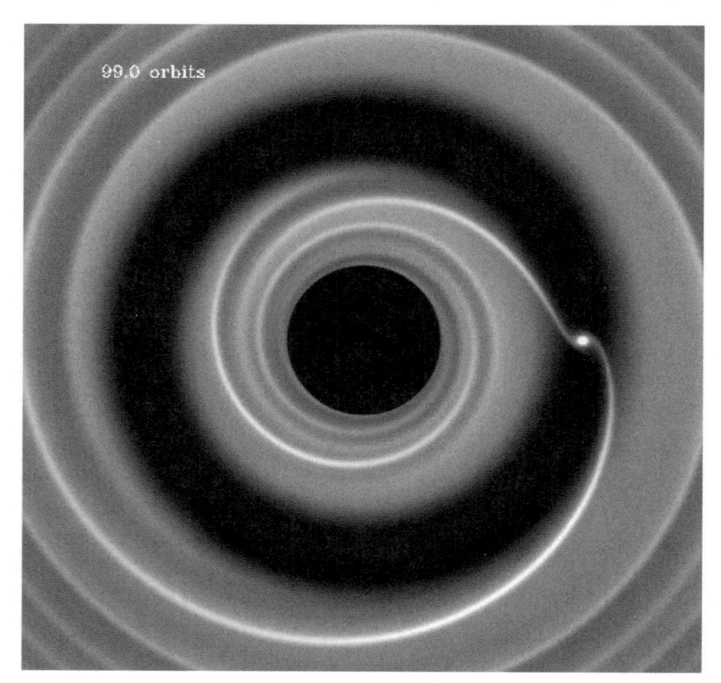

Figure 11.6 A giant kernel opens up a gap in its circumstellar disc. The 'black hole' at the centre is an artefact.

Frederic Masset.

one stage, there is no kernel build-up, and so it is likely that there will be no Type I migration – the giant planet enters the story at the point of a fully-fledged Type II migration.

Migration needs to be halted, otherwise all the giant planets in every system will end up in the star. Either the disc must be removed, or there must be counter-effects. The disc will be removed partly through gradual infall to the star and partly through bursts of activity that young stars are observed to undergo, when outflowing stellar winds and intense UV radiation somehow dissipate the disc (T-Tauri phase). Observations of young stars suggest that the disc lasts 1–10 Ma. This will be too long in some cases for the giant planet to survive, because the Type II migration time, depending on the disc properties and other parameters, can be less than this. There are however, several ways in which counter-effects can appear, including:

- tidal interactions between the planet and the star;
- the effect of mass loss from distended young giant planets;
- magnetic interactions between the star and the disc that halt the inward migration of the disc, and consequently Type II migration too, because in Type II migration the planet and disc migrate together; and
- evaporation by stellar radiation of a narrow zone in the dish-shaped disc at a few AU, thus creating a barrier.

These mechanisms (and others) are rather subtle and are peripheral to the story, so they are not detailed here, but note that they could save otherwise doomed planets.

So far we have considered a single giant planet in a disc, yet the Solar System and at least a few of the known exoplanetary systems have more than one giant, and it is presumed that most of the others do too. Computer models of discs with two giants have shown that interactions between the giants can slow and even reverse Type II migration, and this is one way in which we could have ended up with a Solar System in which Jupiter and Saturn are still beyond the ice line. It is also possible under special choices of circumstellar disc parameters to have limited migration without invoking giant–giant interactions.

The multiple-giant case can also explain giants in highly eccentric orbits. Figure 11.7 shows the eccentricities e versus semimajor axis a, with Jupiter and Saturn for comparison. Some of these eccentricities are very high, particularly at larger a, and the inset in Figure 11.7 shows one example (16 Cygni Bb). It is difficult to produce such large values by formation from the circumstellar disc, without something extra. One possibility arises when the orbital periods of the two giants are in a simple ratio, such as 1 : 2 or 1 : 3 – mean motion resonances (Section 4.3.3), perhaps as a result of migration. In such a case the gravitational interaction between the giant planets can disturb the orbits in a cumulative way, resulting in large orbital changes. An everyday analogue is the pushing of a child on a swing – if you time your pushes correctly a large amplitude builds up.

Another way to achieve large eccentricities is through close encounters between giants. Simulations show that a common outcome is for one giant to be flung into the

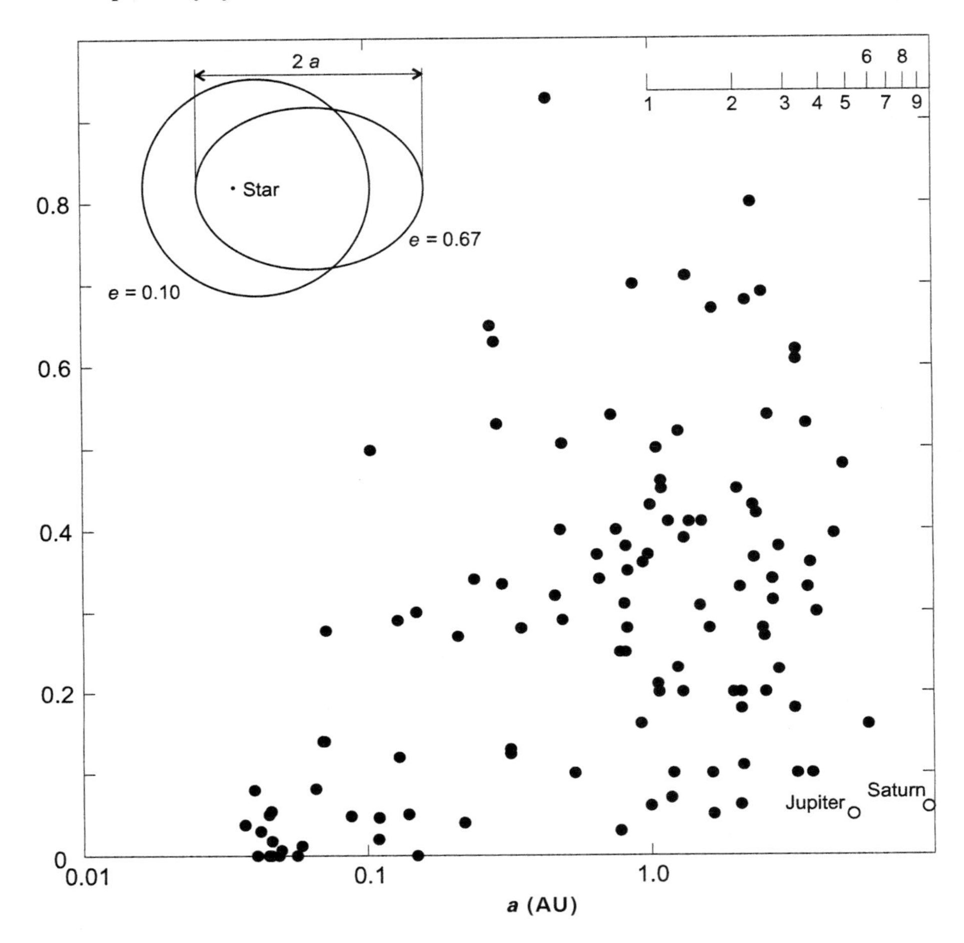

Figure 11.7 Eccentricities e versus semimajor axes a for the known exoplanets. Inset (top left) shows orbits with eccentricities of 0.1 and 0.67 (16 Cygni Bb) with the same semimajor axis a.

cold of interstellar space and for the other to be retained in a high e orbit. These scenarios are also the basis for an additional explanation of giants close to the star. If the surviving giant is in a high e orbit with a small periastron distance, then tidal interactions with the star, perhaps aided by residual disc gas, will reduce e and we can end up with a giant in a small, low e orbit. If the other giant avoided ejection it will be in a large, high e orbit.

Another migration mechanism is possible after the disc has cleared, and this is through the scattering of planetesimals by giant planets. In this way an inner giant planet can migrate inwards, and an outer giant outwards.

In conclusion, the system parameters specifying the star, disc, and the giant planets, are sufficient in number and sufficiently variable, that giants with a great range of masses and orbital semimajor axes can result. We certainly have plausible explanations of the observed exoplanetary systems. The giants in the Solar System emerge as just one of many possible types of outcome. However, plausibility does not mean that it actually happened that way. There is no guarantee that the exoplanetary systems came into existence in the ways described above, though they are the best current explanations that we have.

But what about Earth-mass planets? Could these form and survive, particularly in the HZs of the known exoplanetary systems? If not then the chances of finding potential habitats among the known exoplanetary systems are much reduced.

11.3.2 Giant planet migration and the formation of Earth-mass planets in HZs

A major problem with the speedy Type I migration is that it carries Earth-mass planets rapidly to the star. Only by careful choice of disc parameters is it possible for planets with a mass of about 1 m_E to migrate sufficiently slowly to outlast the disc and thus survive. This can seem rather contrived. Fortunately for our understanding, it has been discovered that the growth of Earth-mass planets interior to the ice line, which is where the HZ will lie, is likely to be rather slow, much slower than the growth of kernels beyond the ice line. Migration during the 1–10 Ma lifetime of the circumstellar disc of gas is then slight, and when the disc disperses it is quite possible that this inner region will contain many embryos with masses $\leq 0.1\, m_E$, plus a disc of planetesimals, from each of which further growth can occur.

Meanwhile, growth has been more rapid beyond the ice line. In the two-stage process this is because the abundance of condensable water provides a swarm of planetesimals with low relative speeds. In the one-stage process it is because the disc instabilities occur early on. Subsequently, the region interior to the ice line might or might not be traversed by a (growing) giant. If, as in the Solar System, the migration of a Jupiter-twin is, at most, slight, the terrestrial planet region will not be greatly disrupted by migration. In other systems, with more extensive migration, even if the giant(s) stops short of the HZ, embryos and planetesimals will be scattered as the giant moves inwards, sweeping orbital resonances across the HZ. If the HZ is traversed by a migrating giant then embryos will certainly be scattered. So, could Earth-mass planets form after migration has ceased, and could any such planets be present in the HZ today? Consider formation first.

It is rather loose to refer to *the* HZ – you have seen that the HZ migrates outwards during the main sequence lifetime (e.g., Figure 8.5). Of relevance to the search for life is whether an Earth-mass planet could have been in the HZ for the past billion or so years. This time span is based on the Earth, where it took about two billion years for life to have a detectable effect on the atmosphere (Section 3.2.3), including the 700 Ma of the heavy bombardment (Section 3.3.6). The question therefore is whether an Earth-mass planet could have formed at a distance from its star that has placed it in the HZ for at least the past billion or so years, and so given any life there the opportunity to become detectable by us. I shall refer to this range of distances as HZ(recent). It is narrower than the HZ at any particular instant. Remember that it is the HZ that moves, not necessarily the planet.

The few computer studies so far made have shown it to be possible for Earth-mass planets to form in the range of distances corresponding to HZ(recent) after giant migration is complete. In cases where the giant comes to reside beyond HZ(recent), sufficient embryos and planetesimals remain provided that the giant planet never comes close to the HZ(recent) outer boundary. Even in cases where the giant traverses HZ(recent) and becomes a hot Jupiter, well inside the interior boundary of HZ(recent), there could be sufficient planetesimals left, and additionally there could be sufficient dust in the disk to form a new generation of planetesimals that lead to embryos and to an Earth-mass planet. If the only systems ruled out are those with giant planets that came to rest in or close to HZ(recent), then perhaps half of the known exoplanetary systems could have had Earth-mass planets form in what is now HZ(recent). If the hot Jupiter systems are ruled out then the proportion falls towards 10%. It must be emphasised that these are crude estimates.

11.3.3 Earth-mass planets in HZs

If, in any system, an Earth-mass planet somehow formed in what is now HZ(recent), the next question is whether it could have been there for at least the past billion or so years. This is not the same question as that of formation, which takes less that 100 Ma, perhaps a lot less. By contrast, survival is against the gravitational buffeting by the giant(s) for the age of the star, which is typically many thousands of millions of years. Computer studies have been made of representative systems. On the basis of these it can be concluded that in perhaps a half of the known exoplanetary systems, Earth-mass planets, always provided that they could have formed, could have been in HZ(recent) for at least the past billion or so years. In some cases the planet could be anywhere in HZ(recent), and in other cases it could only be in variously restricted regions. If, in the studies, an Earth-mass planet has been removed from HZ(recent) its usual fate was to be flung into the cold of interstellar space, or have a collision with its star.

Figure 11.8 shows the sort of regions in various cases where Earth-mass planets could have long been present in HZ(recent). Figure 11.8(a) shows a giant planet in a low eccentricity orbit near its star, well interior to HZ(recent). Planets could be present anywhere within HZ(recent). Figure 11.8(b) shows a giant in a more eccentric orbit not far outside HZ(recent). Planets could be present only in the inner region of HZ(recent). Figure 11.8(c) shows a system like ours, where the giants, Jupiter and Saturn, are well beyond HZ(recent). In such systems the whole HZ, like ours, would be a safe harbour for Earth-mass planets. There are, as yet, no other known systems quite like ours, though they very probably await discovery (Section 11.4.2).

Systems like that in Figure 11.8(a) are the hot Jupiters, and there are many examples of these. A typical one is Rho Coronae Borealis, a main sequence star with a mass about 95% that of the Sun, and rather older. The inner boundary of HZ(recent) is beyond about 0.8 AU. The giant planet, minimum mass 1.04 m_J, is in an orbit with an eccentricity of 0.04 and a semimajor axis of 0.22 AU, placing the giant safely interior to the HZ throughout the star's lifetime. HD72659 typifies a

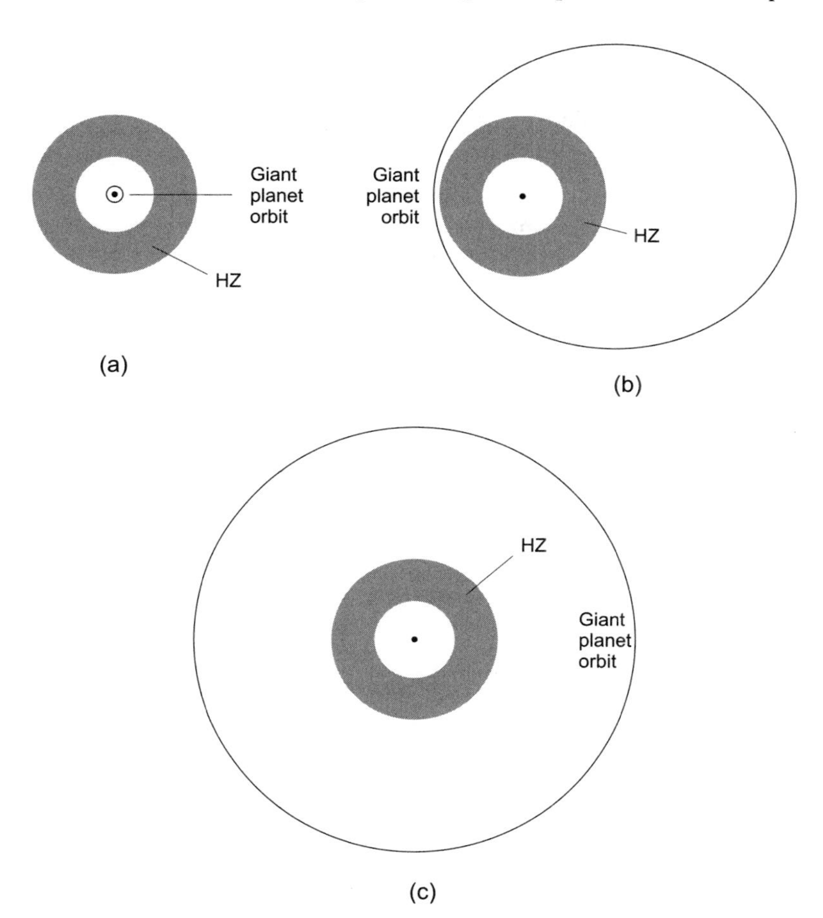

Figure 11.8 Survivable orbits for Earth-mass planets in a HZ(recent) (shaded) where: (a) the giant is very much interior to HZ(recent); (b) the giant is not far outside HZ(recent); and (c) the giant is well outside HZ(recent) as is the case with Jupiter in the Solar System.

main sequence star with a giant planet beyond the HZ. The star, also with a mass 95% that of the Sun, is somewhat younger, and HZ(recent) extends to about 1.5 AU. The planet has a minimum mass of 2.55 m_J and is in a modestly eccentric orbit with $e = 0.18$. With a semimajor axis of 3.24 AU the giant has been safely beyond the HZ throughout the star's lifetime, even if the giant were 1.5 times its minimum mass. The system resembles Figure 11.8(c) but with the giant planet in a more eccentric orbit and somewhat closer to HZ(recent).

The other place where Earth-mass planetary bodies could be in a safe harbour is as massive satellites of giant planets that orbit within HZ(recent). One example is HD23079, a main sequence star 10% more massive than the Sun, and about 3000 Ma old. Its giant planet, minimum mass 2.61 m_J, is in a low eccentricity orbit with $a = 1.65$ AU, which puts it in the middle of HZ(recent). If such satellites are common, then this would increase considerably the number of systems that could have detectable life.

11.4 THE UNDISCOVERED EXOPLANETS

11.4.1 The known exoplanetary systems – a summary

At the time of writing (September 2003), Doppler spectroscopy has made all but one of the 117 confirmed discoveries of non-pulsar planets, in 102 systems, 13 of which are known to be multiple-planet systems (see Schneider website in Resources). The one exception so far, OGLE-TR-56b, was discovered in transit, and later confirmed by Doppler spectroscopy. Subsequent to their discovery, HD209458b has been observed in transit, and Gliese 876b has been detected astrometrically.

The properties of these exoplanetary systems can be summarised as follows:

- The furthest planet discovered by Doppler spectroscopy is around the star HD47536, at 401 ly. OGLE-TR-56 is at about 5000 ly.
- Roughly 10% of nearby solar-type stars have giant planets within 4 AU of the star.
- Stars with high metallicity might be favoured for having planetary systems.
- The planets range in minimum mass ($m \sin(i_0)$) from 0.12 m_J (a bit less than half the mass of Saturn) to the brown dwarf limit, 13 m_J, with a preponderance at the lower end of the mass range.
- The planets are thought to be rich in hydrogen and helium, though only for HD409258b and OGLE-TR-56b is there direct observational evidence for this, from the mean densities.
- About one-third of the planets orbit closer to the star than Mercury does to the Sun, and of the remainder only three planets (perhaps four) are beyond 4 AU, in different systems.
- Three-quarters of the planets have orbital eccentricities greater than 0.1.
- Seven stars with planets are in binary stellar systems, the planet(s) orbiting just one of the two stars.

Gravitational-lensing surveys have shown that further afield, fewer than about one-third of the stars towards the nuclear bulge of the Galaxy have Jupiter-mass planets at projected distances in the range 1.5–4 AU from the star, and fewer than about one-half have rather more massive planets in the range 1–7 AU. Note that the typical star observed is an M dwarf. Transit surveys have failed to detect (Jupiter-size) planets in some star clusters in spite of expectations based on extrapolation from the proportions in the solar neighbourhood. The reasons for this are uncertain.

11.4.2 What planets await discovery and when might we discover them?

Exoplanets continue to be discovered, and it is clear that we have not discovered all of them, even those within a few hundred light years of the Sun. We have not even acquired a representative sample. In particular, *observational selection effects* are playing a major role. Consider those associated with the various techniques, their consequences, and future prospects.

Doppler spectroscopy

The selection effects and several of their consequences have been introduced earlier, in Sections 10.1.4, 10.1.5, and 11.2.1. They can be summarised as follows:

- Though the Doppler signal v_{rA} (the radial velocity amplitude) is independent of distance to the star, nearby stars provide greater radiation flux at our telescopes, thereby making it easier to measure the Doppler shifts in their spectral lines.
- The larger the ratio m/M of the mass of the planet to the mass of the star, and the greater the value of the orbital inclination i_0, the greater the Doppler signal ($v_{rA} \propto m\sin(i_0)/M^{2/3}$).
- The smaller the orbital period P of the planet the greater the Doppler signal, and the shorter the span of time required to detect the orbital motion ($v_{rA} \propto P^{-1/3}$).
- Solar-type stars have sharp spectral lines, good surface stability, are fairly luminous, and are not rare, so there are plenty of bright examples.

Consequently, searches with Doppler spectroscopy have concentrated on nearby solar-type stars. So far, most of those within a few hundred light years of the Sun have been observed by Doppler spectroscopy long enough to reveal any planets of the order of Jupiter's mass in orbits within about 4 AU of the star. To the proportion of about 10% known to have such planets, only a further small proportion might await discovery, undetected so far because the orbit is nearly face-on to us, thus reducing the Doppler signal. A much larger proportion might have giant planets beyond 4 AU, where the Doppler signal v_{rA} is weaker and the orbital period P is longer. These could be detected with current technology, and will surely be found soon, as observational time spans increase. Moreover, for solar-type stars, Doppler spectroscopy can reach well beyond a few hundred light years, as shown by the post-discovery detection of OGLE-TR-56b at 5000 ly. Therefore huge numbers of giant planets might be revealed by Doppler spectroscopy in the next decade. Unfortunately, Earth-mass planets will probably remain undetectable by this technique around solar-type stars.

A small proportion of stars less like the Sun have been observed, particularly the abundant M dwarfs. Many planets might be discovered as these stars are more comprehensively surveyed, though not to any great range, because of the M dwarfs' low luminosity. On the other hand Doppler spectroscopy could reveal Earth-mass planets close to low-mass M dwarfs, where the HZ also happens to lie. We get the orbit of the planet, but only its minimum mass unless it is observed in some other way.

Astrometry

For astrometry, the selection effects and some consequences were introduced in Sections 10.1.2 and 10.1.3, and are summarised as follows:

- The astrometric signal β (the angular size of the star's orbit), decreases as the distance d to the star decreases ($\beta \propto 1/d$).
- The larger the ratio m/M the greater the astrometric signal ($\beta \propto m/M$).
- The larger the semimajor axis a of the planet's orbit (and so the longer its period P) the greater the astrometric signal ($\beta \propto a$).

As yet this technique has only detected one exoplanet – Gl876b by the HST, some years after its discovery by Doppler spectroscopy. This detection was facilitated by the low mass of Gl876, about one-third the mass of the Sun, and the proximity of the star to us, 15.4 ly.

This technique is ideal for detecting planets in large orbits and will surely do so when it acquires the required sensitivity, and as soon as observations have been made for long enough. Developments planned at the VLT could by 2005 give it the capability to detect Jupiter-twins to about 800 ly, and Earth-mass planets a few AU from nearby, low-mass M dwarfs. The Keck telescopes might acquire a comparable capability.

The ESA space telescope Gaia, a sky survey planned for around 2010, will have Jupiter-twins within its capabilities out to about 350 light years, and planets of a few Earth masses 1 AU from M dwarfs out to a few tens of light years. NASA's SIM (also planned for around 2010) will have a better capability through being a targeting mission rather than a sky survey. We get the mass of the planet and details of its orbit.

Transit photometry

The transit technique (Section 10.2) can only detect planets for which the orbital inclination i_0 is within a few degrees of 90° so that the orbit is nearly edge-on to us. So far, just one planet (OGLE-TR-56b) has been discovered in this way. On the other hand it is possible to detect Earth-sized planets around M dwarfs by this technique right now. Earth-twins have to await the space missions COROT (2004), Kepler (2006), and perhaps Eddington. The latter two could detect Earth-twins even at many thousands of light years. We get the size of the planet and its orbit, but not the mass of the planet.

Gravitational microlensing

Gravitational microlensing (Section 10.3) can already detect Earth-mass planets, and can reach out to tens of thousands of light years. We get the mass of the planet if we can see the lensing star, but we learn little about the planet's orbit.

Direct detection

Direct imaging was described in Sections 9.4 and 9.5. The VLT in a few years should be able to see Jupiter-twins to about 15 ly, and the James Head Space Telescope (2010) should have this capability to about 25 ly. Extremely large ground-based telescopes, such as a 100 m OWL (2015?), would be capable of imaging Jupiter-twins to hundreds of light years, and perhaps Earth-twins to a few tens of light years. Darwin and TPF (2015) would image Earth-twins to several tens of light years, and be able to investigate them for life.

Earth-mass planets

Though we have an increasing capability to detect Earth-mass planets, we will, of course, only detect them if they are there. Recall that theoretical studies have shown the following:

- Migration of giant planets makes it less likely that Earth-mass planets could form in what is now HZ(recent) of the star, though 10–50% of the known exoplanetary systems could have such planets.
- Perhaps in half of the known exoplanetary systems, Earth-mass planets, *provided that they could have formed in these systems*, can have been in the HZ for at least the past billion years or so (i.e. in HZ(recent)). This is long enough for any life to have had an effect we could observe, if the Earth is any guide.

We can thus expect to find Earth-mass planets in HZ(recent) in some exoplanetary systems. We can also expect to find such planets outside HZ(recent), such as analogues of Venus in the Solar System, or beyond HZ(recent), where the Solar System has no Earth-mass planets.

To estimate our chances of finding evidence of life beyond the Solar System we need a figure for the proportion of stars in our region of the Galaxy that have Earth-mass planets in HZ(recent). Unfortunately, our census of exoplanetary systems is so incomplete that we have little idea. The theoretical studies of the known systems indicate that the proportion could be high. It could be even higher, given that observational selection effects discriminate against Earth-mass planets and against systems like ours where the giant planets are well beyond the HZ. But we could be disappointed.

To summarise, within a decade we expect the proportion of stars known to have planets to rise considerably. Among these new discoveries there should be lower mass planets, down to Earth-mass, and planets of all masses in larger orbits. By about 2015 we will have space telescopes such as Darwin and TPF capable of seeing the planets themselves, even Earth-mass planets. Huge ground-based telescopes such as OWL might also have this capability in favourable cases.

At present the Solar System looks unique with its giants in large, low-eccentricity orbits, and its whole HZ consequently a particularly secure abode for the formation and survival of Earth-mass planets. We expect systems resembling the Solar System to be discovered, and perhaps even to be common – they would not have had to face the problems of giant planet migration and the gravitational buffeting of close giants. By 2015 we should know how common they really are.

11.4.3 A note on evidence from circumstellar discs

Among the planets awaiting discovery, it is expected that some more will be around those stars known to be surrounded by gas and dust. Many young stars are known to have discs of gas plus some dust, and a few older stars are known to have discs or rings of dust, the gas having been dissipated by stellar activity, notably during the T Tauri phase. As noted earlier, Epsilon Eridani is known to have a dust ring *and* (probably) a planet. The planet is interior to the ring and might be too far from it to cause gravitational distortions in the ring. The ring however is non-uniform, and the cause might be planets near its inner boundary. Distortion is also present in the dust disc around Beta Pictoris (Figure 11.9). The disc is presented to us nearly edge-on, and it is clearly warped. This warp could well be due to the gravitational influence of a giant planet. These are just two examples. Moreover, the very existence of dusty discs or rings around older main sequence stars calls for a mechanism to replenish the dust that otherwise would have been long gone. A likely mechanism is collisions

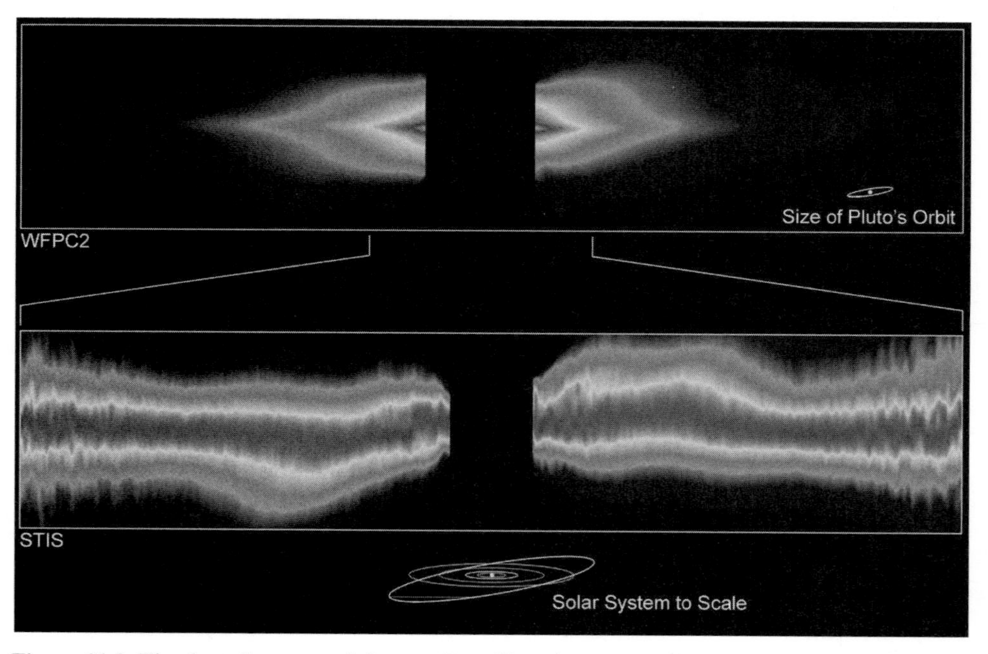

Figure 11.9 The dust disc around the star Beta Pictoris, presented nearly edge-on. The density decreases from the mid-plane. The gap in the middle is part of the imaging system, and is about 30 AU across in the lower image.

Sally Heap (GSFC/NASA).

between asteroids, perhaps comets too in an analogue of the Edgeworth–Kuiper belt in the Solar System (Section 1.1.1). If there are such bodies, there may be planets too.

11.5 STARS, PLANETS, AND LIFE FORMS

We end this Chapter with a few speculative remarks on what extraterrestrial life might be like, which will lead us into Chapter 12, where we discuss how we might discover life on exoplanets. The biochemical basis of the local version of life is not of central concern here – this will be discussed briefly in the next chapter. Of more concern is the effect on the biosphere of the type of star and the type of planet. We apply the principle that we expect life to adapt itself to its environment, including its external form and its sense organs.

For example, eyes are likely to be developed by mobile creatures, and these will use the available light. M dwarfs have a red spectrum peaking in the near-infrared (Figure 8.7), so eyes on a planet of an M dwarf are likely to be sensitive in the near-infrared, and insensitive at green and blue wavelengths. If there is photosynthesis then it will use the longer wavelengths characteristic of M dwarfs. A hot star will invoke corresponding adaptations, and also, for surface life, adaptation to high levels of UV radiation, perhaps through a thick protective skin. If otherwise habitable planets form only at the extremes of the HZ then low or high temperature adaptations will also occur.

A planet's gravity will also have an effect. In the high gravity of a massive rocky planet squat life forms might be common, or life might be confined to oceans and other bodies of liquid water that provide support. 'Birds' might be unknown and 'plants' could hug the surface. There would also be an advantage in small life forms, to reduce differential gravitational forces. In low gravity we expect soaring structures, perhaps very large, and air thick with flying creatures. Low cliffs will not threaten the survival of 'animals' as they would in a high-gravity world.

These are just a few examples of how the basic properties of the star and planet could mould a biosphere. We now turn to how we could detect biospheres, regardless of the form the inhabitants might take.

11.6 SUMMARY

- As of September 2003, Doppler spectroscopy has made all but one of the 117 confirmed discoveries of non-pulsar planets, in 102 systems, 13 of which are known to be multiple-planet systems. The one exception so far, OGLE-TR-56b, was discovered in transit, and later confirmed by Doppler spectroscopy. See the Schneider website in Resources for details.
- All the known exoplanets are giants. About one-third are in orbits closer to the star than Mercury is to the Sun, and of the remainder only three planets (perhaps four) are beyond 4 AU, in different systems. High orbital eccentricities are common, particularly in the larger orbits, with values exceeding 0.6.

- All the giant planets are thought to be rich in hydrogen and helium, like Jupiter, rather than silicate–iron like the Earth. They probably formed beyond the ice line, in which case those now inside the ice line must have got there through migration. Migration through gravitational interaction with the circumstellar disc has a plausible theoretical foundation, and can also explain high orbital eccentricity.
- About 10% of the nearby solar-type stars have giant planets within 4 AU of the star.
- Observational selection effects have so far disfavoured the detection of giant planets in orbits larger than a few AU, and the discovery of Earth-mass planets.
- It is plausible that Earth-mass planets have been present for at least the past billion years or so in HZ(recent) of about half of the known exoplanetary systems. In the systems yet to be discovered there could be a significant proportion that resemble the Solar System, with the giant planets well beyond the HZ and Earth-mass planets secure within the HZ. By 2015 we should know, thanks to a variety of ground-based and space-based telescopes.
- The nature of a biosphere is expected to be influenced by the type of star and the type of planet.
- We will not have the capability to investigate exoplanets to see if they are inhabited until the Darwin or TPF missions are launched in about 2015.

11.7 QUESTIONS

Answers are given at the back of the book.

Question 11.1

Suppose that a planet has $m\sin(i_0) = 1.9\,m_J$. It is observed in transit, which gives a radius $1.2\,R_J$. Show how it can be concluded that this is probably a hydrogen–helium giant, and that it has probably not spent much of its life close to its star.

Question 11.2

A few of the points in Figure 11.4 are labelled with the name of the exoplanet. A system that is less unlike the Solar System than most of the others is 47 UMa (Ursae Majoris), which has two planets, 47 UMa b and 47 UMa c.

(i) Draw the orbits of its giants with the orbits of Jupiter and the Earth to the same scale. You can approximate all the orbits as circles with the star at the centre. Label each giant orbit with the minimum mass of its giant in units of m_J.

(ii) Add the present-day HZ boundaries for 47 UMa. It is similar to the Sun in mass, though 47 UMa is a bit older, 7000 Ma, and thus its HZ is further out – calculations show that it extends from about 1.05 AU to about 2.1 AU.

Figure 12.1 A large satellite around a giant in the HZ.
Julian Baum (Take 27 Ltd).

when the various new ground-based and space telescopes described in Chapter 9 become available. It will also be possible even further into the future when we send probes to the stars. But whether our instruments are on a probe approaching a planet, or orbiting the Sun or Earth, or even at the Earth's surface, we need to ask – how could remote observations like these find life out there?

12.2 DETECTING BIOSPHERES FROM A DISTANCE

12.2.1 Is there life on Earth?

In 1989 Galileo Orbiter was launched by NASA (Figure 12.2). Its primary mission was to study Jupiter and its satellites, and in Section 6.2 you saw how it has boosted the idea that there might be an ocean and hence life on Europa. It reached Jupiter in December 1995, but before that, in December 1990 and again in December 1992, it came within about 6×10^6 km of the Earth. This was in order to gain kinetic energy through gravitational interaction with the Earth, thus enabling

Figure 12.2 The Galileo spacecraft, en route to Jupiter.
NASA/Caltech.

Galileo's modest rockets to raise the massive payload to faraway Jupiter. Advantage was taken of close encounters to see whether life could be detected on the Earth and the Moon. In 1990 the Earth was the object of study, and in 1992 the Moon. Even though the instruments on board were not designed specifically to detect life, it was hoped that the outcome would help astronomers design ways of detecting life elsewhere.

The answer that Galileo Orbiter gave to the question: 'Is there life on Earth?' was a resounding 'Yes!'. The conclusion itself came as no surprise, but it was encouraging that Galileo's instruments could reach it. There were three instruments that provided the evidence.

First, there was the infrared spectrometer called NIMS. This enabled astronomers to identify atmospheric substances through the imprint they made on the spectrum of infrared radiation emitted by the Earth. The substances detected included O_3 and CH_4. Recall from Section 3.2.3 that one of the major effects of the Earth's biosphere is that it sustains molecular oxygen (O_2) as a major component of the atmosphere. O_2 has only a weak spectral signature in the infrared, but through the action of solar UV radiation, O_2 gives rise to an appreciable trace of O_3, and this has such a strong spectral signature in the infrared that it is readily detected (Section 12.2.2). It is difficult to envisage any process other than photosynthesis that could generate sufficient O_2 to yield the amount of O_3 seen. But we can't be quite sure of this. The clincher is CH_4. It is very readily oxidised by O_2 to give CO_2 and H_2O, and as a result it only accounts for about 1 molecule in every 600 000 in the Earth's troposphere. This however was sufficient to give NIMS a clear if small infrared signature. The crucial point is that without a huge rate of release of CH_4 into the oxygen-rich atmosphere the quantity of atmospheric CH_4 would be far less, and it would have been undetectable by NIMS. The Earth has insufficient reservoirs to supply the CH_4, and volcanic emissions are also insufficient. The huge rate of release comes from the biosphere, where CH_4 is generated by large organisms and by certain bacteria. The presence of O_2 and CH_4 together, far from chemical equilibrium, puts the existence of a biosphere on Earth beyond reasonable doubt.

Second, Galileo measured the amount of solar radiation that the Earth reflected at various wavelengths – a reflection spectrum. Around 0.8 μm, in the near-infrared, a sharp rise in reflectance was detected, particularly over the continents. The 'red-edge' as it is called, is due to green vegetation, and is associated with the photosynthesising molecule chlorophyll and with biological structures that reject radiation not utilised by chlorophyll. However, though we know how to interpret this spectral feature on Earth, it is not at all certain that photosynthesis in an alien biosphere would look the same. Perhaps the best that can be hoped for is to see widespread absorption that could not be readily accounted for by common minerals. Overall, this is a less certain indicator of life than pairs of atmospheric gases way out of chemical equilibrium, such as O_2 and CH_4.

Third, the radio receiver on Galileo detected strong radiation confined to a set of very narrow wavelength ranges. Moreover, the radiation at each of these wavelengths was not constant but was modulated in an intricate way that could not be explained by natural processes. These were the various terrestrial radio and television transmissions. The modulation was the information that carried the programme content – a steady wave carries no soap operas. This shows that the Earth not only has a biosphere, but that it has evolved in a particular, possibly very rare manner, with the appearance of technological civilisation. Unambiguous images of cities and other artefacts were not obtained by the relatively small Galileo cameras at the rather large distances of this spacecraft. Figure 12.3(a) shows the modest resolution achieved. Figure 12.3(b) shows the far superior resolution achieved by a satellite orbiting close to the Earth, but even in this case cities and artefacts cannot be seen (though distinct areas of vegetation could be seen in colour).

In December 1992 Galileo's instruments were turned to the Moon, with entirely negative results! There were no pairs of atmospheric gases way out of chemical equilibrium, indeed, hardly any gases at all, no characteristic red reflection, no radio or TV broadcasts. If there is life on the Moon it must be deep in the crust, and this is extremely unlikely given the scarcity of water.

We will now look at infrared and visible–near-infrared spectra in more detail, deferring to Chapter 13 the detection of radio transmissions. Spectra are of huge importance because (like radio transmissions) they could reveal a biosphere from a great distance, and could be used to investigate Earth-mass exoplanets with the sort of instruments we will have in the near future. Consider first the infrared spectrum.

(a)

(b)

Figure 12.3 (a) The Galileo spacecraft views the Earth and Moon, 16 December, 1992, from a range of 6.2×10^6 km. (b) NASA's Terra satellite views the eastern Mediterranean in September 2001. The area covered is a few hundred kilometres across.

(a) and (b) NASA/JPL.

12.2.2 The infrared spectrum of the Earth

Figure 12.4 shows the infrared spectrum of the Earth, as seen from space. This is not the spectrum obtained by the Galileo spacecraft but a more detailed one obtained by the Nimbus-4 satellite in the 1970s. This particular spectrum was acquired in daytime above the western Pacific Ocean, and has been chosen because it resembles the sort of spectrum that would be obtained from a cloud-free Earth from a great distance, when the light from the whole planet would enter the spectrometer. Note that the intervals on the wavelength scale get more crowded at longer values. By contrast, in the frequency scale at the top of the frame the intervals are equally spaced. Frequency f and wavelength λ are related via $\lambda = c/f$, where c is the wave speed. Therefore, if the frequency intervals are equally spaced, the wavelength intervals cannot be. It is common in the infrared to use a scale in which the frequency intervals are equally spaced, though we shall refer to the wavelength scale alone. The vertical scale is proportional to the power emitted.

There are several smooth curves, each labelled with a temperature. These correspond to emission from a black body at temperatures equal to those shown (Section 1.1.2). There is also a jagged curve displaying much detail. This is the power emitted by the Earth. It is the detail in this curve that provides the evidence that the Earth is inhabited.

Surface temperature

Between 8 μm and 12 μm, except for the dip around 9.6 μm, the terrestrial spectrum follows closely the black body curve labelled 27°C (300 K). The absence of strong spectral lines, such as would be generated by gases, indicates that the radiation received by Nimbus-4 in this wavelength range has been emitted by the Earth's surface or by clouds. It has in fact been emitted by the surface, the area being cloud-free. The surface temperature is thus close to 27°C, perhaps a little less. At this temperature, if the atmospheric pressure is sufficiently high, water can exist as a liquid (Figure 2.10). Water exists as a liquid from about 0°C to a higher temperature that depends on the atmospheric pressure. At the surface of the Earth this pressure (about 10^5 Pa) enables water to be liquid up to about 100°C. So, as the Earth has a rather substantial atmosphere, we can conclude that liquid water could exist at its surface. For an exoplanet we would not know whether clouds or the surface were responsible for the 8–12 μm spectrum. If the cloud cover were variable, or if there were spectral features that could be linked to cloud particles, then probably we could tell.

The other important inference from the surface temperature is that it is well within the range for complex carbon compounds to exist. Compounds such as proteins and DNA break up at temperatures above about 160°C, so most of the Earth's surface is safely cool. On the other hand it is not so cold that biochemical reaction rates, which decrease rapidly as temperature falls, are too low to sustain life.

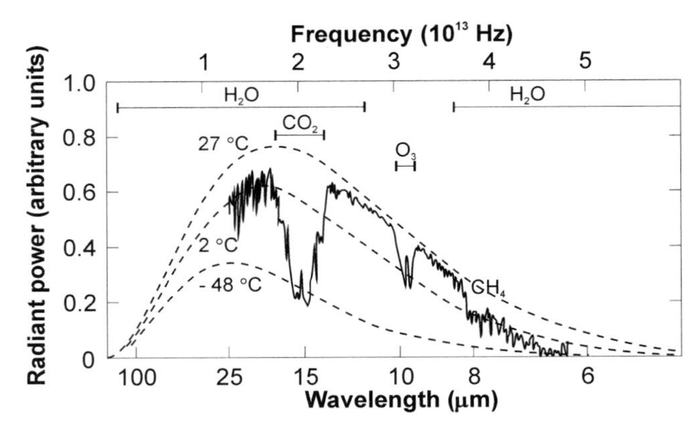

Figure 12.4 Earth's infrared spectrum, as obtained in daytime by the Nimbus-4 satellite over a cloud-free part of the western Pacific Ocean in the 1970s. The radiant power is per unit frequency interval.

Water

That water is *actually* present on Earth, at least as vapour in the atmosphere, is indicated by much of the fine structure in the spectrum in Figure 12.4. The H_2O molecule has a great many narrow absorption lines in the infrared spectrum, so many that they overlap and blend together to form absorption bands. It is these bands that are seen in abundance in Figure 12.4 rather than the individual lines. At the ends of the wavelength range in Figure 12.4 the smooth curve with a temperature of 2°C fits the spectrum fairly well. This temperature is too low for the surface of the western Pacific Ocean. At these wavelengths, the water vapour has absorbed all the radiation emitted from the surface, and has re-emitted it at the lower temperature of its own location.

We can thus infer that the atmospheric temperature at the general altitude of the water vapour is lower than at the surface. The actual altitude of 2°C cannot be obtained from Figure 12.4. To obtain it we need the variation of atmospheric temperature with altitude above the Earth, and this is shown in Figure 12.5, obtained from direct measurements and averaged over the Earth's surface. You can see that 2°C occurs at an altitude of only a few kilometres. Therefore, there must be enough water vapour below this altitude to hide the ground from space at these wavelengths. This is in accord with direct measurements that show water vapour to be heavily concentrated in the lower few kilometres of the Earth's atmosphere.

Carbon dioxide

Around 15 μm the Earth's spectrum in Figure 12.4 has a large dip corresponding to a deep and broad set of overlapping absorption bands due to CO_2. The heaviest absorption corresponds to a temperature of about −50°C, and Figure 12.5 shows that this value occurs in the upper troposphere. This is not because CO_2 is concen-

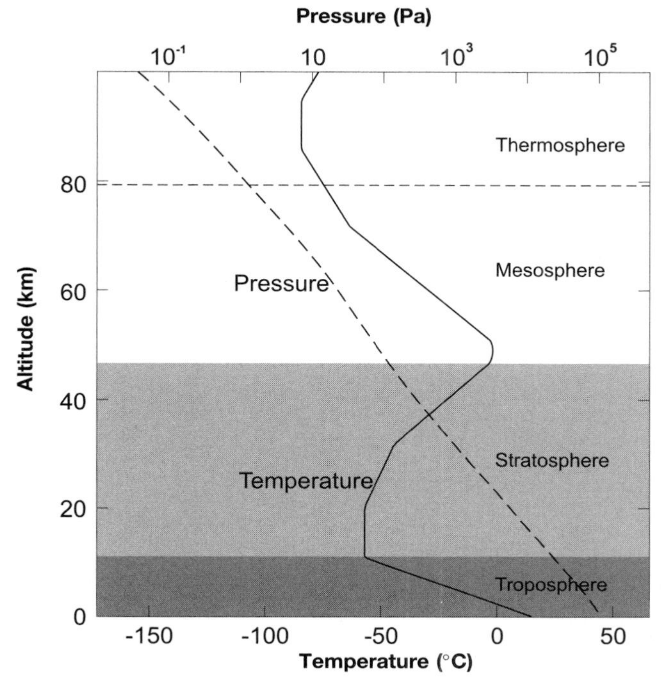

Figure 12.5 The variation of atmospheric temperature and pressure with altitude above the Earth's surface.

trated there – it is not – but because this is as deep as we can see at 15 μm into an atmosphere in which there is sufficient CO_2 at deeper levels to screen them from view. From Figure 12.4 we can conclude that CO_2 is present and therefore that the planet has carbon, which is essential for biomolecules. The actual amount is about 0.035% (as a proportion of all molecules).

Oxygen and methane

The dip around 9.6 μm in Figure 12.4 is due to O_3. This is derived from O_2 by the action of solar UV radiation, and the size of the O_3 feature in Figure 12.4 shows that O_2 must be present in considerable quantity in the Earth's atmosphere. O_2 is not directly detectable in the infrared because it absorbs infrared radiation only very weakly. This is because it consists of two identical atoms (i.e. it is a homonuclear diatomic molecule). In such molecules, roughly speaking, the distribution of electric charge is symmetrical around its centre. As a result, infrared radiation cannot exert much net force on the molecule and so the absorption of photons in the infrared is weak. The O_3 molecule has a different symmetry from O_2, and the net force is correspondingly stronger. It is also stronger in heteronuclear molecules, diatomic or otherwise, such as CO, CO_2, and H_2O, and so these also have strong infrared absorption.

The temperature around the centre of the O_3 absorption is about 0°C. Therefore, Figure 12.5 places the O_3 that is radiating directly to space either in the lower troposphere or in the upper stratosphere. In fact, O_3 is concentrated in the stratosphere so it is the upper stratosphere from where the infrared radiation in Figure 12.4 is coming. The solar UV radiation thus creates O_3 well above the Earth's surface, and it screens the lower atmosphere from the UV that has produced the O_3.

The other question that arises from the presence of O_3 is the source of the O_2 that gave it birth – is it a biosphere? An abiogenic origin is the photolysis of water (Section 3.2.3), in which a UV photon splits H_2O into OH and H. The hydrogen, being of low mass, is lost to space. Reactions of OH lead to the formation of O_2. Photolysis of water has always been generating some O_2 on Earth. However, the O_2 oxidises surface rocks and volcanic gases, and consequently the quantity of O_2 in the atmosphere from photolysis alone would be far less than it is. This indicates that most of the O_2 in the Earth's atmosphere is sustained biogenically, in particular through oxygenic photosynthesis, and though a significant O_3 absorption can result from a modest amount of O_2, the Earth's O_3 absorption does point to abundant O_2.

There are, however, conditions under which photolysis of water could give rise to O_2 in abundance. First, it could do so if the photolytic rate of generation from water were far higher. This will be the case in a few billion years when the luminosity of the Sun will have increased to the point where the Earth's upper atmosphere will be warm enough to hold a lot more water vapour than it does now. More water means more photolysis means more oxygen. However, this supply cannot last long. Photolysis destroys water, and in a few million years all the water will be lost, and the Earth will be dry. In a comparable time the O_2 produced by photolysis will be removed through oxidation of surface rocks and volcanic gases. Thus, unless the Earth were caught in the act of losing its water, photolysis could not account for the high O_2 abundance. Venus seems to have lost its water in this way. This would have happened early in its history because it is closer to the Sun than the Earth.

Second, photolysis of water could give rise to a lot of oxygen if it were being removed geologically at a very low rate. It would then build up slowly over hundreds of millions of years. In this regard, size matters. Large rocky planets like the Earth sustain considerable geological activity (with plate tectonics at its heart in the Earth's case). Small planets are unlikely to sustain such activity. Therefore, if a small planet is sufficiently near its star so that its atmosphere is damp, it could gradually build up a lot of O_2 from the photolysis of water, though in the long term much of this too could be lost to surface oxidation.

Thus, for the Earth, a combination of size and the unlikelihood of catching our planet at the moment it loses nearly all its water, would make it seem likely to an observer in space that oxygenic photosynthesis was at work. As discussed in Section 12.2.1, the detection of CH_4 and O_2 *together* in the Earth's atmosphere puts the existence of a biosphere beyond reasonable doubt!

What about planets beyond the Earth?

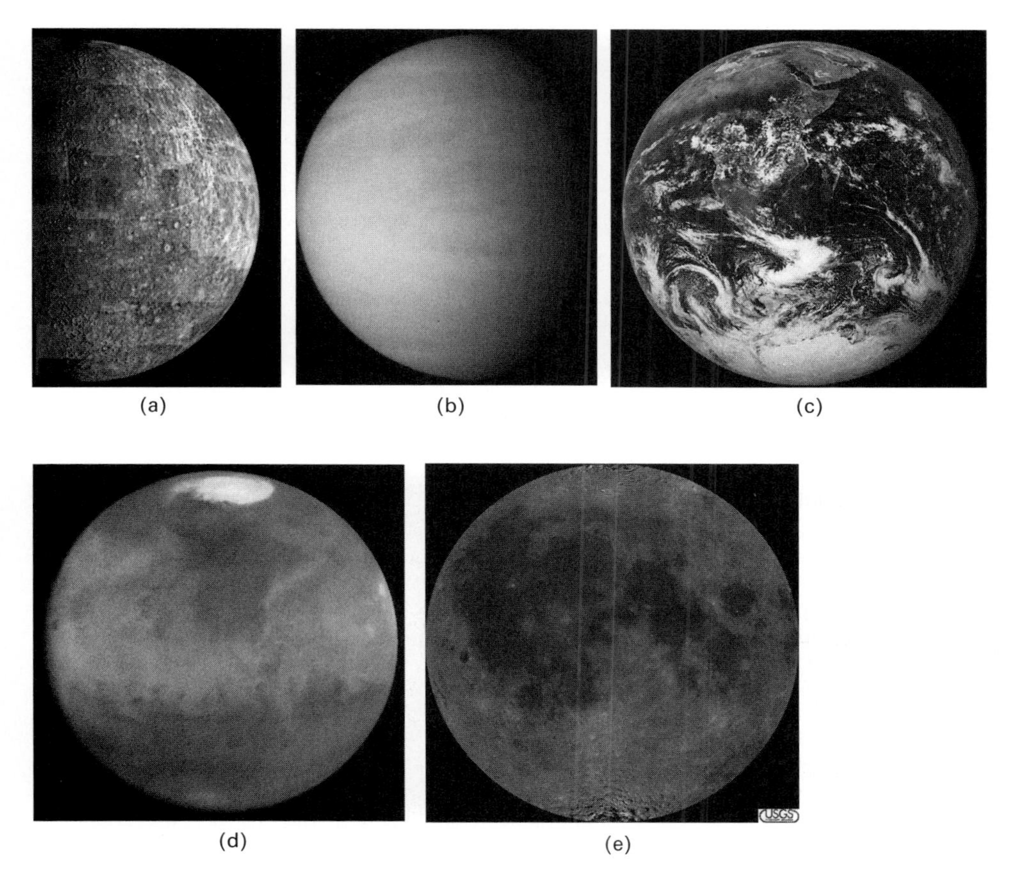

Figure 12.9 Visible-light images of (a) Mercury (b) Venus (c) the Earth (d) Mars (e) the Moon. (a)–(c) NASA/JPL. (d) HST/STScI. (e) USGS.

up to a factor of two. The variation would be even greater if, instead of the total light reflected, light in certain wavebands were isolated. Also, there are longer term terrestrial changes of up to 20%, due to variation in cloud cover, and these too are larger than for the other planets in the Solar System. Such variations would not prove the existence of a biosphere, but they would be indicative of habitability, and combined with other observations, such as the infrared spectrum, they could help build a convincing case.

As well as the reflection spectrum of the planet, we also have the absorption spectrum of its atmosphere at visible and near-infrared wavelengths. This is the outcome of the light reflected by the planet's surface being absorbed at certain wavelengths by the atmospheric constituents. Like the infrared spectrum, the visible–near-infrared absorption spectrum can reveal the atmospheric constituents. O_2 can now be detected directly, because it has a strong spectral signature, particularly at 0.76 μm, unlike the weak spectral signature that it has in the infrared where

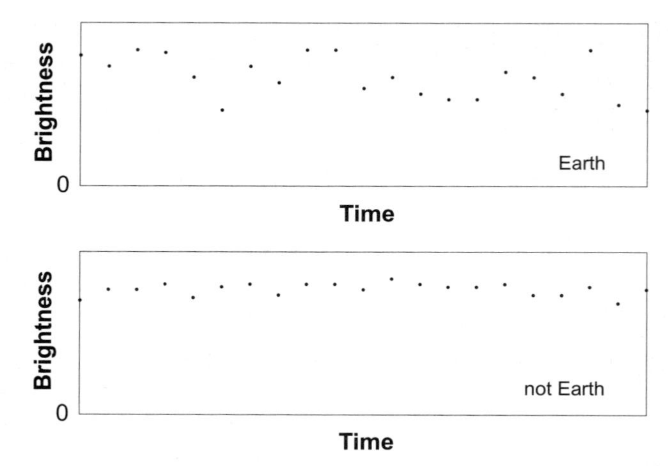

Figure 12.10 Notional light curves showing the brightness of the sunlit hemisphere of the Earth and one of the other terrestrial bodies in the Solar System.

we have to detect it by proxy through O_3. The feasibility of this approach has been demonstrated by examining the Earth's visible spectrum reflected off the dark side of the Moon (Earthshine), though to apply it to Earth-twins we will again need the OWL/Darwin/TPF generation.

As an aside, you might wonder why O_2 has a strong spectral signature at visible wavelengths, given that it is the symmetry of the molecule that prevents it having a strong infrared signature. This is because in the infrared it is the atoms of oxygen as a whole that are influenced by the radiation, and so the symmetry is sensed. At the much shorter visible wavelengths a single electron in one of the atoms is involved and the molecular symmetry is less apparent.

12.2.6 Interstellar probes

The obvious advantage with an interstellar probe (Figure 12.11) is that you can get much nearer the exoplanet. Once there, it would be much easier and quicker to gather spectral data than from a remote Earth-based vantage point, and it also makes it possible to gather spectral information that would be too weak to get from Earth. If the probe got really close, entered the atmosphere and landed on the surface, then direct sampling and close-up imaging would be possible, and any but the most deeply buried biospheres would be detected. So, as we have sent probes to the outer Solar System, why don't we send them to exoplanets? Consider sending a probe to the nearest solar-type star, Alpha Centauri A, at 4.40 ly. The Sun is 1.6×10^{-5} ly away, and so Alpha Centauri A is about 275 000 times further than the Sun, and that makes it difficult to reach. At an average speed of 10% of the speed of light it would take 44 years for a probe to reach Alpha Centauri A. This is rather a long journey. Moreover, the energy required to accelerate a probe to 10% of the speed of light is huge (Section 13.4.1), and would need a propulsion method other than chemical.

Figure 12.11 An artist's impression of an interstellar probe.
Artwork copyright 2003 by Don Dixon cosmographica.com.

What have we achieved with space probes so far? Since our first steps into space in the 1950s many probes have been launched, and though none of them has been aimed at the stars, there are four NASA spacecraft that will leave the Solar System, all of them having completed their missions to the outer planets. These are Pioneer 10, Pioneer 11, Voyager 1, and Voyager 2. Though Pioneer 10 was the first of these to be launched, in 1972, it is Voyager 1, launched in 1977, that is now furthest away. In 2003, it was about twice as far as Pluto, though this puts it only about one-thousandth of a light year away! It has been slowed by the Sun's gravity to a few kilometres per second, and so it will be tens of thousands of years before it gets among the nearer stars, and very much longer before it accidentally gets close to one. Nevertheless, the possibility that it will be found by aliens has not been ignored. Voyager 1 carries a grooved copper disc, bearing sounds and images of Earth. Whether any aliens could play and understand it is a matter of debate.

It is likely to be the end of this century before we find a feasible and affordable way to launch a probe that could achieve an average speed of 10% of the speed of light. Even if we could, there is still the 44-year travel time to Alpha Centauri A, plus the 4.4 years needed for information from the probe to reach us at the speed of light. There is also the problem of slowing the probe down so that it does not whiz past any planet in a few seconds. One possibility is to use radiation pressure from solar photons acting on huge sails to accelerate the craft out of the Solar System, perhaps supplemented by a laser beam, and then slow the craft down with the photons from its destination star. However, it is for a distant successor to this book to give interstellar probes more space. Fortunately, developments in spectroscopy from ground-based and space telescopes give us a real prospect of finding life beyond the Solar System in the next few decades.

12.3 SUMMARY

- We are particularly concerned with carbon–liquid water life that we could detect from afar, which probably restricts us to planets with habitable surfaces that have been in the HZ of their star for at least the past billion years or so.
- To determine if an exoplanet has a biosphere we could perform direct sampling and imaging at its surface – if we were able to land a probe on the planet. That possibility is a century or more off. In the foreseeable future we have to rely on observations from within the Solar System, and the first requirement is then that we can analyse the electromagnetic radiation that we receive from the planet.
- The infrared spectrum of an exoplanet can reveal the temperature at its surface if there are spectral windows to the surface, and within the atmosphere otherwise. It can also reveal the atmospheric composition. If evidence of water vapour and CO_2 is found, and if the surface temperatures are within the range for liquid water and complex carbon compounds, then the conditions would probably be suitable for carbon–liquid water surface life.
- That a biosphere was actually present would be indicated by a strong O_3 absorption feature, because this would point to considerable quantities of atmospheric oxygen, such as could be sustained readily by oxygenic photosynthesis. Photolysis of water as a source of abundant O_2 would be possible if the planet had long been geologically inactive, or if it were caught in a brief period of water loss.
- Much stronger atmospheric evidence of a biosphere would be provided by redox pairs far from chemical equilibrium with each other, such as O_2 (as indicated by O_3) and CH_4.
- Lack of abundant O_2 in the atmosphere would not necessarily imply the absence of a biosphere.
- The visible and near-infrared spectrum of an exoplanet might also reveal habitability or even a biosphere, by detecting appropriate gases, or the effects associated with chlorophyll or other biological substances, or by revealing a variable light curve.
- It will require telescopes like OWL, Darwin, and TPF to provide the spectra that would enable us to determine whether an exoplanet were inhabited.

12.4 QUESTIONS

Answers are given at the back of the book.

Question 12.1

Suppose that an otherwise Earth-like planet was entirely covered in cloud, and that its effect was to block all radiation to space from altitudes less than 10 km. Discuss briefly whether its infrared spectrum would reveal whether is was inhabited. You will find the pressure curve in Figure 12.5 useful for part of your answer.

Question 12.2

Explain how it can be deduced from Figure 12.6 that:

(i) the surface temperature at mid-latitudes in daytime on Mars is about 7°C;
(ii) there is a considerable proportion of CO_2 in the martian atmosphere; and
(iii) there is very little water vapour in the martian atmosphere.

Question 12.3

By referring to Figure 12.4, describe the changes that would take place in the infrared spectrum of an Earth-twin with carbon–liquid water life, from just before the biosphere has had much effect on the composition of the atmosphere, to when the biosphere is at its most active, to when the biosphere has retreated to a few areas near the poles. The retreat is caused by the increase in its star's luminosity as it ages.

Question 12.4

Discuss the possibility of detecting an aquatic biosphere on a Europa-like satellite of an exoplanet, through infrared and visible–near-infrared observations made from the Earth.

13

Extraterrestrial intelligence

There is one way in which we could discover extraterrestrial life now, with no more than the instruments we already have, and this is by life telling us that it is there. This could be by transmitting radio or laser signals that we receive, or by sending probes into the Solar System that we find, or in any other way that a technological intelligence could make its existence known. And indeed it is technological intelligence that we will have discovered. Regardless of the likelihood of such a discovery, it would be an astonishing finding. We would not only have learned that there is life in at least one other place in the Universe, not only that it had evolved into multicellular species, and not only that one of these species had evolved intelligence, but also that that intelligence was technological. We would have discovered what is usually called extraterrestrial intelligence (ETI). Whatever they looked like, these creatures would be thinking like us – they would be trying to understand the Universe and manipulate their environment in the same way that we do. What are the chances of such a discovery, and how might it be made?

13.1 THE NUMBER OF TECHNOLOGICAL INTELLIGENCES IN THE GALAXY

The Galaxy contains about 2×10^9 stars. There are billions of other galaxies in the observable Universe, of various types, some containing more stars, others fewer (Section 1.3.2). Overall, the observable Universe contains such a huge number of stars that it is beyond reasonable doubt that it contains at least a few other technological intelligences. However, discovering any beyond our own Galaxy is unlikely given the vast distances involved. It is therefore usual to focus the search for extraterrestrial intelligence (SETI, pronounced 'seh-tee') in our own Galaxy.

Consider the number N of planets in the Galaxy that are the sources of attempts at interstellar communication today. There is an equation for N well known to astronomers. This is the *Drake equation*, named after the US astronomer and SETI pioneer Frank Drake who presented it at a meeting in the USA in 1961. There are various forms of this equation. The one used here is:

$$N = R_* f_p n_E f_l f_i f_c L \qquad (13.1)$$

where R_* is the rate at which suitable stars are born in the Galaxy; f_p is fraction of these stars that have planets; n_E is the mean number of habitable planets per planetary system; f_l is the fraction of the habitable planets that actually have a biosphere; f_i is the fraction of biospheres in which species evolve that are intelligent in some sense; f_c is the fraction of intelligent species that become technological (technological species i.e. have the potential to be detected over interstellar distances); and L is the average time for which a planet harbouring technological species can be detected over interstellar distances through technological activity. It is implicit in the Drake equation that none of these factors has varied since the Galaxy was born. This is not the case, but in obtaining an order of magnitude estimate of N it will suffice.

To see that Equation (13.1) gives an estimate of N consider it in the form:

$$N = R_c L \qquad (13.2)$$

Here, R_c is the (assumed constant) rate at which planets harbouring technological species appear in the Galaxy – this is clear from the equation $R_c = R_* f_p n_E f_l f_i f_c$. Figure 13.1 shows such planets appearing at the rate R_c. In the time L the number that have appeared is $R_c L$. Thereafter, for each new one, another planet comes to the end of its detectable lifetime L, and so $R_c L$ is the steady state value of N.

An alternative meaning for N is obtained if we regard L as the average time for which a planet has technological species, rather than the time for which they are detectable. N is then the number of planets with technological species in the Galaxy today. If technological species are typically detectable for nearly the whole time for which a planet has such species, then the two values of N are similar.

Whichever of these two meanings we attach to N, in order to estimate its value we have to estimate the values of all the contributory factors. $R_* f_p n_E$ is the product of three astronomical factors. R_* includes suitable stars only, and thus we exclude the short-lived O, B, and A stars (Section 8.2.1). These are in any case rare and so we are left with about 99% of the stars in the Galaxy. Some astronomers exclude M dwarfs, which are very common, for the reasons outlined in Section 8.2.3. An estimate of

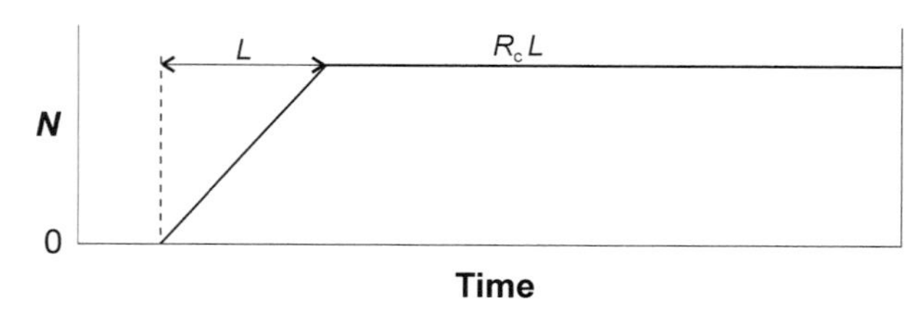

Figure 13.1 The build-up of the number of technological civilisations to a steady state, assuming a constant rate of appearance R_c, and that all have a lifetime L.

$R_* \sim 1$ per year is then arrived at. If we include M dwarfs then R_* is about five times larger. The value of f_p is not negligible, and might approach 0.5 (Section 11.4). In estimating n_E note that we must include any location on or in a planetary body from which ETI could make its existence known to us. Thus we must not only include planets or large satellites with habitable surfaces, but potential habitats outside the HZ, such as planetary interiors. We might estimate that, on average, there would be one location per planetary system that was inhabitable in this broad sense. In this case $R_* f_p n_E \leq 0.5$, if brown dwarfs are excluded, otherwise a few times larger, but given the uncertainties these values could easily be 10 or 100 times smaller. If that seems disconcertingly imprecise, the imprecision pales into insignificance alongside those in most of the remaining factors!

The next three factors, f_l, f_i, and f_c are biological. It is widely believed among astrobiologists that if a planet is habitable then it will probably have a biosphere, though our ignorance about the way life started on Earth makes some scientists dubious. Taking a slightly more conservative view let's set $f_l \sim 0.5$. Now we come to the really uncertain factors. There is very little agreement among biologists about the chance of an intelligent species evolving from a biosphere, and even less agreement about the chance of an intelligent species becoming technological. Therefore, estimates of the product $f_i f_c$ vary from close to zero to close to 1.

The final factor L is also highly uncertain. Consider its interpretation as a detectable time. We have been detectable through accidental leakages of radio, TV, and radar for a few decades. Deliberate attempts at serious interstellar communication have been confined to a handful of cases since 1974. Will we be detectable (through leakages or through deliberate attempts) for another 10^2, 10^4, 10^6 years, or far longer? We have no idea. Adopting the other interpretation of L, as the average time for which a planet harbours technological species, note that *Homo sapiens* has been around for about 100 000 years. But we have no idea for how much longer we will be here. Will it be 10^3, 10^6, 10^9, or many more years? Will there be other technological species to succeed us?

The uncertainty in L, coupled with those in f_i, and f_c, means that, for either meaning of L the corresponding value of N could be anything from 1 (which is us alone) to hundreds of millions. Therefore, the Drake equation is not useful for calculating N. Its use is that it displays the important factors and reminds us that the factors are multiplied. The safe conclusion is that SETI is, of essence, an experimental science – if we want to know we have to search.

13.2 SEARCHING FOR ETI

There are three ways in which we could detect ETI. We could:

- detect radiation that travels from them to us;
- detect their spacecraft or other artefacts in the Solar System (even the aliens themselves!); and/or

- detect features of their planet, or of their environment in general, that reveal technological modification.

In considering each of these we shall often use our own technological capability as a point of reference. We have only recently become detectable across interstellar distances, and our ability to communicate with the stars is slight. It is likely that another intelligence will be older and therefore more advanced than us, so if a means of communication looks feasible to us it is likely to be at least feasible to them, and very probably something they could implement with great facility.

Though there are other types of radiation (such as gravitational waves), electromagnetic radiation seems best for interstellar communication. This is because:

- it is relatively easy to generate, to beam at a selected target, and to detect;
- there are wavelength ranges that are only weakly absorbed or deflected by the interstellar medium, and only weakly absorbed by planetary atmospheres; and
- information can be imprinted on the radiation with low energy cost – the photon energy hf can be low.

The one problem with electromagnetic radiation is that it takes a long time for messages to travel over interstellar distances. It travels at the speed of light c, and though by terrestrial standards this is an enormous $300\,000\,\text{km s}^{-1}$, it still takes 4.2 years to travel the distance to the nearest star. The current laws of physics do not rule out faster-than-light travel (superluminal travel), so could we use speedier entities?

One possibility is particles called tachyons. These could only travel faster than light. Just as ordinary matter cannot be accelerated to speeds as great as c, tachyons cannot be slowed down to c. Unfortunately they remain only a theoretical possibility, as yet unobserved.

Another possibility relies on the phenomenon of quantum entanglement. In quantum theory it seems possible that two particles can be inextricably linked, so that if something happens to one, then something happens to the other instantaneously, no matter how far away it is. If this could be utilised then there could be instantaneous communication, though for the moment it is just another theoretical suggestion. For practical purposes, electromagnetic waves offer the fastest means of communication that we know of at present.

Figure 13.2 shows the electromagnetic spectrum. It covers a huge range of wavelengths (or frequencies), and so the question arises which parts of this spectrum are best for interstellar communication. In the late 1950s, which mark the start of the modern era of SETI, the answer was 'microwaves' and 'optical', and this remains the answer today, though for the first few decades it was microwave searches that dominated SETI. It was only in the mid-1990s that searches began to proliferate in the optical part of the spectrum. Consider first, searches at microwavelengths, then those at optical wavelengths.

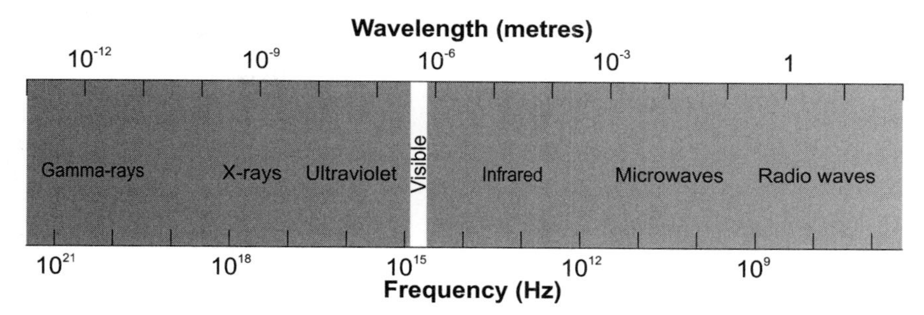

Figure 13.2 The electromagnetic spectrum.

13.3 MICROWAVE AND OPTICAL SEARCHES

13.3.1 Microwave searches

Figure 13.3 illustrates some of the reasons why microwaves have been favoured. Several distinct curves are shown. One curve shows the power (in arbitrary units) in the radiation that we typically receive from the Galaxy. This is due to what is called synchrotron radiation from electrons as they move through the magnetic fields that pervade our Galaxy. It declines steeply as frequency rises, so at frequencies greater than about 10^3 MHz a larger source is the microwave radiation left over from the Big Bang. At even higher frequencies, absorption in the Earth's atmosphere raises the power threshold for the detection of extraterrestrial sources, as illustrated by a third curve in Figure 13.3. The solid curve shows these three effects combined. A favourable region emerges in the lower frequency part of the microwave range.

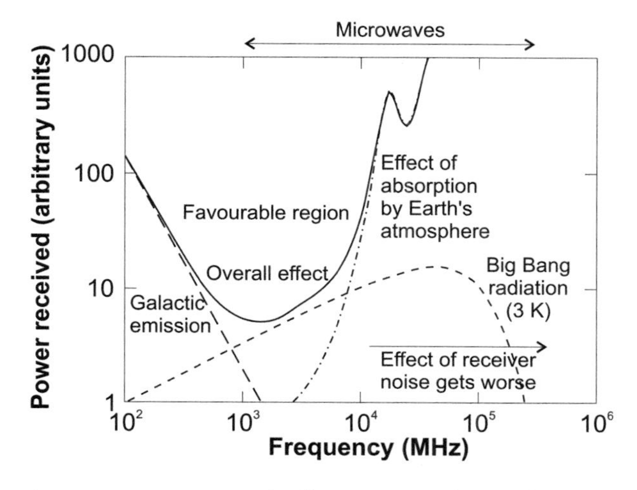

Figure 13.3 The microwave spectrum, and adjacent parts of the infrared and radio ranges of the electromagnetic spectrum, showing the microwave 'window' for SETI.

In addition, there is an increase in microwave receiver noise as frequency increases, and this exceeds the effect of the Big Bang radiation at frequencies greater than about 10^4 MHz, though it does not rival atmospheric absorption.

Overall, the region around 10^3 MHz is best – there is little background (except for terrestrial broadcasts in some frequency ranges), and little absorption in the Earth's atmosphere. In addition the photon energy hf is extremely low at such low frequencies, and so information can be imprinted on the radiation with low energy cost. Consequently, the very first SETI in the modern era was made at these microwave frequencies. This was Project Ozma in 1960, led by Frank Drake, and carried out with a radiotelescope at Green Bank, West Virginia, with a 25 m diameter dish. A total time of 150 hours were spent looking at the nearby solar-type stars Tau Ceti and Epsilon Eridani at frequencies around 1420 MHz. No signals from ETI were found.

The choice of 1420 MHz was not accidental. But why make a choice at all? Why not look at all the frequencies in the microwave window in a single detector? The reason is illustrated in Figure 13.4, which shows the advantage of ETI confining its transmission to a narrow range of frequencies. This helps the transmission to 'outshine' the natural background. If neither background nor signal fluctuated this would be less of a requirement. But they do, and so we get the sort of signal-to-noise problem outlined in Section 9.2.1. The range of frequencies that the signal is spread over and that the receiver detects is called the bandwidth. In Project Ozma the receiver bandwidth was 100 Hz. The 'window' around 1000 MHz is about 10 000 MHz wide, and so the bandwidth is the tiny fraction 10^{-8} of the window. Even so, you might wonder if it would have been possible to scan the receiver across the window. This was not practical, if only because of the enormous time it would have taken to search 10^8 frequency bands. Consequently, the receiver was scanned only in the vicinity of 1420 MHz.

This particular vicinity was chosen because radiation at 1420.4 MHz is emitted by a hydrogen atom when it undergoes a transition in which the spins of the electron and the nucleus (a proton) go from having their rotation axes parallel (pointing in the same direction) to antiparallel (pointing in opposite directions). The rationale was that this spectral line is one of the most clear and widespread in the interstellar medium, thus providing ETI with a basis for selection from among the host of other microwave frequencies. Communication would probably be *near* this hydrogen line rather than on it, to facilitate discrimination from the natural sources.

Current searches

Since Project Ozma there have been a few hundred microwave searches, some of them lasting many years. Several searches are in operation today. Bandwidths less than 1 Hz are being used in some cases, coupled with receiving systems that can examine tens or hundreds of millions of frequency channels simultaneously. One such is the Billion Channel Extraterrestrial Assay (BETA), masterminded by the US physicist Paul Horowitz, and running on the 26 m radiotelescope at Oak Ridge, USA since 1995. BETA uses a channel width of 0.5 Hz over the range

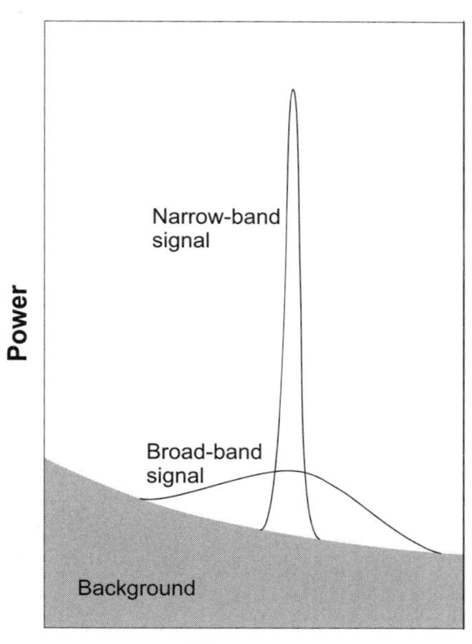

Figure 13.4 The advantage of a narrow bandwidth transmission in outshining the background.

1400–1720 MHz (so doesn't quite reach a billion channels). In 1998 the privately funded SETI Institute in California continued Project Phoenix, which had earlier risen from the ashes of cancelled NASA searches. It is now using small proportions of the observing time of the largest single-dish radiotelescope in the world – the 305 m dish at Arecibo in Puerto Rico – and the 76 m Lovell Radiotelescope at Jodrell Bank in the UK (Figures 13.5 and 13.6). The Search for Extraterrestrial Radio Emissions from Nearby Developed Independent Populations (SERENDIP IV) is a project in which anyone with a computer connected to the internet can take part. A detection system is permanently connected to the Arecibo radiotelescope. Without interfering appreciably with other observations, it taps the (1420 ± 1.25) MHz signal at 0.6 Hz resolution, from whatever part of the sky the telescope is looking at while devoted to other projects. A screensaver called SETI@home takes a small portion of the data, analyses it, and returns the analysis to the main SERENDIP computer. Details can be obtained via the website listed in Resources.

The sensitivity of these searches is sufficient to detect the sort of microwave transmissions that we could send to ourselves from a maximum range that is a substantial fraction of the diameter of the Galaxy. For example, the Arecibo radiotelescope could communicate with a similar radiotelescope far beyond the solar neighbourhood. Quite how far depends on the power fed into the transmission, its bandwidth, and the frequency of transmission. The large range owes much to the

Figure 13.5 The 305 m dish of the Arecibo radiotelescope in Puerto Rico.
Courtesy of the NAIC-Arecibo Observatory, a facility of the NSF.

tight angular confinement of the transmitted beam. The beam has an angular structure similar to the point spread function for imaging discussed in Section 9.1, with a central lobe angular diameter about λ/D radians, into which most of the power is concentrated. With $D = 305\,\mathrm{m}$ and $\lambda = 0.03\,\mathrm{m}$ (10 000 MHz), $\lambda/D = 20$ arcsec, so the beam would be tightly collimated. If the beam were as strong in all directions then it would have what is called its equivalent isotropically radiated power (EIRP). In this case, if the large power 10^6 W were fed into the beam this would be equivalent to an EIRP of order 10^{15} W. ETI could probably transmit even greater powers. So, if they are targeting us, we should have the capability to detect their signals.

13.3.2 What a microwave signal from ETI might be like

A signal is sought that is clearly not of natural or terrestrial origin. A terrestrial origin can be ruled out if the signal comes from a particular location among the stars. To see whether this is the case requires either that the signal is observed with at least two widely separated telescopes, or that the signal lasts long enough for the

Figure 13.6 The 76 m Lovell Radiotelescope at Jodrell Bank, UK. Part of its observing time is devoted to the SETI Project Phoenix.
Ian Morison.

Earth's rotation to change the direction to a source in the stars. The Earth rotates at 15° per hour, so the motion of a celestial source should be apparent in an hour.

To rule out a natural origin we need to be aware of pulsars (Section 11.1). These constitute a subclass of neutron stars, which are remnants of massive stars. A neutron star can emit a beam of microwaves (and other wavelengths) at an angle to its rotation axis, as illustrated in Figure 13.7(a). If, as the pulsar rotates, the beam sweeps across the Solar System, then we can observe a regular series of pulses as in Figure 13.7(b), and the neutron star is seen as a pulsar. Each pulse consists of a short train of microwaves. Pulsar rotation periods range from a few seconds to about a millisecond, so this is the range of pulse spacings. If we did not know about rotating neutron stars then we might fall into the trap of thinking we had detected a transmission from ETI. Indeed, when the first pulsar was discovered in 1967 this possibility was considered, though only briefly. The explanation in terms of a rotating neutron star was soon at hand.

A telltale absence from a pulsar signal is *information* – the pulses carry no message. They could do so if the spacing of the pulses varied in accord with some code, but they beat as constantly as a highly accurate clock. One way to imprint a

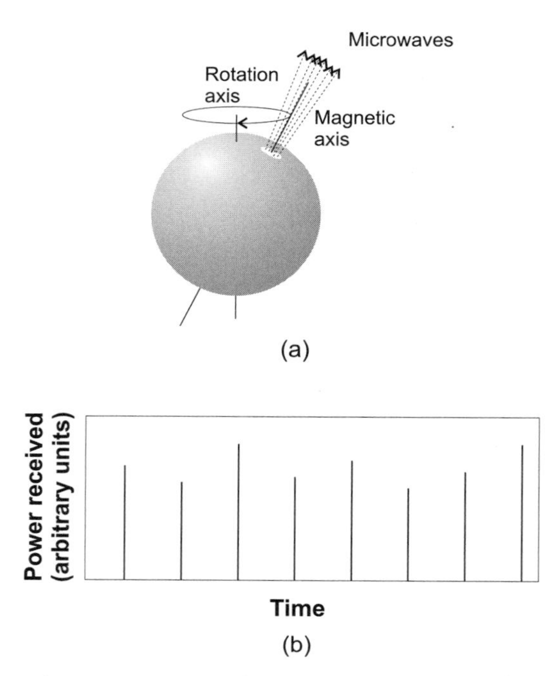

Figure 13.7 (a) A rotating neutron star emitting beams of microwaves. (b) The train of pulses that could be observed on Earth.

message on a signal is in a binary code. This is any code made up of two symbols, one of which can be regarded as 0 (zero), the other as 1. For example, the frequency of microwaves could be switched between two adjacent values, as in Figure 13.8, in which case the one frequency could represent 0 and the other 1. The great advantages of a binary code are that it is readily recognisable, and that the imprint on microwaves is comparatively well preserved across interstellar distances. Though some natural process might generate microwaves that alternate between two frequencies, it would not generate a meaningful sequence of binary digits 0 and 1. Our first guess might be that they constituted a string of numbers rather than symbols in some alien script. Thus, in decimal terms, 0 and 1 have their usual meanings, but 10 is 2, 11 is 3, 100 is 4, and so on (or possibly 01 and 001 for 2 and 4 if we are meant to read from right to left). Proceeding on this assumption, suppose that the string could then be interpreted (in decimal form) as the digits 31415927 repeated several times. These are the first 8 digits in π, the ratio of the circumference of a circle to its diameter. It would then be beyond reasonable doubt that we had detected a signal from ETI.

To summarise, we expect a microwave signal from ETI to be narrow band, and to be coded in a comparatively recognisable way (probably binary) with a message that would be distinct from the imprint of any natural process (e.g. the digits of π). Whether we could understand any content beyond universal mathematics is a matter of debate, but remember that if ETI is trying to communicate then it will make the

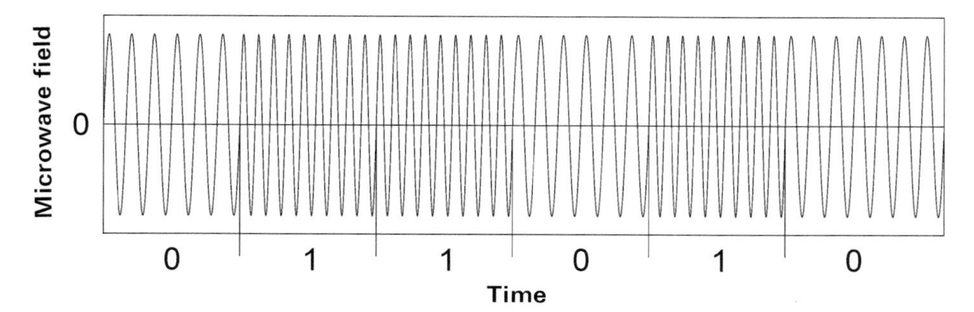

Figure 13.8 Microwaves coded in binary through the frequency of the wave varying between two adjacent values, and the corresponding sequence of zeroes and ones. The frequency difference is exaggerated here, and there would be far more cycles between frequency changes.

message as transparent as possible. Quite what we might learn, and what the effect would be on us, is unknown, but would surely be profound. First, however, we must detect a signal from ETI.

13.3.3 The outcome of microwave searches and their future

Though some peculiar signals have been observed, all have been brief and unrepeated, and none has borne the imprint of artificial origin. Therefore, there is no evidence for ETI in the microwave searches to date. One explanation is that we are alone. Another is that ETI transmissions are too weak or too broadband. This is unlikely to be the case for deliberate attempts at communication (Section 13.3.1), but probably would be the case if the transmissions were intended only to be detectable within their planetary system or between neighbouring systems. This then puts us in the eavesdropping mode of detection. It has been possible for ETI to eavesdrop on us ever since we began to make radio and television broadcasts, and to use radar. Ignoring the early, very weak transmissions, a shell of microwaves and radiowaves has travelled nearly 100 ly from the Earth, encompassing 20 000 stars. These transmissions are however comparatively weak, and as they spread out the inverse square law makes them even weaker. Moreover, the strongest of our transmissions, radar, have bandwidths of about 1000 Hz. Consequently, the *snr* is poor even within the Solar System. Even from the nearest stars, ETI would have to have receivers far more sensitive than ours to detect our leakages, and the large bandwidths might cause ETI to fail to recognise them as artificial.

But if ETI is out there and is trying to communicate, then the likely reasons for failure include the limited number of narrow frequency bands that we have searched, and the limited number of stars that we have so far targeted. There is also the limited time for which we have targeted each star, assuming that ETI sweeps a tightly collimated beam around the sky, and therefore that the beam dwells on us for only a small proportion of the time.

Searches continue, and there are plans for much larger ones. These include the Allen Telescope Array (ATA) under construction at Hat Creek Observatory, USA. There will be 350 dishes, each 6.1 m in diameter, which will give it a collecting area of about 10 000 m^2. This is only about one-seventh that of the Arecibo telescope, but the ATA will simultaneously cover multiple targets over the range 500–11 000 MHz, and much of its time will be dedicated to SETI. The ATA should be fully operational by 2005. A smaller array is being constructed by the SETI League in New Jersey, USA. If these and other radiotelescopes fail to discover ETI by, say, 2025, then we could conclude either that ETI is extremely rare in the Galaxy, or that ETI rarely attempts microwave interstellar communication.

13.3.4 Searches at optical wavelengths (OSETI)

Optical wavelengths span not only visible radiation, but also UV and infrared radiation (Figure 13.2). However, in OSETI it is visible radiation that has so far dominated the searches, and the following discussion is therefore restricted to 'visible' SETI.

At first sight visible wavelengths do not seem promising. Though the Earth's atmosphere is transparent at visible wavelengths, there are copious natural sources in the form of stars and glowing interstellar gas. Moreover, any planets would be near to stars. The device that makes OSETI worthwhile is the high-power laser. Laser light has many remarkable properties, but the ones of most relevance here are that extremely tightly collimated beams can be produced, with angular diameters considerably less than an arcsec, and that extremely high powers can be achieved. Even by the early 1990s powers had reached the order of 10^{13} W within well-separated pulses each lasting a few nanoseconds (1 ns $= 10^{-9}$ s). This, with collimation of order 0.1 arcsec, gives an EIRP in each pulse of the order of the Sun's total power of 3.85×10^{26} W! It is thus easy to envisage ETI building pulsed lasers that, within the angular spread of the beam, could outshine their star's total output by several orders of magnitude.

The type of laser signal that ETI might transmit is shown in Figure 13.9, and most OSETI is based on it. The huge pulses have peak power of order 10^{18} W, but are only 1 ns in duration and are spaced by 1 s. Thus the mean power is 'only' 10^9 W, about the output of a large power station, so could be regarded as well within the capacity of ETI not much in advance of us. If the beam collimation of the 10^{18} W giant pulses was of order 0.01 arcsec (which is feasible), then the EIRP in the pulse would be roughly 10^{34} W, nearly 10^8 times the luminosity of a solar-type star. In this case, a 10 m optical telescope on Earth would easily detect the short bursts of photons in the pulses at a range up to a few hundred light years. Moreover, there is no problem of which frequency to select. Even though each nanosecond pulse contains about half a million visible wavelengths, and thus has a well-defined frequency, the photon signal-to-noise ratio, even at 300 ly, is adequate without frequency filtering.

These giant pulses, if regularly spaced, would, like those from pulsars, carry no message. Even if the spacing was varied the message would arrive very slowly. The

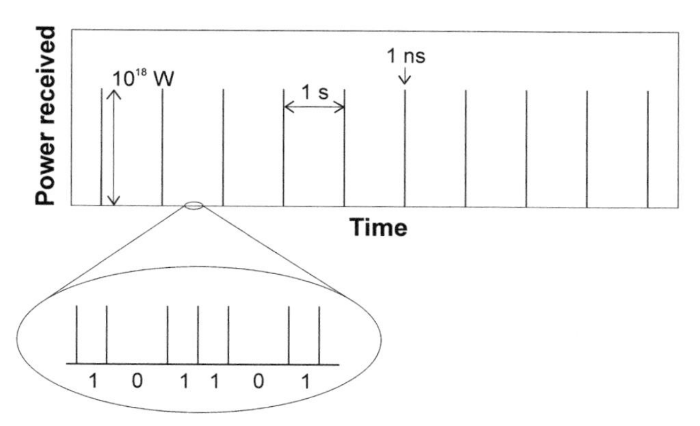

Figure 13.9 A possible form of optical transmission from ETI. Note that each giant nanosecond pulse would contain about half a million wavelengths of visible light. It is the weaker pulses that deliver a message.

huge pulses do, however, grab our attention, and ETI could have included weaker pulses that arrive at a faster rate (Figure 13.9). We would now notice these pulses, and variations in their spacing would deliver a message quickly. The weaker message pulses would benefit from frequency filtering, though this could be broadband.

Within 300 ly of the Sun there are about half a million stars. Not all of these will be accessible to OSETI, because, unlike microwaves, visible radiation is absorbed and scattered by dust. Therefore, dusty regions will be inaccessible, and this is an important restriction at greater ranges, where there are dusty regions such as the Orion Nebula and the central regions of the Galaxy.

Another consideration for OSETI arises from the very narrowness of the laser beams. If ETI is targeting a particular star, they have to know where the star will be when their optical signal reaches it. With a 0.01 arcsec collimation, the beam is only 1 AU across at a range of 300 ly. Therefore, for a star 300 ly away, they have to know where it will be in 300 years, and they have to know it to an accuracy of about 1 AU. They thus have to know proper motions very well. Only in recent years have we acquired sufficiently accurate data, with Hipparcos (Section 10.1.3).

Only a small amount of effort has been devoted to OSETI, but in recent years the activity has been growing, in part following the development of high-power pulsed lasers. An early pioneer is the British engineer Stuart Kingsley, who has operated the Columbus OSETI Observatory in the USA since the early 1990s. It is privately owned and run, and based on a 0.25 m telescope. Paul Horowitz and others have operated an OSETI observatory since 1998. It is based on a 1.5 m telescope at Oak Ridge in the USA, and piggybacks on searches for exoplanets. Both these searches are for pulses, which will remain the dominant type of search. The number of optical searches is growing, and many more are planned, though as yet the largest telescope planned will have a modest 1.8 m aperture, and is being built by Horowitz and others. In addition, it is to be hoped that at least a small proportion of the time on the world's largest telescopes will be devoted to OSETI. That will increase the chance of the first detection of ETI.

13.4 SPACECRAFT AND OTHER ARTEFACTS FROM ETI

There is no evidence that the Solar System has ever been visited by ETI, either in person or robotically. Is this because interstellar travel is extremely difficult? Is it so difficult that we are unlikely ever to find ETI in this way? To answer this question, consider how difficult it would be for *us* to travel to the stars.

13.4.1 Interstellar travel

The fundamental requirement is for the spacecraft to acquire sufficient energy for it to reach the stars in a reasonable time. Consider first the energy needed to combat gravity. The gravitational energy E_g needed to lift a body with a mass m from a distance r to a distance r' from a body with a mass M is:

$$E_g = GMm\left(\frac{1}{r} - \frac{1}{r'}\right) \tag{13.3}$$

This equation requires either that m and M have spherically symmetrical mass distributions or, as in the present case, that they are separated by a distance much larger than their own dimensions. Per unit mass lifted we have:

$$\frac{E_g}{m} = GM\left(\frac{1}{r} - \frac{1}{r'}\right) \tag{13.4}$$

To get away from the Earth to interstellar space most of the energy needed is to combat the Sun's gravity, and amounts to about $10^9\,\mathrm{J\,kg^{-1}}$. Lifting 1 kg from the Earth's surface to a height of 1 m requires just under 10 J, so you can see that a *lot* of energy is required to escape the Solar System. Moreover, to this we must add the energy needed to lift the rocket fuel, which initially accounts for nearly all of the mass of the spacecraft.

If we give the spacecraft just enough energy to reach interstellar space it will arrive there travelling very slowly. We must do better than this. About the best we could hope to do at present would be to deliver the spacecraft with a speed v of about $100\,\mathrm{km\,s^{-1}}$, which is 0.033% of the speed of light. The kinetic energy E_k is $mv^2/2$, which, per unit mass, is:

$$\frac{E_k}{m} = \frac{v^2}{2} \tag{13.5}$$

This corresponds to $5 \times 10^9\,\mathrm{J\,kg^{-1}}$ – about five times E_g/m. Again we must add the kinetic energy that was needed to carry the fuel. The speed of about $100\,\mathrm{km\,s^{-1}}$ is based on chemical fuels and gravity boosts involving the Sun and Jupiter. In a gravity boost the spacecraft encounters a body in such a way that the kinetic energy of the spacecraft is increased at the expense of the kinetic energy of the body. (The mass of Jupiter and the Sun is so great that their change in orbital speed is negligible.)

At $100\,\mathrm{km\,s^{-1}}$ it would take about 12 600 years to reach the nearest star. At 10% of the speed of light the time reduces to 42 years, but E_k/m has risen to $4.5 \times 10^{14}\,\mathrm{J\,kg^{-1}}$! This means that, to deliver just 1 kg to interstellar space at 10% of the speed of light, not including the energy needed to accelerate the fuel, the equivalent of the whole energy output of a huge 10^9 W electrical power station for five days is required.

To approach 10% of the speed of light we need to substitute chemical fuels with something far more potent. One possibility is nuclear fusion involving ^{2}H and either ^{3}H or the very scarce helium isotope ^{3}He. Per kg of these isotopes, about 10^7 times more energy is released than in chemical reactions, and this release need not be explosive. Even more energy, about 200 times more, is released from the annihilation of matter with antimatter. Antimatter differs from matter in that certain of its properties are reversed. Like matter it has positive mass, but, for example, opposite electric charge, so that the antiproton is negatively charged, the antielectron (positron) is positively charged, and so on. When a particle meets its antiparticle both are annihilated and their total masses are converted into energy, in accord with Einstein's equation $E = mc^2$ where c is the speed of light. This gives $9.0 \times 10^{16}\,\mathrm{J\,kg^{-1}}$. Antimatter is rare in the Solar System, and probably throughout the observable Universe. It appears in various nuclear reactions, but it is difficult to store – if it meets ordinary matter it is, of course, annihilated.

A completely different approach is the solar sail (Figure 13.10), in which the spacecraft is pushed along by the pressure on a huge sail of the photon flow that constitutes electromagnetic radiation. Within the Solar System the Sun's radiation would be the source of photons, but in interstellar space it would be better to blow the craft along with a laser beam. This could still be regarded as solar sailing if the laser was solar powered, such as by photovoltaic cells.

There are more exotic possibilities. Wormholes are short cuts through space-time, which is the four dimensional construct in relativity consisting of the three space dimensions and time. A wormhole is a channel that permits travel from one point in space-time to another without us having to travel through ordinary space. This permits superluminal travel. It also permits travel back in time, which would lead to all sorts of problems – any slight alteration in history made by a time traveller could alter the world they travelled from, including the possibility that they did not now exist and so could not have travelled back in time to make the alteration! Therefore, though general relativity permits the existence of wormholes, there would surely have to be some rule that prevents backward time travel. Wormholes would be unstable, so to stabilise them they would need to be threaded by matter with the curious property of negative energy density. Quantum physics permits the existence of such matter, but so far, like wormholes themselves, none is known.

Another possibility is warp drive, in which a space-craft warps space-time in its vicinity in such a way that the space-craft moves superluminally from one point in

Figure 13.10 Solar sailing to the stars.
Illustration courtesy of the Planetary Society.

space-time to another. As with wormholes, time travel would need to be ruled out, and matter with negative energy density would be needed to maintain stability. According to Star Trek, warp drive lies a few centuries off. In fact, we cannot be sure it is feasible at all. On the other hand, we know that our physics is incomplete, so we have no firm idea what the future holds.

In conclusion, with our present technology, we are far short of being able to travel to the nearest stars with reasonable journey times on a human timescale. You might therefore be surprised to learn that we could embark on accomplishing the far grander exploration of the whole Galaxy. If newcomers to space technology like us are on this threshold, then maybe ETI is already well advanced in this enterprise.

13.4.2 Galactic exploration

In the 1940s and 1950s the Hungarian-born US mathematician John von Neumann developed the idea of the self-replicating probe. This has the capability of making a copy of itself from materials in its environment. As applied to galactic exploration any such probe sent from Earth to a nearby star would, on arrival, make more than one copy of itself from local materials. These copies would then be launched towards other stars. Meanwhile, the original probe would have transmitted information to

Earth about its own destination star and any planets. When the second generation probes arrived at their stars the process would be repeated. This exponential growth in the number of probes would enable the whole Galaxy to be explored in the order of 100 Ma even with our present chemical rocket technology. This is only about 1% of the age of the Galaxy. The one thing we lack at present is the will to devote the huge resources needed to make such a probe.

More fancifully, a probe could also colonise a planetary system. It could carry the instructions to make human beings, and it could make a life-supporting environment for them to inhabit. These would take the form of huge space stations, thus creating O'Neill colonies, named after the US physicist Gerard K. O'Neill, who showed in 1974 how large space colonies could be built.

If we could complete our exploration of the Galaxy by a time from now that is only 1% of its age, then any ETI that started exploration a relatively short time in the past could have largely or fully completed the exploration already. If so, then there might already be ETI artefacts in the Solar System. Some scientists have speculated that the asteroid belt would be a good place to look. Are all of those tiny asteroids really just lumps of rock?

Overall, it might not be worth making a huge effort to find alien artefacts, but we should at least be aware of the possibility of accidental discovery.

13.5 TECHNOLOGICAL MODIFICATIONS BY ETI OF THEIR COSMIC ENVIRONMENT

In 1964 the Russian astronomer Nikolai Kardashev classified civilisations into three types, based on their rate of energy consumption. Type I civilisations utilise the energy resources of their planet, including the stellar radiation it receives. We are a Type I civilisation. Type II utilise a large proportion of the whole energy output of the host star, not just the tiny fraction that falls on the planet. Type III utilise a large proportion of the energy output of their galaxy. Each type could be detected from afar through its technological modification of its environment.

Type I would be hard to detect. You have already seen in Section 13.3.3 that eavesdropping on communications is unlikely to be successful, with defence radar in our own case offering the best prospect to ETI. Another possibility is the gamma radiation from nuclear fission or fusion, though if we could detect such radiation then it would be likely to be at deadly levels for ETI, perhaps as a result of nuclear conflict – a depressing thought.

Type II civilisations are more detectable. In 1959 the British physicist Freeman J. Dyson envisaged ETI surrounding its star with a swarm of artefacts that divert a large proportion of the star's energy for use (Figure 13.11). Such a Dyson sphere (which is *not* a solid, complete sphere), like any energy device, does not consume energy, nor does the energy build up. Instead, the energy flows through various energy conversion devices, is utilised, and then emerges in some final form. Some fraction of the star's radiation, which is predominantly at visible and near-infrared wavelengths, will thus be utilised and emerge at mid- and far-infrared wavelengths.

Figure 13.11 A Dyson sphere, diverting a large proportion of a star's energy for use by a Type II civilisation.
Anders Sandberg.

The star would seem to have a huge infrared excess. Such stars are known, but there is always a natural explanation, such as a star surrounded by a dusty envelope that absorbs the stellar radiation, which warms the dust, which then radiates at infrared wavelengths in accord with its temperature.

A Type III civilisation might be detected through its reorganisation of a galaxy. But though there is quite a menagerie of galaxies, none as yet requires such an extraordinary explanation. The galaxy-wide communications that a Type III civilisation presumably needs, might be powerful enough to be detectable in eavesdropping mode, unless electromagnetic radiation or any other type of radiation that we could detect, was no longer the basis of their communications.

13.6 THE FERMI PARADOX

As yet there is no observational evidence for ETI. Many people conclude that this leaves the existence of ETI open. We might have failed to search for microwave or optical transmissions at the right time, or from the right star, or at the right frequencies. We might have failed to find artefacts in the Solar System, even though they could be there. We might have failed to detect Dyson spheres, even though they could be there too. Others argue that the failure to detect ETI is simply because ETI does not exist! Around 1950 the Italian physicist Enrico Fermi posed this failure as the paradox that now carries his name. *Fermi's paradox* states that if

ETI exists, then it must be widespread, in which case why aren't they among us? The assumption is that there has been plenty of time for ETI to have emerged and to have colonised the Galaxy – you saw in Section 13.4.2 that this could take a very small fraction of the age of the Galaxy. An obvious solution to the paradox is that there are no other technological species in the Galaxy.

Many and varied are the refutations of what many find to be this bleak conclusion. One extraordinary hypothesis is that ETI is among us, looking human! This is difficult to test, or refute. Another hypothesis is that Galaxy-wide ETI will not make its existence known to us until we are sufficiently advanced to join the galactic community. Frank Drake offers the less provocative hypothesis that it makes far better sense for technological species to colonise their own planetary system than endure the costs and hazards of going to other stars. Another possibility is that ETI is strongly disinclined to launch self-replicating probes, because the software controlling replication would certainly develop faults, and who knows what rogue behaviour could result?

It is also possible that ETI has only just emerged in the Galaxy, and that we are among the first technological civilisations. This puts us in a rather special place in Galactic history, but it cannot be ruled out. Even if we are not among the very first, ETI could still be thinly dispersed, particularly if the imperative to explore and colonise is weak. There might even be a weak imperative to communicate. On Earth we have seen a considerable reluctance to signal our presence across interstellar space (see Section 13.7). Perhaps, therefore, our first contact will be with a doomed civilisation that has little to lose by making its presence known.

There are many more refutations to the conclusion that we are the only technologically intelligent civilisation in the Galaxy (see Resources), not the least of which is that we have not yet searched hard enough or long enough. Certainly, if ETI is trying to make its communications obvious even to recently emerged species like ours, then we have some hope of success.

SETI must be treated as an observational quest. If we want to know, we must search, though how much scientific effort should be devoted to the search is another question.

13.7 COMMUNICATING WITH EXTRATERRESTRIAL INTELLIGENCE (CETI)

We have not yet much considered any deliberate attempts that *we* have made at interstellar communication. Such attempts come under the subject called communication with ETI, or CETI (pronounced 'keh-tee'). An early proposal for CETI was in 1826, by the German mathematician, astronomer, and physicist Carl Friedrich Gauss. At the time it was thought that the Moon might be inhabited, and so Gauss suggested that we could signal to the 'Selenites' by cutting down areas of Siberian forest to illustrate Pythagoras's theorem for right-angled triangles, as in Figure 13.12. It was never attempted.

Figure 13.12 A way of signalling to ETI on the Moon through illustrating Pythagoras's theorem.

In the modern era the first major attempt at CETI started in 1972 with the launch of the spacecraft Pioneer 10. Under the inspiration of Frank Drake and the US astronomer Carl Sagan, it carried a metal plaque etched as in Figure 13.13. The man and woman between them show characteristics of all races, and their size is given by the drawing of Pioneer 10 behind them. Centre left are lines drawn from the Solar System to various pulsars, with the pulsar period encoded along the line. This gives our position in the Galaxy. At the top is a diagram illustrating the 1420 MHz emission of hydrogen atoms (Section 13.3.1), and at the bottom is a diagram of the Solar System. Pioneer 10's main mission was a fly-by of Jupiter, which it successfully achieved in 1973. It is now (2003) about 80 AU from the Sun and will escape from the Solar System, though it is not heading towards any nearby stars. Pioneer 11, launched in 1973, carries a message with the same content. It too, after fly-bys of Jupiter and Saturn, is heading out of the Solar System.

Voyagers 1 and 2, launched in 1977, also carry messages. In this case there is a copper disc with images and sounds of Earth. This was before the era of the CD, so the disc is grooved and a stylus has to be run along it to recover the information. The container of the disc is etched with some of the information in Figure 13.13, plus diagrams illustrating how the disc has to be played. Voyager 1 flew by Jupiter and Saturn, and is heading for interstellar space, as is Voyager 2, which also visited Uranus and Neptune.

The probability of ETI coming across any of these spacecraft is extremely small, and even if ETI did find one, they might not understand the messages. The first deliberate attempt at microwave communication had the merit of being targeted at

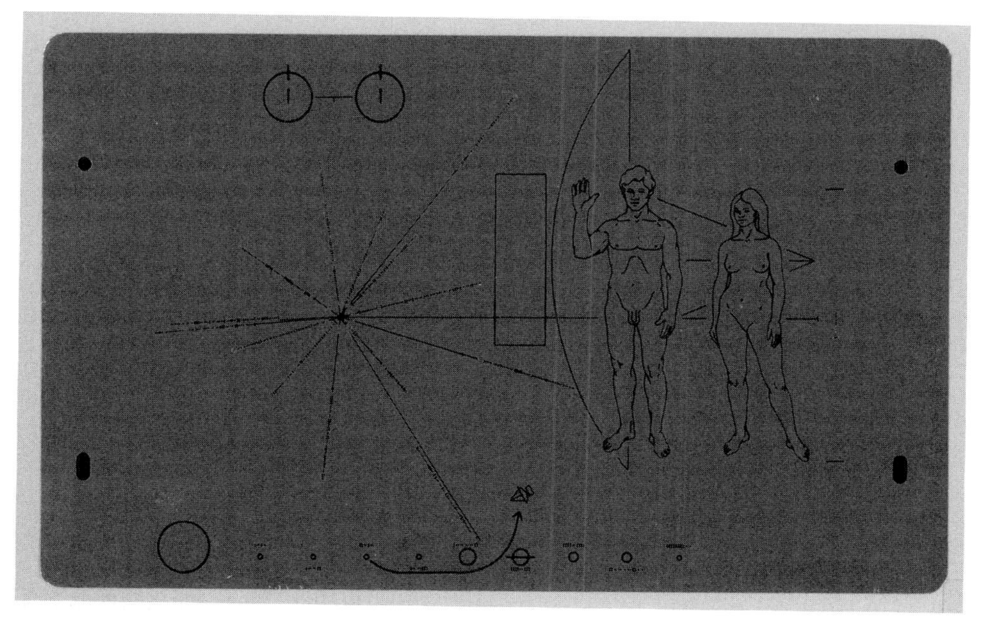

Figure 13.13 The message plaque on board Pioneer 10 and Pioneer 11, now leaving the Solar System.
NASA.

stars. This was in 1974, when at the opening ceremony of the refurbished Arecibo radiotelescope the telescope was used to send a message to a globular cluster of about 300 000 stars called M13, about 21 000 ly away. The message will thus get there in about 21 000 years. Over the bandwidth, centred on 2400 MHz, the message outshines the Sun. It is in binary code, a total of 1679 bits, and it took three minutes to transmit. If correctly decoded it shows a 2-D blocky image that includes (among other things) an illustration of binary numbers, a simple drawing of the double helix of DNA, the outline of a human, and an indication that we live on the third planet from our star. No location of the Sun is given, though this could be roughly determined by the direction from which the transmission came.

Given the large number of stars that will receive this short message in 21 000 years, there is some chance that we will have given away our existence far more effectively than by leakages of our radio, TV, and radar transmissions. This was the basis of some protest about the Arecibo message at the time, though the low metallicity of the stars in globular clusters probably makes rocky planets rare, and the proximity of stars in globular clusters might destabilise planetary orbits (Section 8.3). Nevertheless, radioastronomers have not deliberately targeted any stars subsequently. Also, except possibly for defence radar, we have probably not leaked anything detectable to stellar distances. It might seem silly to be cautious about giving away our existence, but even though the chances of contact are slight the consequences could be huge, so it deserves some consideration.

This raises the question of whether we should reply if we do pick up a signal from ETI. This is still a matter of debate, but I suspect we would find the temptation irresistible. The decision might depend on what we learn about ETI from the contact, and on the way that humanity reacts to the momentous discovery that it is not the only technological species in the Galaxy.

13.8 SUMMARY

- The discovery of another technological intelligence would show that, whatever their physical form, there was at least one other species in the Universe that thought rather like we did. This would be a momentous discovery.
- The likelihood of technological species evolving on a planetary body, and the time for which such species exist, are so uncertain that the number of technological intelligences in the Galaxy today could be anything from 1 (us alone) to hundreds of millions.
- There are three ways in which we could detect extraterrestrial intelligence. We could:
  - detect radiation that travels from them to us, particularly at microwave and optical frequencies;
  - detect their spacecraft or other artefacts in the Solar System; and/or
  - detect features of their planet, or of their environment in general, that revealed technological modification.
- Microwave searches have been made since 1960, and have been the main technique employed in SETI, though since the mid-1990s there has been a significant rise in optical searches (OSETI). No convincing signals have yet been detected, but there are many more targets and frequencies to search, and previous targets and frequencies must be re-examined in case the transmission is intermittent.
- We are on the threshold of being able to embark on exploration of the Galaxy, so it is possible (if unlikely) that alien spacecraft or other artefacts are either in the Solar System today or could arrive here at any time.
- Of the three types of Kardashev civilisations, a Type I (like us) will be very hard to detect through their technological modification of their planet and its environment. Type II might be detectable if they construct Dyson spheres. Type III might be detectable if they reorganise their galaxy, or, in eavesdropping mode, if their interstellar communications give powerful leakage.
- Though there have been attempts, based on the Fermi paradox, to show that SETI is futile, SETI is best treated as an observational quest, with a very uncertain outcome.
- CETI has had little resource devoted to it, and is controversial – should we signal our presence to the cosmos?

13.9 QUESTIONS

Answers are given at the back of the book.

Question 13.1

One of the factors in the Drake equation has been described as ‘sociological’. Which factor do you think this is? Give reasons for your choice. Place reasonable upper and lower limits on the value of this sociological factor.

Question 13.2

Suppose that a radiotelescope is equipped with a detector that can search simultaneously 10^8 adjacent channels each 1.5 Hz wide. It is to search the whole of the favourable microwave region, and will target 1000 nearby stars. It will spend one hour on each star, and cycle through them so that each star is visited 10 times.

(i) Calculate the overall bandwidth of the detector at any particular instant.
(ii) Calculate the number of hours between visits to a given star.
(iii) Assuming that the search is uninterrupted, calculate how many months in total the search will take.
(iv) Calculate for how many minutes each star will have been observed at any particular frequency.

Question 13.3

Describe any way other than frequency switching in which a binary code could be imprinted on a microwave transmission. Describe any advantages or disadvantages over frequency switching that the method of your choice might have.

Question 13.4

Outline how you could detect laser pulses from ETI with the unaided eye.

Question 13.5

Describe one big advantage that a Type II civilisation has over a Type I civilisation in sending a spacecraft to another star.

Glossary

Adaptive optics An optical system for partially correcting the effects of atmospheric turbulence on a telescopic image.

Amino acids The 20 or so building blocks of biological proteins.

Aphelion The point in the orbit of a body when it is furthest from the Sun.

Archaea One of the three domains of life, the other two being the Bacteria and the Eukarya. Bacteria and Archaea are prokaryotes; Eukarya are eukaryotes.

Arcsec One second of arc (i.e. 1/3600 degrees of an arc).

Astrometry The accurate positional measurement of a celestial object such as a star.

Astronomical unit (AU) The semimajor axis of the Earth's orbit around the Sun. It has a value of 1.496×10^{11} m. It is a measure of the average distance of the Earth from the Sun.

ATP and ADP Small organic molecules that play a central role in energy storage and transfer within a cell.

Bacteria One of the three domains of life, the other two being the Archaea and the Eukarya. Bacteria and Archaea are prokaryotes; Eukarya are eukaryotes.

Biosphere On the Earth, or on any other planet, the assemblage of all things living and their remains.

Black body An ideal thermal source of electromagnetic radiation. The spectrum, which is broad, depends only on the temperature of the source. The higher the temperature the more the radiation is shifted to shorter wavelengths.

Brown dwarf A luminous object, made largely of hydrogen and helium, with a mass between about 13 and 80 times the mass of Jupiter (0.013 and 0.08 times the mass of the Sun). It is too massive to be a giant planet, but not massive enough to have a main sequence phase of core hydrogen fusion.

Carbonate–silicate cycle The geochemical feedback whereby a surface cooling of the Earth reduces the rate at which silicate rocks are weathered to form carbonates. This results in a rise in the carbon dioxide content of the atmosphere, thus offsetting the cooling through an enhanced greenhouse effect. The reverse effect also occurs, to complete the feedback scheme.

Carbon–liquid water life Life based on complex carbon compounds and liquid water. All terrestrial life has this basis.

Chemosynthesis The creation of an energy store in a cell using chemical reactions that do not involve photosynthesis.

Chirality That property of a molecule whereby it cannot be superimposed on its mirror image.

Coacervates Droplets bounded by membranes, in which there are higher concentrations of organic polymers than in the surroundings. These could be related to structures that were precursors of the first true cells.

Cold start A planet that is beyond the habitable zone (HZ) of a star when it is very young, and might remain too cold even when the zone moves out to encompass it.

Column mass The mass per unit area of a surface, such as the atmospheric mass per unit area of a planet's surface.

Diffraction limit A fundamental optical limit to the performance of a telescope that determines how much fine detail is visible in the image.

Doppler spectroscopy The measurement of radial motion by measuring Doppler shifts in the spectral lines of a moving object.

Drake equation An expression that displays the factors that determine the number of planets in the Galaxy that are the sources of attempts at interstellar communication at the present time.

Earth-like planets Rocky, Earth-mass planets endowed with suitable volatiles.

Earth-mass planet A planetary body from a few times the mass of Mars to a few times the mass of the Earth.

Eccentricity A measure of the departure of an ellipse from circular form. The eccentricity $e = FC/a$, where FC is the distance from the centre of the ellipse to either focus, and a is the semimajor axis.

Effective temperature The temperature of a black body that radiates the same power per unit area as some other body.

Ellipse A geometrical shape, like a circle viewed obliquely. The shape of the orbit of one body around another under mutual gravitational attraction, when each body is small compared to their separation. The orbit of a planet around a star, or a satellite around a planet, are approximately elliptical.

Embryos Rocky or icy–rocky bodies each with a mass of the order of a few percent that of the Earth. In the nebular theory, they are an important stage in the formation of the terrestrial planets.

Eukarya One of the three domains of life, the other two being the Archaea and the Bacteria. Bacteria and Archaea are prokaryotes; Eukarya are eukaryotes.

Eukaryotic cell More complicated than the prokaryotic cell, with internal structures called organelles. It came after the prokaryotic cell, and evolved from it. Nearly all multicellular creatures are made of eukaryotic cells.

Evolution by natural selection Evolution of organisms though the occurrence of random mutations that prove to be beneficial to the survival and reproduction of descendents.

Exoplanetary system A planetary system of a star other than the Sun.

Extremophiles Organisms that live in what we regard as extreme environmental conditions.

Faint Sun problem The evidence that when the Sun was young and faint, the Earth was nevertheless unfrozen.

Fermi's paradox This states that if extraterrestrial intelligence exists, then it must be widespread, in which case why isn't it among us?

Foci (ellipse) Two particular locations on the major axis of an ellipse, equidistant from its centre. For a planet orbiting a star, the star is at one of the foci, and the other is empty.

Giant planet (1) A planet consisting largely of hydrogen and helium, several factors of ten more massive than the Earth. (2) A planet (sometimes called a subgiant planet) a few tens of times the mass of the Earth, consisting largely of hydrogen, helium, and water.

Giant star The phase in the lifetime of a medium or low mass star that succeeds the main sequence phase.

Gravitational microlensing An apparent brightening of a background star produced by the gravitational field of a foreground star.

Greenhouse effect The name given to the phenomenon whereby the surface temperature of a planet is raised because the atmosphere absorbs some of the infrared radiation emitted by the surface and reradiates a proportion back to the surface.

Habitable zone (HZ) That range of distances around a star within which the stellar radiation would maintain water in liquid form over at least a substantial fraction of the surface of a sufficiently massive rocky planet.

Heavy bombardment The bombardment of the inner Solar System by planetesimals in the early history of the Solar System, from about 4600–3900 million years ago. A mixture of rocky bodies and volatile-rich bodies was involved.

Heavy elements The chemical elements other than the two lightest elements, hydrogen and helium.

Hertzsprung-Russell diagram A diagram displaying the luminosities versus effective temperatures of stars (or parameters related to these quantities).

Hot Jupiter A giant exoplanet that is close to its star, much closer than Jupiter is to the Sun.

Hydrothermal vents Outpourings of hot water and dissolved gases, typically at places where new planetary crust is being created.

Integration time The time for which photons are accumulated.

Interference fringes A pattern resulting from the combination of waves in and out of phase with each other.

Kepler's laws Three laws that describe the motion of the planets around the Sun. They can be generalised to cover the motions of planets around stars.

Kernels Large planetary embryos, rich in water, that could have been an essential stage in the formation of giant planets, by becoming massive enough to capture large masses of hydrogen and helium from the nebular gas.

Last common ancestor The most recent ancestor from which all life today on Earth originated.

Light year (ly) The distance that light travels in a vacuum in a year – near enough the distance travelled through space in a year. One light year is 9.46×10^{15} m.

Luminosity The total power radiated by a body, over its whole surface, over all electromagnetic wavelengths.

Main sequence phase That phase in the lifetime of a star when thermonuclear fusion in its core is sustaining its energy output by converting hydrogen to helium.

Mass extinctions Relatively short periods of time when large proportions of species became extinct.

Mean motion resonances The orbital periods of two or more bodies are in simple ratios.

Metallicity The proportion of heavy elements in a star (i.e. elements other than hydrogen and helium).

Methanogenesis Chemosynthesis in which methane appears as a byproduct.

Mutations Changes in DNA through a variety of causes.

Neutron star The remnant of a supergiant. A neutron star is about 10 km across, about the mass of the Sun, and made almost entirely of neutrons.

Nucleic acids Complex carbon compounds, comprising RNA and DNA. Both are involved in protein synthesis and in information storage for making proteins. Some forms of RNA also act as catalysts called ribozymes.

Nucleotides Building blocks of RNA and DNA.

Nulling interferometry An interferometric arrangement whereby there is heavily destructive interference on some axis in the optical system.

Observational selection effects Features of the methods of observing objects that favour certain types of object over others.

Opposition (of a planet) As viewed from a planet, the configuration when a second planet lies in the opposite direction in the sky to the star at the centre of the planetary system (e.g. Mars as seen from Earth every 2.14 years on average).

Orbital period The time that a body takes to go around its orbit once.

Oxygenic photosynthesis Photosynthesis in which oxygen is produced as a byproduct.

Parsec The distance to an object that appears to shift in position by one second of arc when our viewpoint shifts by one astronomical unit. One parsec is 3.26 light years.

Perihelion The point in the orbit of a body when it is closest to the Sun.

Photodissociation (photolysis) The break-up of a molecule (or atom) by the action of a photon.

Photons A stream of particles that describes how electromagnetic radiation interacts with matter.

Photosynthesis The manufacture of organic compounds from inorganic compounds (carbon dioxide and water) by means of energy from photons.

Planetary body An icy–rocky or rocky body, either a planet or a large satellite, at least a few hundred kilometres in radius. A giant planet.

Planetesimals Rocky, or icy–rocky bodies, 0.1–10 km across, that are potential building blocks of planets.

Plate tectonics The process that is responsible for most of the refashioning of the

Earth's surface. The rocky surface is divided into a few dozen plates in motion with respect to each other. They are created at some plate boundaries, slide past each other at other boundaries, and are destroyed at boundaries where one plate dives beneath another.

Point spread function (*psf*) The image formed by a telescope of a distant point.

Prokaryotic cell A relatively simple cell with no organelles. This type of cell predates the more complex eukaryotic cell.

Proteins Complex carbon compound that play a variety of roles in cells. Some are structural, and others (as enzymes) catalyse biochemical reactions.

Pulsar A neutron star that we observe by the beam of electromagnetic radiation that it sweeps across us as it rotates, giving us a series of regular pulses.

Radiometric dating The dating of events in the history of a rock by means of radioactive isotopes.

Redox reaction A chemical reaction in which there is a net transfer of electrons from one or more atoms or molecules to one or more others.

Resolving power A measure of the fine detail that can be acquired by a telescope.

Ribozymes RNA that catalyses biochemical reactions.

RNA world A hypothetical stage in the emergence of life on Earth where RNA was used rather than DNA as the repository of genetic information.

Semimajor axis Half the longest dimension of an ellipse.

Signal-to-noise ratio (*snr*) A measure of the facility with which a signal can be extracted from statistical noise.

Solar nebula The circumsolar disc of gas and dust from which the planets grew.

Spectral resolution The minimum interval of wavelength over which spectral features can be discriminated.

Stromatolites Layers of minerals laid down by colonies of oxygenic photosynthetic bacteria – cyanobacteria. There are living examples and fossilised examples.

Synchronous rotation The orbital period of a satellite equals its rotation period, and the two rotations are in the same direction. Applies also to a planet in orbit around a star.

Terrestrial planets Mercury, Venus, Earth, and Mars (i.e. the substantial rocky bodies that dominate the inner Solar System). The term can be extended to include analogues of such planets in other planetary systems.

The Galaxy The assemblage of about two hundred thousand million stars in which we live. Many of these stars we see as the Milky Way.

Tidal heating The heating of a body, such as a planetary body, arising from the gravitational distortion caused by other bodies.

Transit The passage of one celestial body in front of another.

Type I migration A theoretical scheme whereby a planet migrates through a circumstellar disc without opening a gap in the disc.

Type II migration A theoretical scheme at larger planetary mass than in Type I migration, whereby a gap is opened in the circumstellar disc and the rate of migration slows.

Answers to questions

Comments are in brackets [ ] – the content of which is not expected as part of the answer.

CHAPTER 1

Question 1.1

(i) From Section 1.1.1, the distance FC in Figure 1.2 is *ea*. Thus, FC $= 0.9 \times 1.0\,\text{AU} = 0.9\,\text{AU}$. This is also the distance of the Sun from C. The perihelion distance is $a -$ FC $= 0.1\,\text{AU}$. The aphelion distance is $a +$ FC $= 1.9\,\text{AU}$.

(ii) The Earth would thus get much closer to the Sun than torrid Mercury (0.387) and retreat from the Sun to beyond where frigid Mars lies (1.524 AU). It would thus be alternately too hot and too cold for liquid water at its surface, and thus be inhabitable by life that requires liquid water. [In fact, if the atmosphere were sufficiently massive, the extremes might be averaged out, though recall from Kepler's second law that the planet would spend far longer at aphelion than at perihelion, so it might still be too cold.]

Question 1.2

At 0.1 AU the stars are both well interior to the 1 AU orbit where we know the Earth is habitable. In fact, with *two* solar-type stars the HZ is further out, so 0.1 AU will be even more safely removed from the stars. Therefore, provided that the 'Earth' could form in such a binary system, it should be in a stable orbit. At a stellar separation of 1 AU the location where we would like to place an 'Earth' is severely disrupted. There would be no stable orbits. At 100 AU the stars are so far apart that 'Earths' at 1 AU around either star should be stable.

In globular clusters the stars could approach each other so closely that planetary orbits might be disrupted. [In fact, even though globular clusters look crowded, as in Figure 1.14, sufficiently close approaches might be rare. This is discussed briefly in Chapter 11.]

CHAPTER 2

Question 2.1

1 Many more elements than hydrogen, oxygen, and carbon are required. Nitrogen is present in amino acids (Figure 2.3), and in the bases in RNA and DNA (Section 2.2.1). Phosphorus is present in the phosphate groups in RNA and DNA, and in ATP/ADP (Sections 2.2.1 and 2.2.2). NaCl is present in all cells (Section 2.2.2). [Many other elements also occur, such as calcium.]
2 All life needs liquid water, but not necessarily all the time (Section 2.2).
3 Hydrocarbons are not found in cells (Section 2.2.2). Carbon is an essential basis of proteins, RNA, DNA, polysaccharides, and lipids. [Maybe there was confusion here with carbohydrates?]

Question 2.2

(i) These would be anaerobic thermophiles performing chemosynthesis. As such they would be prokaryotes.
(ii) This is within the range of sufficient solar radiation for photosynthesis, and there would be oxygen dissolved in the water. Aerobic, photosynthesising organisms would be present, including prokaryotes and unicellular and multicellular eukaryotes.
(iii) The same as (i). The organisms could be independent of the rest of the biosphere.

Question 2.3

The known hyperthermophiles are prokaryotes, yet the nucleus and mitochondria within the cell suggest it is a eukaryote. If this is the case then this organism must be placed in the domain of the Eukarya. As it is unicellular it could not be placed in the animal or plant kingdoms. It could be a new species of fungus or protoctist, but none of these are hyperthermophiles. Further study is needed – it might even merit being classified as a member of a new, fifth kingdom in the Eukarya.

Question 2.4

The sequence is:

Animalia–Fungi
Archaea–Eukarya
Bacteria–Eukarya.

In Figure 2.13, the distance back from the tips of the 'Animals' and 'Fungi' branches to where they join at the last common ancestor is the shortest. Tracing the branches further back we then come to the branching between Archaea and Eukarya, and finally to that between Bacteria and Eukarya.

CHAPTER 3

Question 3.1

With 24 hours corresponding to 4600 Ma, one hour corresponds to 191.7 Ma, and one minute to 3.19 Ma.

End of heavy bombardment	3900 Ma	03:39
Appearance of prokaryotes	(pre-) 3850–3000? Ma	(pre-) 03:55–08:21
Appearance of eukaryotes	(pre-) 2700 Ma	(pre-) 09:55
Rise of atmospheric oxygen	2300 Ma	12:00
Spread of multicellular organisms	545 Ma	21:09
Permian mass extinction	248 Ma	22:42
Appearance of *Homo sapiens*	0.1 Ma	23:59:58.1
Start of the Common Era	0.002 Ma	23:59:59.96

Question 3.2

If the biosphere died, the Earth's atmosphere would gradually revert back to the composition it had soon after the end of the heavy bombardment (i.e. dominated by CO_2 and N_2, plus some water vapour). There would be more CO_2 as the rate of removal would be lower, and less N_2 as the rate of release would be lower. In the absence of photosynthesis there would be far less O_2 [though probably enough to sustain an effective UV screen in the form of O_3].

Question 3.3

The enormous advantage is that we would have a much longer and more complete geological record in which the evolutionary steps before the last common ancestor would be preserved. By contrast, the oldest rocks on Earth are about the same age as the earlier possible date for the last common ancestor – the heavy bombardment has destroyed the older rocks.

Question 3.4

(a) If RNA of the correct chirality (sugars in the D form) were discovered in a meteorite, then we could bypass the problem of how to make the bases and the ribose needed for the first RNA molecules on Earth. We would then have a route into the RNA world.

(b) This discovery would not necessarily boost our understanding of life in the Solar System because we would probably be no closer to understanding how this meteoritic RNA formed, though there should have been more time (free from the heavy bombardment), more space, and a greater range of environments to accomplish the synthesis.

CHAPTER 4

Question 4.1

The Moon is a small body (Figure 1.7). Therefore, even if you did not know that it is almost devoid of atmosphere, you might expect this to be the case, as it is with Mercury (Section 4.2.1). With little or no atmosphere there is insufficient atmospheric pressure for water to be stable as a liquid. Therefore, even though the Moon has always been in the Sun's HZ, its surface is not habitable. [This does not rule out life in its interior, though this is extremely unlikely – see the answer to Question 4.3.]

Question 4.2

Figure 4.2 shows that the luminosity and temperature increases from zero age to 4 600 Ma have caused the inner boundary of the HZ to migrate outwards from about the orbit of Venus to halfway to the Earth's orbit. The further increase in luminosity to 11 000 Ma is even greater than during the first 4 600 Ma, and moreover the effective temperature will then be lower than it is today, promoting further outward migration. Therefore, it is to be expected that the Sun's HZ at 11 000 Ma will lie beyond the Earth. [Calculations show that the inner boundary of the Sun's HZ at 11 000 Ma will be at around 1.3 AU, 0.3 AU beyond the Earth.]

Question 4.3

(a) The Table gives the mean surface temperature of the Moon as 1°C. This is subject to huge variations, depending on whether it is day or night. But at a few tens of metres, as for Mercury (Section 4.3), the diurnal swing is moderated and so the temperature will not be far from 1°C. At greater depth the temperature will be higher, because of the Moon's internal heat sources, and if there were volatiles present their pressure could exceed 610 Pa. Thus water could be stable as a liquid. [Unfortunately, the Moon is poorly endowed with volatiles, including water, so it is highly unlikely that liquid water is present at any depth.]

(b) Equation (4.6) gives the dependence of the rate of tidal heating of a satellite on various parameters:

$$W_{\text{tidal}} \propto (Mm)^2 R \frac{e}{a^6}$$

where M is the mass of the planet, m that of the satellite, R is the radius of the satellite, e its orbital eccentricity, and a its orbital semimajor axis. Putting in values for Europa from Table 4.2, which is significantly tidally heated by Jupiter gives:

$$W_{\text{Europa}} \propto (318\,M_{\text{E}} \times 48.7 \times 10^{21}\,\text{kg})^2\,1565\,\text{km} \times 0.010/(671\,000\,\text{km})^6$$

For the Moon, the values in the Table in the question give:

$$W_{\text{Moon}} \propto (M_{\text{E}} \times 73.5 \times 10^{21}\,\text{kg})^2\,1738\,\text{km} \times 0.0549/(384\,500\,\text{km})^6$$

Therefore, assuming similar constants of proportionality:

$$W_{\text{Europa}}/W_{\text{Moon}} \approx 260$$

This shows that the tidal heating of the Moon is much feebler than that of Europa, so will not raise temperatures much. [We have ignored the factors representing the composition of the satellites, but these are not so different as to alter the conclusion.]

CHAPTER 5

Question 5.1

Planetary interiors tend to cool, and therefore a particular temperature, such as 0°C, would be reached at shallower depths in the past than in the present. The Hesperian epoch ended over 1800 Ma ago, and therefore in this epoch water should have been liquid at shallower depths than today. This means that carbon–liquid water life could have been present at shallower depths. [At the very earliest times a planetary interior will heat up, until the rate of energy generation inside it balances the rate of energy loss by radiation to space. But this phase would be confined to the Noachian epoch.]

Question 5.2

At several times its actual mass, Mars would then have a mass comparable to that of the Earth. It would then have been able to sustain a more massive atmosphere, because:

- its level of geological activity would not have declined so much, thus less of its atmosphere would have been trapped in its crust;
- leakage to space would have been slower, due to its stronger gravity; and
- it probably would have retained a magnetic field strong enough to ward off the erosive effects on the atmosphere of the impacts of solar wind particles.

With a more massive atmosphere it is likely that water could be stable as a liquid on its surface today, in which case it would be habitable by carbon–liquid water life.

Question 5.3

(a) The question sets 4400 Ma as the time on Mars from which occurred steps towards the origin of life and its subsequent evolution. If conditions suitable for this progress lasted on the martian surface only until the end of the Noachian epoch, then that gives at most (4400 – 3500 Ma) = 900 Ma for its evolution. The comparable starting time on Earth is 3900 Ma. At 3000 Ma (900 Ma later) there were no eukaryotes and no multicellular eukaryotes (Section 3.2.1). Therefore, if deteriorating surface conditions halted evolution on Mars then, on the basis of evolution on Earth, life might well not have got beyond unicellular prokaryotes. If, however, martian life continued to evolve deep in the crust, then not only might there be fossils of eukaryotes and multicellular eukaryotes, but also living organisms of these types.

(b) If life originated early on Mars then it could have been delivered to Earth soon after the heavy bombardment ceased, in the form of martian meteorites – we might all be martians!

CHAPTER 6

Question 6.1

(a) Gravitational interactions between the Galilean satellites, boosted by mean motion resonances, cause changes in their orbital eccentricities (Section 4.3.3).

(b) If the orbital eccentricity of Europa were reduced close to zero for a long time the interior of Europa would cool, having been robbed of tidal heating (Section 4.3.2). Heating by long-lived radioisotopes would be insufficient to maintain a widespread ocean of water, and so it would freeze.

(c) If the orbital eccentricity of Ganymede were increased considerably, then, if this lasted long enough, the enhanced tidal heating might produce a widespread ocean in its water shell (Figure 4.6). This might not be adjacent to the rocky mantle and therefore there might be no hydrothermal vents. Nevertheless, the presence of liquid water plus dissolved redox pairs that require enzyme-type molecules to catalyse the reactions, might be viable energy sources, at least until chemical equilibrium is reached. Life might therefore originate, but die out. If the liquid water was adjacent to hydrothermal vents life might have a much longer history.

Question 6.2

A mission like that in Figure 6.6 should be designed, ideally, to recognise all the following manifestations of life:

- macroscopic, presumably multicellular life forms, that are obviously alive (e.g. fish!);

- unicellular life forms that resemble those on Earth;
- unicellular lifeforms that differ from those on Earth in fundamental structural and biochemical ways;
- biological materials like proteins, RNA, and DNA that are similar to those on Earth;
- biological materials that differ in important ways from those on Earth (e.g. proteins based on different amino acids, or that consist of equal mixtures of D and L forms).

Question 6.3

If the surface of the rocky interior of Europa were at the right sort of temperature for liquid water, there could be pores in the rock where life could exist, just as in the Earth's crust. Deeper in the rocky interior the pressure would close the pores. Indeed, they might even be closed under the 100-km-thick shell of ice.

CHAPTER 7

Question 7.1

Clearly, terrestrial life has to migrate away from the Sun. This would be possible in the form of meteoroids collisionally ejected from Earth and reaching bodies like Mars (the opposite of martian meteorites reaching the Earth), or reaching the large satellites of the giant planets. Large impacts would be needed to eject material through the Earth's more substantial atmosphere and through its stronger gravitational field. There would also have to be viable ecological communities trapped in the material. To survive beyond when the Sun becomes a white dwarf, the material would either have to reach planets around other stars [very unlikely], or reach planetary bodies that somehow come to reside in the habitable zone of a white dwarf [also very unlikely].

The other way is for viable ecological communities to reach habitable planetary bodies via spacecraft. This would require technologically advanced species to be present on Earth at that far distant time [this would be a better prospect that meteoroid ejection].

Question 7.2

From Equation (7.1):

$$\frac{L_{\text{WD}}}{L_{\text{Sun}}} = \left(\frac{R_{\text{WD}}}{R_{\text{Sun}}}\right)^2 \left(\frac{T_{\text{eWD}}}{T_{\text{eSun}}}\right)^4$$

The Sun has a radius of 6.96×10^5 km, and a white dwarf about that of the Earth, 6378 km (Table 1.2). The effective temperature of the Sun is 5780 K (Section 1.1.2), and that of a young white dwarf is given as 2×10^4 K. Therefore:

$$\frac{L_{\mathrm{WD}}}{L_{\mathrm{Sun}}} = \left(\frac{6378}{6.96 \times 10^5}\right)^2 \left(\frac{2 \times 10^4}{5780}\right)^4 = 0.012$$

With such a low luminosity, the HZ will be close to the star, well within 1 AU.

CHAPTER 8

Question 8.1

(i) The main-sequence lifetime of an A3 star is less that 2000 Ma, too short for life to have a detectable effect on its surface or atmosphere.
(ii) The HZ will be within a few AU of the binary system. Therefore, at a separation of only 3 AU the stars would gravitationally disturb any planets in the HZ, so there would be no planets there.
(iii) In a globular cluster the stars are so densely packed that gravitational disruption of a planetary system might be likely.
(iv) A G2V star 1000 Ma old is too young for any life at its surface to have modified the surface or atmosphere to an extent that could be detected from afar.
(v) With M dwarfs not ruled out of consideration, this M0 dwarf is old enough to have distantly observable life, and it is at a suitable distance from the Galactic centre. The only possible problem is the metallicity, which we are not given. This star is in the thick disc, and here the proportion of low metallicity stars is high. On the other hand, this is not a very old star, so its metallicity is probably high enough for planets of the order of the mass of the Earth.

Question 8.2

Two of the giant planets in the Solar System, Jupiter and Saturn, are predominantly made of hydrogen and helium. If this is true of the giant planets around other stars [it is] then it could be that whereas giant planets can form at low metallicity, rocky bodies with masses about that of the Earth cannot – there might only be Mars-mass bodies, and these might not be able to support surface life for long enough to be detectable from afar.

CHAPTER 9

Question 9.1

(a) If the Earth were moved closer to the Sun then, if the cloud cover and surface reflectivity remained the same, the Earth would reflect more sunlight because the solar intensity would be greater. The contrast ratio would thus improve at visible and near-infrared wavelengths. The surface would rise in temperature and so the

Earth would emit more infrared radiation, so the contrast ratio would also improve at mid-infrared wavelengths.

It is, however, possible that the cloud cover would increase, like Venus. In this case the outcome for reflected sunlight is not clear – it depends on whether the increase in cloud cover or the increase in solar intensity wins. The outcome is also unclear at infrared wavelengths. It depends on whether the absorption of solar radiation has increased or decreased – the emission to space has to balance this absorption.

(b) If the Earth were moved closer to the Sun it would tend to get easier to detect directly if the contrast ratio improved, otherwise harder. It would certainly tend to get harder to detect through its image lying closer to the core of the Sun's image as spread by the point spread function (*psf*).

Question 9.2

A coronagraph attenuates the *psf* of an object on the telescope axis, such as the star that a planet orbits, so it is useful in a space telescope. Adaptive optics is a (partial) correction for the effects of turbulence in the atmosphere above a telescope. In space there is no such atmosphere, so adaptive optics is not useful.

Question 9.3

(i) The *snr* ratio in a telescope image is given by Equation (9.3), and shows that the greater the area of the mirror A, the larger the *snr*.

(ii) The detail that can be obtained in interferometry increases as the spacing between the telescopes increases (Figure 9.9).

CHAPTER 10

Question 10.1

(a) We use Equation (10.3), rearranged as:

$$d = \left(\frac{m}{M}\right)\left(\frac{a}{\beta}\right)$$

The value of β to be used is about twice the astrometric precision of 10 μas (i.e., 20 μas or 1.0×10^{-10} radians). Putting in the values for a Jupiter-twin:

$$d = \left(\frac{1.9 \times 10^{27}}{2.0 \times 10^{30}}\right)\left(\frac{5.2\,\text{AU}}{1.0 \times 10^{-10}}\right) = 4.9 \times 10^{7}\,\text{AU} = 780\,\text{ly}$$

(b) Applying the same method as in (a):

$$d = \left(\frac{6.0 \times 10^{24}}{2.0 \times 10^{29}}\right)\left(\frac{1.0\,\text{AU}}{1.0 \times 10^{-10}}\right) = 3.0 \times 10^{5}\,\text{AU} = 4.7\,\text{ly}$$

[Wolf 359 at 7.8 ly is the closest star with such a low mass. The Earth-mass planet would thus have to be somewhat further from this star to produce the desired effect.]

Question 10.2

The radial velocity range is $2.5 \times 10^{-7}\,c$, where c is the speed of light (Equation 10.7). The range is thus $75\,\mathrm{m\,s^{-1}}$, and the amplitude v_{rA} is half of this, $37.5\,\mathrm{m\,s^{-1}}$. From Equation (10.12):

$$m\sin(i_0) = 37.5\,\mathrm{m\,s^{-1}}(20 \times 24 \times 3600\,\mathrm{s})^{1/3}(2.0 \times 10^{30}\,\mathrm{kg})^{2/3}/(2\pi\, 6.67 \times 10^{-11}\,\mathrm{N\,m^2\,kg^{-2}})^{1/3}$$
$$= 9.6 \times 10^{26}\,\mathrm{kg} = 160\,m_{\mathrm{E}}$$

This is the minimum mass – we do not know i_0.

The semimajor axis can be deduced from Equation (10.4). As this is a solar mass star this equation takes the form $a = P^{2/3}$ if a is in AU and P is in years (Section 1.1.1). Thus:

$$a = (20/365)^{2/3}\,\mathrm{AU} = 0.14\,\mathrm{AU}$$

The planet is thus a giant, orbiting close to its star. The cyclic changes in radial velocity are sinusoidal, and so the orbit is circular.

There could be a second companion, though it must be producing a much smaller Doppler signal for it to be so far undiscernable in the cyclic radial velocity changes.

Question 10.3

(a) From Equation (10.17), the mass of the companion is given by:

$$m = (t_{\mathrm{Ep}}/t_{\mathrm{E}})^2 M = (0.15/60)^2\, 0.5\, M_\odot = 3.1 \times 10^{-6}\, M_\odot = 1.0\, m_{\mathrm{E}}$$

(b) With such a faint star it will be difficult to make accurate spectral measurements for Doppler spectroscopy and photometric measurements for transit photometry. The latter has the added difficulty of requiring a transit. Astrometry is unlikely to be possible because of the distance to the star that makes it extremely faint in the sky. [There are likely to be many cases of gravitational microlensing where follow-up is impossible for these reasons. Also, a second microlensing event is very unlikely given the chance alignment needed.]

CHAPTER 11

Question 11.1

A planet of radius $1.2\,R_{\mathrm{J}}$ with the same mean density as Jupiter will have a mass $(1.2)^3\,m_{\mathrm{J}}$ (i.e. $1.7\,m_{\mathrm{J}}$ (Equation 11.1)). Because this planet is observed in transit we know that its actual mass is 1.9 m_J. This indicates a mean density slightly greater than that of Jupiter, but this can easily be put down to greater internal pressures at this higher mass. A hydrogen–helium composition is thus probable.

If it had spent much of its life close to its star then it would have cooled so slowly that it would be distended to the extent that it would have a much lower mean density than that of Jupiter, and hence have a far larger radius for its mass.

Question 11.2

[This table of data is derived from Figure 11.4. Values for $m\sin(i_0)$ in the range 2.5–2.6 m_J for 47 UMa b are OK, and 0.74–0.78 m_J for 47 UMa c. For a, acceptable ranges are 2.05–2.13 AU and 3.65–3.81 AU respectively.

	$m\sin(i_0)$ (m_J)	a (AU)
47 UMa b	2.54	2.09
47 UMa c	0.76	3.73

Figure Q11.2 is based on this Table and on the data given in the question.]

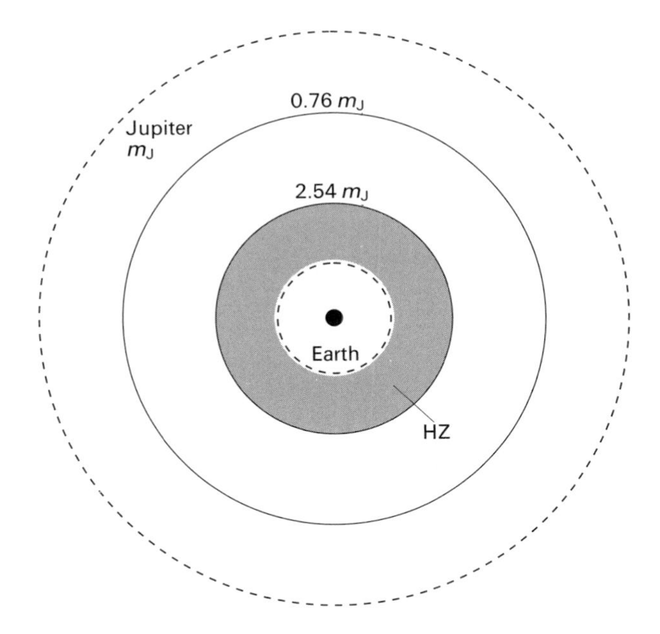

Figure Q11.2 47 Ursae Majoris, with HZ (now) and the orbits of Jupiter and the Earth.

Question 11.3

Epsilon Eridani

This is a system in which the giant is in a fairly large, eccentric orbit. When the embryo was Earth-mass it was undergoing Type I migration, and thus spiralling inwards rapidly. When it had grown to several Earth-masses the migration became Type II and the inward rate consequently slowed. It then grew to become a giant. Subsequently it must have either entered a mean motion resonance with a second giant or had a close encounter with one. It thus acquired its eccentric orbit. The other giant has either been flung out of the system or is in a large orbit that has so far escaped definite detection.

HD168746

This is a system in which the giant is in a small, low-eccentricity orbit. When the embryo was Earth-mass it was undergoing Type I migration, and thus spiralling inwards rapidly. When it had grown to several Earth-masses the migration became Type II and the inward rate consequently slowed. It then grew to become a giant, though if it is not much greater than its minimum mass of $0.23\,m_J$ its growth was rather stunted. It came to rest in its present orbit when either the circumstellar disc dissipated, or when tidal interactions with the star counteracted the inward push of the disc, or through other counter-effects.

Question 11.4

(a) With a giant within the HZ there is little chance that an Earth-mass planet would have been there. Such a planet could survive near the star, well interior to the HZ, and also far from the star, well outside the HZ.

(b) A planetary body that is actually habitable could be a large satellite of the giant – the satellite as well as the giant would be within the HZ regardless of any tidal heating. Another possibility is a planet with the order of the mass of the Earth outside the HZ, provided that its interior is still warm – subterranean life is then a possibility [as discussed in Section 4.3].

Question 11.5

(a) These values can be read from the P and a scales in either Figure 10.2 or 10.10.

A $P = 200$ days for a solar-mass star corresponds to a semimajor axis $a = 0.67\,\text{AU}$.

B $P = 2000$ days for a star of mass half that of the Sun corresponds to $a = 2.5\,\text{AU}$.

(b) *The radial-velocity amplitudes*

A With $P = 200$ days, Figure 10.10 shows that $v_{rA} = 35\,\text{m}\,\text{s}^{-1}$ for a solar-mass star with $m\sin(i_0) = m_J$.
With $i_0 = 89.9°$ and $m = 0.0030\,m_J$ we have $m\sin(i_0) = 0.0030\,m_J$.

Given that v_{rA} is proportional to $m_p \sin(i_0)$, $v_{rA} = 35\,\mathrm{m\,s^{-1}} \times 0.0030 = 0.11\,\mathrm{m\,s^{-1}}$.

B With $P = 2000$ days, Figure 10.10 shows that $v_{rA} = 26\,\mathrm{m\,s^{-1}}$ for a half solar-mass star with $m \sin(i_0) = m_J$. With $i_0 = 30°$ and $m = 2.0\,m_J$ we have $m \sin(i_0) = 1.0\,m_J$. Thus, v_{rA} remains $26\,\mathrm{m\,s^{-1}}$.

(c) *The astrometric motions*

A With $P = 200$ days, Figure 10.2 shows that $\beta d = 2.0 \times 10^{-3}$ arcsec ly for a solar-mass star with $m = m_J$. Given that βd is proportional to m_p, $\beta d = 2.0 \times 10^{-3}$ arcsec ly $\times\, 0.0030 = 6.0 \times 10^{-6}$ arcsec ly. Now $d = 25$ ly light years, so $\beta = 6.0 \times 10^{-6}$ arcsec ly/25 ly $= 2.4 \times 10^{-7}$ arcsec.

B With $P = 2000$ days, Figure 10.2 shows that $\beta d = 1.6 \times 10^{-2}$ arcsec ly for a half solar-mass star with $m = m_J$. Given that βd is proportional to m, $\beta d = 1.6 \times^{-2}$ arcsec ly $\times\, 2.0 = 3.2 \times 10^{-2}$ arcsec ly. Now $d = 250$ ly, so $\beta = 3.2 \times 10^{-2}$ arcsec ly/250 ly $= 1.3 \times 10^{-4}$ arcsec.

(d) System A has an Earth-mass planet in a small orbit. The values of v_{rA} and β are below the thresholds given in the question so it is not detectable by these means. It is however in an orbit presented almost edge-on and so it will transit its star. However, as it is an Earth-mass planet in front of a solar-mass main-sequence star it will cause a dip in apparent luminosity of about 0.01%. Therefore, it will be below the given threshold of detection.

System B has a Jupiter-mass planet in a fairly large orbit around a low-mass star. The value of v_{rA} is above the threshold given in the question so it is detectable by Doppler spectroscopy. The value of β is below the threshold so it is not detectable astrometrically. Its inclination shows that it will not transit its star.

(e) In System A, the HZ is now exterior to the planet (Figure 8.5), so the planet is not habitable. The planet in System B would be outside the HZ of a solar-mass star, so is even further outside the HZ of a lower mass star (Figure 8.5). The planet is thus not habitable. Moreover it is a giant, so is an unlikely place for life even were it inside the HZ. [Remember that the HZ is that range of distances from a star in which the stellar radiation on an Earth-like planet would sustain water as a liquid on at least parts of the surface. A planet or satellite can be habitable outside this zone if some other source of heat, such as tidal heating, is available.]

CHAPTER 12

Question 12.1

Most of the O_3 is in the stratosphere, above the cloud, and so it will be detectable in the infrared spectrum [as an emission line, but I don't want to go into that]. Turning to methane, the atmospheric pressure at 10 km is only about one-tenth that at the ground (Figure 12.5), and a reasonable assumption is that it is not concentrated at high altitudes [which is correct]. Therefore, the detection of methane is unlikely. We thus only have O_3, leaving uncertainty about habitation.

Question 12.2

(i) Except for the absorption band, the general shape of the spectrum fits quite well the 7°C curve in Figure 12.6. Figure 12.6 is for a cloud-free part of Mars, and so the atmosphere is likely to be transparent outside the absorption band, in which case this must be the temperature of the surface.

(ii) The absorption band centred around 15 μm closely resembles the CO_2 absorption band in the Earth's atmosphere (Figure 12.4), so must be due to CO_2 in a comparable quantity to that in the Earth's atmosphere. The martian atmosphere is so much thinner than the Earth's atmosphere that CO_2 thus constitutes a considerable proportion of it.

(iii) There is no sign of the numerous small absorption bands due to water vapour, yet the fine structure in the CO_2 absorption band indicates that the spectrum has sufficient resolution to have detected the water bands. The atmosphere must therefore be fairly dry.

Question 12.3

Initially, the infrared spectrum would resemble that in Figure 12.4 except that there would be no O_3 absorption band. [The surface temperature, as indicated by the general shape of the spectrum, might also be a bit different.] The O_3 feature would then grow, provided that the carbon–liquid water life involved copious oxygen-releasing photosynthesis. As the biosphere shrinks the O_3 feature would diminish. Given that the biosphere shrinks due to an increase in the star's luminosity, the spectrum (outside the CO_2 absorption band) would shift to shorter wavelengths to fit a higher surface temperature. [The strengths of the CO_2 and H_2O bands are also likely to vary.]

Question 12.4

The first difficulty is the need for super-resolution, so that the satellite is seen distinctly from its giant planet. Second, an aquatic biosphere on a 'Europa' would be beneath a thick crust of ice. Above the ice there would be little or no atmosphere, and in any case a biosphere beneath a thick ice crust is very likely to have a negligible influence on any atmosphere. If water escaping through cracks bore organic substances that stained the ice then it might be possible to detect features in the visible–near-infrared reflection spectrum (cf. chlorophyll on Earth).

CHAPTER 13

Question 13.1

The 'sociological' factor is L, interpreted as the time for which a planet harbouring a technological species can be detected over interstellar distances through technological activity. This is sociological in the sense that it depends on the goals and ambitions of the societies that presumably constitute a technological civilisation, and on how long its social structures enable it to survive with advanced technology.

We have been detectable for about 100 years, so that is a defensible lower limit on L. An upper limit is obtained from the assumption that detectability only ceases when the planetary system becomes uninhabitable. In the case of the Solar System this will be a few billion years from now [Section 7.2].

Question 13.2

(i) The overall bandwidth at any instant is $1.5\,\text{Hz} \times 10^8 = 1.5 \times 10^8\,\text{Hz}$, or 150 MHz.
(ii) With 1000 stars, the interval between one hour observations is 1000 hours, which is about 1.4 months.
(iii) For 1000 stars each visited 10 times, an uninterrupted search will last 10 000 hours, or about 14 months.
(iv) The favourable microwave region is about 10 000 MHz wide [arguably about half of this]. With an overall bandwidth of 150 MHz it takes $10\,000/150 = 67$ moves to cover the whole of this for one star. Each star is observed for 10 hours, so each frequency is observed for 600 minutes/$67 = 9$ minutes.

Question 13.3

One way would be to switch the radio transmission on or off, or from high power to low power [this is called amplitude modulation]. This has the disadvantage that large powers have to be switched rapidly. [There are several other, quite technical, possibilities. One is to switch the beam on and off target, not by moving the radio dish, which would require large power, but by controlling the direction from the output of an array of radiotelescopes, operated interferometrically. Another possibility is to use the fact that electromagnetic waves are transverse (Figure 9.8). The planes in which the electric and magnetic fields lie can be made to rotate around the direction in which the wave travels, and this circular polarisation (Section 3.3.4) can, like a corkscrew, be clockwise or anticlockwise. By switching between these two possibilities a binary message can be encoded.]

Question 13.4

One would need to choose a particular laser frequency and use a narrow-band frequency filter centred on this frequency. It might then be possible to see the pulses from a star as a series of flashes, the star being apparently brighter during each pulse. The pulses would need spacings wide enough for the eye to resolve [1 s is plenty wide enough]. Visibility would be enhanced if the pulses were relatively long in their duration, and if the star were nearby. [It would actually be possible to see suitable pulses in daytime – the number of photons from the blue sky in a very narrow band of frequencies would be far less than from a laser pulse from a nearby star.]

Question 13.5

The one big advantage is the energy that the Type II civilisation has available. By utilising a large proportion of the energy output of their star, the energy needed to send a spacecraft to a star would be a far smaller proportion of the energy available than in the case of a Type I civilisation.

Resources

PHYSICAL CONSTANTS

Many of these are known to more significant figures than are given here.

Astronomical unit (AU)	1.496×10^{11} m
Light year (ly)	9.461×10^{15} m
Parsec (pc)	3.086×10^{16} m
Avogadro constant (N_A)	6.022×10^{23} mol^{-1}
Gravitational constant (G)	6.674×10^{-11} N m^2 kg^{-2}
Planck's constant (h)	6.626×10^{-34} J s
Second of arc (arcsec)	1/3600 of a degree of arc
Speed of light (c)	2.998×10^{8} m s^{-1}
Stefan-Boltzmann constant (σ)	5.670×10^{-8} W m^{-2} K^{-4}
Mass of Earth (m_E)	5.974×10^{24} kg
Mass of Jupiter (m_J)	1.899×10^{27} kg
Mass of Sun ($M_\odot$)	1.989×10^{30} kg

BOOKS

Physics, optics

Physics for Scientists and Engineers, R Serway and J W Jewett, 2003, Brooks Cole. (A compendium. Undergraduate level.)

Predicting Motion (The Physical World, Book 3), R J A Lambourne (editor), 2000, Institute of Physics Publishing. (Newtonian mechanics, etc. Undergraduate level.)

Optics, E Hecht, 2001, Addison Wesley. (Undergraduate level.)

Principles of Optics: Electromagnetic Theory of Propagation, Interference and Diffraction of Light, M Born and E Wolf, 1999, Cambridge University Press. (Advanced undergraduate/graduate level.)

Optics, F G Smith and J H Thomson, 1988, Wiley. (Undergraduate level.)

Statistics (for physical sciences)

Data Reduction and Error Analysis for the Physical Sciences, P R Bevington and D K Robinson, 2002, McGraw-Hill. (Undergraduate level.)

Statistics for Nuclear and Particle Physicists, L Lyons, 1992, Cambridge University Press. (Undergraduate level.)

Astronomy, celestial mechanics

The Sun and Stars, S Green and M Jones (editors), 2004, The Open University and Cambridge University Press. (Introductory text in stellar astronomy. Undergraduate level.)

Galaxies and Cosmology, M Jones and R J A Lambourne (editors), 2004, The Open University and Cambridge University Press. (Introductory text in galactic astronomy and cosmology. Undergraduate level.)

Foundations of Astronomy, M Seeds, 2003, Thompson. (Introductory, broad ranging text. Lower undergraduate level.)

The Solar Corona, L Gloub and J M Pasachoff, 1997, Cambridge University Press. (Includes coronagraphy. Undergraduate level.)

Astrophysics of the Sun, H Zarin, 1988, Cambridge University Press. (Includes coronagraphy. Undergraduate level.)

Solar System Dynamics, C D Murray and S F Dermott, 1999, Cambridge University Press. (Moderately mathematical text in celestial mechanics, applying to planetary systems in general.)

Planetary systems (mainly Solar System)

The Solar System, N McBride and I Gilmour (editors), 2004, The Open University and Cambridge University Press. (Introduction to the Solar System, with a geological emphasis. Undergraduate level.)

The Earth in Context: A Guide to the Solar System, D M Harland, 2001, Springer-Praxis. (Introduction to the Solar System, with a geological emphasis. Undergraduate level.)

Planetary Science: The Science of Planets around Stars, G H A Cole and M M Woolfson, 2002, Institute of Physics Publishing. (Broad-ranging topics in planetary science, with a physical science emphasis. Undergraduate level.)

Discovering the Solar System, B W Jones, 1999, John Wiley and Sons. (Introduction to all aspects of the Solar System, in a broadly thematic approach. Undergraduate level.)

Scientific Frontiers in Research on Extrasolar Planets, D Deming and S Seager (editors), 2003, Astronomical Society of the Pacific. (A collection of research papers presented at a conference in Washington, DC in 2002.)

Distant Wanderers, B Dorminey, 2002, Copernicus Books. (A popular account, including the history of extrasolar planet discoveries and the scientists involved.)

Extrasolar Planets: The Search for New Worlds, S Clark, 1998, Wiley-Praxis. (Extrasolar planets and their context – the early discoveries. General readership.)

Astrobiology

Life in the Universe: The Science of Astrobiology, I Gilmour and M A Sephton (editors), 2004, The Open University and Cambridge University Press. (Broad coverage. Undergraduate level)

Life in the Universe, J Bennett, S Shostak, and B Jakosky, 2003, Addison Wesley (Broad coverage. Non-specialist undergraduate level.)

The Origin of Life, P Davies, 2003, Penguin. (General readership.)

The Search for Life in the Universe, D Goldsmith and T Owen, 2001, University Science Books. (Introduction to the search for extraterrestrial life. Undergraduate level.)

Earth, Life and the Universe, K Tritton, 2001, Curved Air Publications. (Broad coverage. General readership.)

Search for Life, M Grady, 2001, The Natural History Museum London. (Brief overview of the search for extraterrestrial life. General readership.)

Rare Earth: Why Complex Life is Uncommon in the Universe, P D Ward and D Brownlee, 2000, Copernicus Springer-Verlag. (A particular, controversial viewpoint that complex life in the Universe is rare. General readership.)

Life on Other Worlds and How to Find it, S Clark, 2000, Springer-Praxis. (Introduction to the search for extraterrestrial life. General readership.)

Bioastronomy '99: A New Era in Astronomy, G A Lemarchand and K J Meech (editors), 2000, Astronomical Society of the Pacific. (A collection of research papers presented at a conference in Hawaii in 1999.)

Earth: Evolution of a Habitable World, J I Lunine, 1999, Cambridge University Press. (A survey of life on Earth and its origin. Undergraduate level.)

The Biological Universe, S J Dick, 1996, Cambridge University Press. (Broad coverage, if a bit dated in parts. General readership.)

SETI

Where is Everybody? S Webb, 2002, Copernicus-Praxis. (An examination of the chances of us finding ETI. The Fermi paradox. General readership.)

Is Anyone Out There? The Scientific Search for Extraterrestrial Intelligence, Frank Drake and Dava Sobel, 1992, Souvenir Press. (A popular account of SETI by the pioneer, Frank Drake. General readership.)

WEB SITES

Extrasolar planets

Catalogue maintained by J Schneider
`http://www.obspm.fr/encycl/catalog.html`

Space agencies

European Southern Observatory
`http://www.eso.org/`

European Space Agency
`http://www.esa.int/` `http://sci.esa.int/`
`http://www.rsd.esa.int/`

NASA
http://www.nas.gov/
SETI Institute
http://www.seti.org/

Telescopes (ground based)

Keck
http://www2.keck.hawaii.edu/
VLT
http://www.eso.org/projects/vlt/
OverWhelmingly Large Telescope (OWL)
http://www.eso.org/projects/owl/

Telescopes (space based) and space missions

Solar System exploration
http://sse.jpl.nasa.gov/
Mars exploration
http://mpfwww.jpl.nasa.gov/ http://mars.jpl.nas.gov/
Mars Express and Beagle 2
http://sci.esa.int/marsexpress/
http://beagle2.open.ac.uk/index.htm
Mars Exploration Rovers
http://mars.jpl.nasa.gov/mer
Galileo mission to Jupiter
http://galileo.jpl.nasa.gov
Europa Orbiter
http://www.jpl.nasa.gov/europaorbiter/
Cassini-Huygens mission to the saturnian system
http://ciclops.lpl.arizona.edu/ http://sci.esa.int/huygens/
COROT
http://www.astrsp-mrs.fr/projets/corot/
Eddington
http://astro.estec.esa.nl/Eddington/
Kepler
http://www.kepler.arc.nasa.gov/
James Webb Space Telescope
http://ngst.gsfc.nasa.gov
Space Interferometry Mission
http://sim.jpl.nasa.gov
Gaia
http://sci.esa.int/home/gaia http://www.rssd.esa.int/GAIA/

Darwin

http://sci.esa.int/home/darwin/

Terrestrial Planet Finder (TPF)

http://tpf.jpl.nasa.gov

Images

European Southern Observatory

http://www.eso.org/

Hubble Space Telescope

http://www.stsci.edu/resources

NASA

http://photojournal.jpl.nasa.gov

http://nssdc.gsfc.nasa.gov/planetary/

Beagle 2

http://www.beagle2.com/resources/photo-album.htm

SETI

SETI@home screensaver, for participating in SETI

http://setiathome.ssl.berkeley.edu/

SETI at optical wavelengths (OSETI)

http://www.coseti.org

Index